Future City

Volume 11

Future City Description

As of 2008, for the first time in human history, half of the world's population now live in cities. And with concerns about issues such as climate change, energy supply and environmental health receiving increasing political attention, interest in the sustainable development of our future cities has grown dramatically.

Yet despite a wealth of literature on green architecture, evidence-based design and sustainable planning, only a fraction of the current literature successfully integrates the necessary theory and practice from across the full range of relevant disciplines.

Springer's *Future City* series combines expertise from designers, and from natural and social scientists, to discuss the wide range of issues facing the architects, planners, developers and inhabitants of the world's future cities. Its aim is to encourage the integration of ecological theory into the aesthetic, social and practical realities of contemporary urban development.

More information about this series at http://www.springer.com/series/8178

Niki Frantzeskaki • Katharina Hölscher
Matthew Bach • Flor Avelino
Editors

Co-creating Sustainable Urban Futures

A Primer on Applying Transition Management in Cities

Springer

Editors
Niki Frantzeskaki
DRIFT Faculty of Social Sciences
Erasmus University
Rotterdam, The Netherlands

Katharina Hölscher
DRIFT Faculty of Social Sciences
Erasmus University
Rotterdam, The Netherlands

Matthew Bach
DRIFT Faculty of Social Sciences
Erasmus University
Rotterdam, The Netherlands

Flor Avelino
DRIFT Faculty of Social Sciences
Erasmus University
Rotterdam, The Netherlands

ISSN 1876-0899 ISSN 1876-0880 (electronic)
Future City
ISBN 978-3-319-69271-5 ISBN 978-3-319-69273-9 (eBook)
https://doi.org/10.1007/978-3-319-69273-9

Library of Congress Control Number: 2017963359

Printed on acid-free paper

This Springer imprint is published by Springer Nature
The registered company is Springer International Publishing AG
The registered company address is: Gewerbestrasse 11, 6330 Cham, Switzerland

My Experience with Transition Management

My first observation being part of a transition management process within the city of Rotterdam was the fact that it provided a 'good excuse' to create the much-needed space within a city organisation for creativity and innovation. We have seen the transition management process functioning as a strong shield towards risk-averse forces. Exactly what was needed!

Especially when the crisis hit hard, the standard reaction of cities was not to start anything new and performing the city tasks with the minimum possible effort. But what cities desperately needed to tackle wicked urban problems was the exact opposite: new ways of working and new ways of tackling the problems. The previous ways of working certainly proofed that just doing 'business as usual' will not make a difference for our citizens. We feel it is our obligation to find ways of working that can make a real difference.

We have seen transition management gently forcing cities out of their comfort zone by involving in its policy processes the outside world, especially those with a real drive and track record for change. This can feel 'risky' for urban policy officers and will make them try to find a way out of this 'new thing'. Luckily, we have also seen that transition management can offer the much-needed firm handhold to cling on to and provides assurance: 'this can really work'.

On the one hand, reassurances come from the fact that transition management activities are being part of a scientifically well-elaborated method. On the other hand, the DRIFT action researchers can provide first-hand examples from cities that applied the method, leading to real action and, more importantly, real change in the urban environment. This is convincing evidence for other cities.

We first-hand experienced that being in the middle of the first steps of framing and envisioning the process can be overwhelming. Dealing with the overload of information and taking on board the opinions of frontrunners and unusual suspects are challenging. Just like not knowing what the future of this process will bring. The bravery of the cities involved in the project I have in my mind – MUSIC project – was impressive. The city case studies enriched the understanding and fine-tuning of the transition management method for cities. These brave cities followed all steps of the transition management process from setting the scene towards real action plans, producing actions that changed their cities, to more sustainable and innovative ones.

Urban Planning Department, Strategy and Investment
Municipality of Rotterdam, Rotterdam,
The Netherlands

Cleo Pouw

Contents

innovators even in incremental steps (Bulkeley and Betsill 2010; Frantzeskaki et al. 2016b). At the same time, in most of the cities, an active citizenry puts in place new ways of doing and organizing, innovates with new solutions and alters uses of places in cities (Frantzeskaki et al. 2016a). The local governments are becoming more curious and open to people's voice and creativity, rethinking planning as open to citizen-based innovations by putting citizens as partners in planning. This comes more in line to the 'self-organisation' mode of planning (Alfasi and Portugali 2004) and to polycentric multi-level governance approaches (Bulkeley and Betsill 2010) that recognize multiple-actors and multiple-spheres of influence coexist in planning processes and outcomes. For such modes of planning to become less of an 'isolated experiment' and more of a governance experimentation (Bulkeley and Castan Broto 2013; Loorbach et al. 2015; Van der Heijden 2016), there is a need for new approaches that can facilitate the uptake, mainstreaming and even embedding of 'governance experimentation' to urban planning and governance practice.

Finding the right approaches to tap into the potentials of cities and reroute their development pathways towards sustainability and resilience is paramount. The governance of urban sustainability transitions requires new strategic environmental planning processes that integrate diverse sources of knowledge and perspectives, make 'glocal' connections between local urban challenges and broader global, national and regional macro-developments and enable pragmatic decisions in a context of complexity and uncertainty (Newman 1999; Scott 2001). This challenges urban practitioners and scientists alike to take up a process-content thinking and to reflect on how to navigate societal complexity while mobilizing innovative and transformative societal potential towards action for sustainability. At the same time, tactical urbanism (manifested in action-based networks, civil society initiatives and planning actions) is a way to localize global solutions but also to create and design solutions at human scale from cities to deal with climate change. For those solutions to be integrated in the uses of space and to be socially and spatially acceptable and effective, a social production of these solutions needs to be encountered. Transition management is one governance approach that facilitates a co-creation arena as part of the social production of these solutions (Loorbach et al. 2010; Frantzeskaki et al. 2012; Nevens et al. 2013).

In this book we present transition management as one approach suitable for strategic environmental planning that can bring together multiple change agents (frontrunners) and mediating innovative solutions to planning processes. In practice, transition management comes down to a combination of developing around a common understanding of a transition challenge and a shared ambition to drive it toward sustainability. Using the transition perspective as a lens, transition teams develop an analysis of a persistent problem in a complex societal system. This could be a geographical area (from neighborhood to city, country, or region), a sector or societal issue. Based on the preliminary analysis, the transition team can identify regime and niche-actors that are potential contributors to a desired transition. These are brought together in so-called transition arenas or experiments. In and around these arenas and experiments, a shared discourse, ambition, and agenda are developed in such a way that it empowers participants and enables them to translate it to

Transition Management in and for Cities: Introducing a New Governance Approach to Address Urban Challenges

Niki Frantzeskaki, Katharina Hölscher, Julia M. Wittmayer, Flor Avelino, and Matthew Bach

Abstract For introducing our book, this chapter provides the arguments on why transition management as a governance approach is suitable for cities, as a new strategic planning approach. It also includes a thorough literature review of transition management applications from 2001 to 2017 that shows its spread as a heuristic, operational and theoretical model for transition governance in multiple sectors and across local, regional and national levels. From the literature review and the book contributions we derive ten directions for future research and development of transition management that can benefit not only its operational applications as a process methodology but also its theoretical and heuristic strengths, pointing at theoretical and conceptual deepening and broadening of its tenets and principles. Next, the chapter provides an overview of the book, its four parts and summaries of all chapter contributions. Last, we provide a 'guide' on how to read this book, and how to best use the knowledge and experience shared in readers' research and urban practice.

Introduction

Cities always face challenges and have the role to be fast paced problem solvers. Solutions are tested and trialed in cities. With more combined challenges at hand ranging from climate change, to urban poverty, cities search for new ways to address them, also questing approaches that help them understand how problems are created as well as for searching solutions. Local governments become more and more equipped to not only employ new approaches in search of systemic solutions but also to recognize needs for innovative ways in engaging with urban change agents and to broaden their role from 'planners' and 'regulators' to facilitators and

N. Frantzeskaki (✉) · K. Hölscher · J.M. Wittmayer · F. Avelino · M. Bach
Dutch Research Institute For Transitions (DRIFT), Faculty of Social Sciences, Erasmus University Rotterdam, Rotterdam, The Netherlands
e-mail: n.frantzeskaki@drift.eur.nl; hoelscher@drift.eur.nl; wittmayer@drift.eur.nl; avelino@drift.eur.nl; bach@drift.eur.nl

N. Frantzeskaki et al. (eds.), *Co-creating Sustainable Urban Futures*, Future City 11, https://doi.org/10.1007/978-3-319-69273-9_1

innovators even in incremental steps (Bulkeley and Betsill 2010; Frantzeskaki et al. 2016b). At the same time, in most of the cities, an active citizenry puts in place new ways of doing and organizing, innovates with new solutions and alters uses of places in cities (Frantzeskaki et al. 2016a). The local governments are becoming more curious and open to people's voice and creativity, rethinking planning as open to citizen-based innovations by putting citizens as partners in planning. This comes more in line to the 'self-organisation' mode of planning (Alfasi and Portugali 2004) and to polycentric multi-level governance approaches (Bulkeley and Betsill 2010) that recognize multiple-actors and multiple-spheres of influence coexist in planning processes and outcomes. For such modes of planning to become less of an 'isolated experiment' and more of a governance experimentation (Bulkeley and Castan Broto 2013; Loorbach et al. 2015; Van der Heijden 2016), there is a need for new approaches that can facilitate the uptake, mainstreaming and even embedding of 'governance experimentation' to urban planning and governance practice.

Finding the right approaches to tap into the potentials of cities and reroute their development pathways towards sustainability and resilience is paramount. The governance of urban sustainability transitions requires new strategic environmental planning processes that integrate diverse sources of knowledge and perspectives, make 'glocal' connections between local urban challenges and broader global, national and regional macro-developments and enable pragmatic decisions in a context of complexity and uncertainty (Newman 1999; Scott 2001). This challenges urban practitioners and scientists alike to take up a process-content thinking and to reflect on how to navigate societal complexity while mobilizing innovative and transformative societal potential towards action for sustainability. At the same time, tactical urbanism (manifested in action-based networks, civil society initiatives and planning actions) is a way to localize global solutions but also to create and design solutions at human scale from cities to deal with climate change. For those solutions to be integrated in the uses of space and to be socially and spatially acceptable and effective, a social production of these solutions needs to be encountered. Transition management is one governance approach that facilitates a co-creation arena as part of the social production of these solutions (Loorbach et al. 2010; Frantzeskaki et al. 2012; Nevens et al. 2013).

In this book we present transition management as one approach suitable for strategic environmental planning that can bring together multiple change agents (frontrunners) and mediating innovative solutions to planning processes. In practice, transition management comes down to a combination of developing around a common understanding of a transition challenge and a shared ambition to drive it toward sustainability. Using the transition perspective as a lens, transition teams develop an analysis of a persistent problem in a complex societal system. This could be a geographical area (from neighborhood to city, country, or region), a sector or societal issue. Based on the preliminary analysis, the transition team can identify regime and niche-actors that are potential contributors to a desired transition. These are brought together in so-called transition arenas or experiments. In and around these arenas and experiments, a shared discourse, ambition, and agenda are developed in such a way that it empowers participants and enables them to translate it to

Transition Management in and for Cities: Introducing a New Governance Approach to Address Urban Challenges

Niki Frantzeskaki, Katharina Hölscher, Julia M. Wittmayer, Flor Avelino, and Matthew Bach

Abstract For introducing our book, this chapter provides the arguments on why transition management as a governance approach is suitable for cities, as a new strategic planning approach. It also includes a thorough literature review of transition management applications from 2001 to 2017 that shows its spread as a heuristic, operational and theoretical model for transition governance in multiple sectors and across local, regional and national levels. From the literature review and the book contributions we derive ten directions for future research and development of transition management that can benefit not only its operational applications as a process methodology but also its theoretical and heuristic strengths, pointing at theoretical and conceptual deepening and broadening of its tenets and principles. Next, the chapter provides an overview of the book, its four parts and summaries of all chapter contributions. Last, we provide a 'guide' on how to read this book, and how to best use the knowledge and experience shared in readers' research and urban practice.

Introduction

Cities always face challenges and have the role to be fast paced problem solvers. Solutions are tested and trialed in cities. With more combined challenges at hand ranging from climate change, to urban poverty, cities search for new ways to address them, also questing approaches that help them understand how problems are created as well as for searching solutions. Local governments become more and more equipped to not only employ new approaches in search of systemic solutions but also to recognize needs for innovative ways in engaging with urban change agents and to broaden their role from 'planners' and 'regulators' to facilitators and

N. Frantzeskaki (✉) · K. Hölscher · J.M. Wittmayer · F. Avelino · M. Bach
Dutch Research Institute For Transitions (DRIFT), Faculty of Social Sciences, Erasmus University Rotterdam, Rotterdam, The Netherlands
e-mail: n.frantzeskaki@drift.eur.nl; hoelscher@drift.eur.nl; wittmayer@drift.eur.nl; avelino@drift.eur.nl; bach@drift.eur.nl

N. Frantzeskaki et al. (eds.), *Co-creating Sustainable Urban Futures*, Future City 11, https://doi.org/10.1007/978-3-319-69273-9_1

their own daily environments. Through such a social learning process, individuals start to relate to a broader common context and ambitions, creating the conditions for diffusion, self-organizational processes, and emergent innovation (Hölscher et al. 2017). Transition management thus proposes to develop informal networks in which individuals and, later, organizations and networks are provided the mental, social, and physical space to develop new ideas, common language, and ambitions, well as new joint projects. In doing so over a longer period of time, participants will increasingly translate the transition perspective and ideas into their own operating context, while understanding the complexities, unpredictability and non-linear causalities involved. Transition management processes therewith indirectly influence and change regular policies in government, business, research, and civil society. It is therefore important to evaluate and plan strategically which actors are involved in transition management processes and how their co-creation of strategic agendas influences the role they will take up in sustainability transitions in the making.

This book comes to provide an insight view of transition management as practice-oriented process methodology to develop orientations for and actions to influence sustainability transitions in cities. We bring together the theory, practice and critical reflection of transition management and its applications in cities and for cities. The outcomes of transition management in cities range from new initiatives and projects that can be impulses for sustainability transitions (Frantzeskaki et al. 2014b; Ernst et al. 2016), establishing new connections between urban actors and change urban actor-networks (Wittmayer et al. 2016) as well as reconfiguring institutional rules and organisation. The new knowledge co-produced in transition management arenas of engagement catalyses and mediates the setting of new urban institutions, innovative solutions and the ways they can be aligned and integrated in existing structures, cultures and practices in cities (Wittmayer et al. 2016; Loorbach et al. 2015). Wolfram in this book (chapter "Urban Planning and Transition Management: Rationalities, Instruments and Dialectics", this volume) also points at the performance of transition management as an approach for collective knowledge production.

How Has Transition Management Been Applied?

Transition management is a theoretical and methodological framework that offers operational guidance on how to set up a strategic co-creation processes for change-oriented interventions in complex adaptive societal (sub-)systems (Rotmans et al. 2001; Loorbach 2010; Frantzeskaki et al. 2012; Nevens et al. 2013). Transition management builds upon Integrated Sustainability Assessment framework (Jaeger 2009) and is differentiated by taking on board a consistent focus on transformations and strategic development for transformative change. With the focus on the persistent structural and cultural constraints underlying shifts towards more sustainable lifestyles, it highlights the need for long-term fundamental change in existing socio-technical systems and provides a tool for understanding how governance, cultural systems, infrastructures and social practices co-evolve creating lock-ins and radical changes (Kemp et al. 2007a, b).

Transition management has in the past 18 years been applied to diverse sustainability questions, policy contexts and geographical scales (Loorbach et al. 2015; Frantzeskaki et al. 2014b, 2017) and has become one of the most prevalent approaches that is currently used in parts of Europe to scientifically ground the governance of sustainability transitions. The applications have shown that the approach is able to support governance of and for sustainability transitions (Nevens and Roorda 2013; Frantzeskaki et al. 2014a; Loorbach et al. 2015). It provides a portfolio of tools that have as a common objective to enable change in practices and structures (institutions) directed towards sustainable development targets. In response to criticism, a fruitful dialogue has started to advance the approach.

We conducted a systematic literature review in journal papers on the documented applications of transition management in order to identify: (a) whether the framework is suitable for application across sectors and across different geo-political contexts, (b) whether researchers who had applied the framework indicated needs for adaptations or further development, and (c) limitations and methodological pitfalls. This section reports on our findings from a systematic and extensive literature review that identified more than 900 papers and through consistent screening reviewed a total of 115 papers (until 01.07.2017). Table 1 is presenting an overview of the literature review.

A paramount lesson we derive from the literature review is that transition management is not anymore a 'Dutch' approach. It has been applied in many countries and in a variety of sectors. From the literature review we discovered that Transition Management approach is suitable for a cross-sectoral application and different geo-political contexts given that it has been applied in a great variety of sectors, in different countries across Europe and outside Europe with a flexible way (theoretical, heuristic and operational applications of the framework have been reported). This literature review as well as the entire book refute the statement of Chang et al. (2017) that transition management has not been tested in real life situations.

A frequent misconception in all the reviewed papers that did not apply transition management but only referenced it, is that many authors *confuse* the multi-level perspective developed by Geels (2005) with transition management developed by Loorbach (2007) (e.g. Marquardt et al. 2016 that refers to the multi-level perspective as transition management throughout the paper). We read in many papers the two being used and misused interchangeably. We hope that this book, together with the growing research on transition management as our literature review reveals, to outcast these misconceptions.

Since its inception, transition management has been practiced in a variety of policy fields, sectors and at different geographical scales in the Netherlands and beyond (Loorbach et al. 2015; Frantzeskaki et al. 2014b, 2017; Frantzeskaki and Shiroyama 2016). Initially, the transition (management) approach was mainly deployed in research and empirical experience at national levels and mainly

Table 1 Applications of transition management identified by a literature review (papers read and critically reviewed, published from 2001 to 2017)

Theoretical applications
Focus: Position and advocate transition management as a new governance approach for transformative change
Contribution: *Informing a new conceptualization of governance:*
*Theoretical contributions and questions of politics and agency
*Thematic focus on innovation and technology
*Multiple sectors, multiple scale applications
Scientific papers: Avelino (2009), Avelino and Grin (2017), Beers et al. (2010), Brones et al. (2017), de Bruijne et al. (2010), Campos et al. (2016), de Graaf and van der Brugge (2010), Eames and Egmose (2011), Elzen et al. (2004), Elzen and Wieczorek (2005), Foxon et al. (2009, 2010), Frantzeskaki and Loorbach (2010), Frantzeskaki et al. (2012), Genus and Coles (2008), Gössling et al. (2012), Grin (2012), Heiskanen et al. (2009), Hendriks and Grin (2007), Hendriks (2009), Holtz et al. (2008), Huétink (2010), Hurlbert et al. (2011), Kemp et al. (2005, 2007a, b), Kemp and Rotmans (2009), Lachman (2013), Loorbach (2010), Loorbach and Frantzeskaki (2012), Loorbach and Rotmans (2010), Loorbach et al. (2010, 2011), Meadowcroft (2009a, b), Nill and Kemp (2009), Raven et al. (2010), Rotmans and Loorbach (2010), Rotmans and Loorbach (2009), Nevens et al. (2013), Park et al. (2012), Scuttari et al. (2016), Shove and Walker (2007), Smith and Stirling (2010), Sondeijker et al. (2006), Späth and Rohracher (2010), Stephens and Graham (2010), Tukker and Butter (2007), Van der Brugge and van Raak (2007), Vergragt (2011), Vollenbroek (2002), Voß and Bornemann (2011), Voß et al. (2009), Walker and Shove (2007), Westley et al. (2011), and Wiek et al. (2006)
Operational applications
Focus: Adopt process tools (e.g. the transition arena) in setting up participatory processes for scenario development and strategy formulation
Contribution: *Setting up processes of co-imagining, and co-creating:*
*Process adoption for scenarios and strategy development
*Sectoral focus on water, energy and mobility
*Lessons to improve process methodology and context fitness are lacking
Scientific papers: Alkemade et al. (2009), Beers et al. (2010), Bos et al. (2013a, b), Brones et al. (2017), Cramer (2013), Eames and Egmose (2011), Frantzeskaki and Loorbach (2010), Frantzeskaki et al. (2012), Foxon et al. (2010), Gössling et al. (2012), Grin (2012), Hendriks and Grin (2007), Huétink (2010), Johansen and van den Bosch (2017), Loorbach and Rotmans (2010), Loorbach et al. (2010), Kemp et al. (2005, 2007a), Kemp and Rotmans (2009), Kern and Smith (2008), Köves et al. (2013), Lopes et al. (2012), Luiten and van Sandick (2007), Nevens et al. (2013), Raven et al. (2010), Rotmans and Loorbach (2009), Park et al. (2012), Schapke et al. (2017), Sondeijker et al. (2006), van der Brugge et al. (2005), van de Kerkhof and Wieczorek (2005), Wittmayer et al. (2016), van der Voorn et al. (2012), Vreugdenhil et al. (2012), and Wiek et al. (2006)
Heuristic applications
Focus: Employ it as a descriptive and/or diagnostic lens to understand and explain dynamics of on-going governance processes
Contribution: *Analysing and rethinking on-going or past governance processes:*
*Critical turn on transition management
*Strong focus on critical infrastructure (including health sector)
*National scale preference
Scientific Papers: Avelino (2009), Beers et al. (2010), Bos et al. (2013a, b), Bosman and Rotmans (2016), Caron-Flinterman et al. (2007), Cramer (2013), de Gooyert et al. (2016), de

(continued)

Table 1 (continued)

Graaf and van der Brugge (2010), Eames and Egmose (2011), Elzen et al. (2004), Foxon et al. (2010), Frantzeskaki and Loorbach (2010), Gössling et al. (2012), Grin (2012), Heiskanen et al. (2009), Hekkert et al. (2007), Hendriks and Grin (2007), Huétink (2010), Hurlbert et al. (2011), Kemp et al. (2005, 2007a, b), Kemp and Rotmans (2009), Kern and Howlett (2009), Kern and Smith (2008), Loorbach and Frantzeskaki (2012), Loorbach and Rotmans (2010), Loorbach et al. (2010, 2011), Luiten and van Sandick (2007), Meadowcroft (2009a, b), Monaghan (2009), Nevens et al. (2013), Nooteboom (2007), Park et al. (2012), Rotmans et al. (2001), Scarse and Smith (2009), Smith and Stirling (2010),Späth and Rohracher (2010), Stephens and Graham (2010), Tukker and Butter (2007), van der Brugge et al. (2005), van der Brugge and van Raak (2007), Vollenbroek (2002), Voß and Bornemann (2011), Vreugdenhil et al. (2012), Walker and Shove (2007), and Westley et al. (2011)

sectoral policy transformations (e.g. energy, water, mobility, building and living, material use) (Avelino 2011; Loorbach and Rotmans 2010; Verbong and Loorbach 2012). The approach was increasingly applied to regional and local scales (Roorda et al. 2014; Nevens et al. 2013; Nevens and Roorda 2013; Wittmayer et al. 2011, 2014; Frantzeskaki et al. 2010) and few to cases outside Europe (Poustie et al. 2016; Ferguson et al. 2013a, b; Silvestri and Frantzeskaki, chapter "Transition Manage ment for Local Sustainability: A Case Study from La Botija Protected Area, San Marcos de Colón, Honduras" this volume). However, applications have still largely focused empirically on Western affluent societies (Frantzeskaki and Shiroyama 2016).

The transition management framework has shown analytical robustness in different types of applications: *theoretical* applications position and advocate transition management as a new governance framework for transformative change, *operational* applications adopt transition management process tools (e.g. the transition arena, transition experiments) in setting up and realising participatory processes for scenario development and strategy formulation, and *heuristic* applications employ transition management as a descriptive and/or diagnostic lens to understand and explain dynamics of on-going governance processes (Frantzeskaki et al. 2014b) (see Table 1).

Theoretical and heuristic applications of transition management take up an interpretative way to analyse and examine past or contemporary transition dynamics. As such they contribute to the understanding of the governance *of* sustainability transitions (Frantzeskaki and Shiroyama 2016). Governance of sustainability transitions refers to issues that are relevant to the debate about types of interventions and how they play out in changing the direction and the pace of sustainability transitions. The operational applications of transition management provide knowledge for explaining the 'successes' or fitness of transition management as a process method that is instrumental to the governance *for* sustainability transitions. Governance for sustainability transitions is the prescriptive dimension: the methods, instruments and frameworks that are promoting, enabling or triggering sustainability transitions. Despite its contribution to governance for and of sustainability transitions, empirical cases open-up many new questions including the role of cultural diversity, actor roles and learning processes (ibid).

Reviewing transition management applications in developed (Western-) European contexts Loorbach et al. (2015) find that the underlying guiding principles and methodological ingredients provide a basic set to develop context sensitive implementations. Transition management showed to add value by triggering explorations of more fundamental change processes and empowering social innovations. This underlines the need for departing from blueprint planning or standard recipes. The authors also acknowledge, however, that transition management is "so far not achieving the aspired large scale systemic changes (Loorbach 2014)" (ibid: 25). They identify several challenges in relation to governance interventions in transitions more generally and transition management specifically, including the question of inclusivity (of e.g. marginalised perspectives), facilitation techniques for social learning and capacity building, and regime persistence. It points to current research directions for transition management, such as explorations of institutional change, top-down strategies, new forms of power, the role of scale and the psychology of transitions.

Over the years, transition management has also been criticised, sprouting a fruitful dialogue between researchers. Transition management invoked a fundamental debate around the question about the extent transitions can actually be managed (Walker and Shove 2007). As a governance framework for transitions that builds on complexity theory (Rotmans et al. 2001; Rotmans and Loorbach 2009), transition management is also often debated as serving a new 'dogma' for incrementalism (Shove and Walker 2007; Frantzeskaki et al. 2012b), while also being at risk of tautologically entrapment in 'explaining complexity' and 'enabling complexity' rather than navigating societal complexity by creating knowledge for interventions that can trigger transitions to sustainability (Vasileiadou and Safarzyńska 2010). Are concerned by the 'double role' of researchers in transition management processes. Against the background of these and other criticisms a productive scientific dialogue emerged addressing these challenges to the methodology (Jhagroe and Loorbach 2015; Avelino 2011; Jhagroe and Frantzeskaki 2012; Wittmayer et al. 2014, Wittmayer and Schäpke 2014).

These criticisms to transition management are still problematic considering that, thus far, the promises of transition management are not rigorously evaluated by drawing on empirical examples. One reason is the still thin empirical application of transition management thus far despite its uptake as a heuristic and theoretical model from a growing number of researchers. Another one is that transition management is still used in an exploratory manner. It is suggested that transition management would greatly benefit from linking to debates on other theories concerned with governance for sustainability, such as institutional theories, policy studies (Kern and Howlett 2009), adaptive management and resilience scholarship (Olsson et al. 2014).

From the literature review and the contributions in this book, we derive overall ten directions for future research and advancement of transition management, that are summarized in Box 1.

Box 1: Future Directions for Developing and Researching Transition Management

#1: Broaden the thematic scope to include environmental, social and economic sustainability dimensions

#2: New process orientations including destabilization and disempowerment next to empowerment and stabilisation

#3: Develop transition management as a visionary proactive approach

#4: Consider the role of agency and the role of individuals in transition processes and in agenda realisation

#5: Consider issues of conflict, politics, power, and democratic legitimacy of transition management processes and outcomes

#6: Theoretically systematize the principles and methodologically the phases of transition management

#7: Identify methods to connect every phase of transition management with on-going urban planning processes and practices beyond communication tools

#8: Consider the spatial perspective when drafting visions, pathways and agendas with transition management in cities

#9: Consider how transition management steers into, or steers clear from contemporary urban politics

#10: Strengthen co-creation and participatory nature of transition management in cities

Looking at the applications and the suitability of transition management for urban planning and governance, the book contributions provide valuable insights to rethink the thematic scope, the normative orientation of the transformative actions/solutions and agenda upon formulation and the response approach (reactive versus proactive) that resonates transition management as a planning approach.

Future Direction #1: Broaden the thematic scope to include environmental, social and economic sustainability dimensions: Transition management is initially applied in a variety of sectors as our literature reviewed shows. At the same time, this volume offers a collection of cases that predominantly apply transition management in environmental sustainability issues: resource management, climate change, sustainable energy. The 'themes' or domains of application of transition management has broadened over time, including themes of social sustainability as social cohesion and inclusion (see chapter "Transition Management in Urban Neighbourhoods: The Case of Carnisse, Rotterdam, The Netherlands", this volume) as well as issues of place making (see chapter "Translating Transitions Thinking and Transition Management into the City Planning World", this volume). With cities broaden their sustainability agendas, it is important to consider social and environmental topics as equally important. Wachsmuth et al. (2016 p. 393) point to this need that "policy makers should treat social equity and ecological effectiveness as mutually reinforcing dynamics in urban sustainability. They should bring the widest range of social movements to the table and see these groups' demands – such as revitalizing rent regulation and public housing – as central. This would entail more frequent meetings of larger groups of stakeholders and different metrics of policy success." It therefore remains for future research to examine the applicability and suitability of transition management for socio-economically persistent problems like urban poverty, urban degeneration and deprivation and gentrification.

Future Direction #2: New process orientations including destabilization and disempowerment next to empowerment and stabilisation: Transition management is a governance experimentation approach (Loorbach et al. 2015). As such it has been set to facilitate dialogues with innovative thinkers and change agents in the cities across sectors that result in strategic and operational interventions that disrupt conventional planning pathways (see chapters "Transition Management in Urban Neighbourhoods: The Case of Carnisse, Rotterdam, The Netherlands", "Translating Transitions Thinking and Transition Management into the City Planning World", and "Transition Management in Urban China: Co-creation of New Homes or a Monopoly Game?", this volume). As such, it is set to 'stabilise' or 'empower' innovative solutions. It remains to be seen whether transition management as a governance experimentation approach can also 'destabilise' or 'disempower' non-sustainable and non-desirable solutions in ways that may prove valuable in preparing cities to deal with discontent, unease and disappointment. Think for example the new deindustrialization that major cities will undergo in the future due to pacing up of the digitalization of the market, automation and new forms of green and sustainable economies that will destabilize current economies (e.g. implying that fossil fuel based jobs, service economy jobs may be lost due to digitalization and automatization of services).

Future Direction #3: Develop transition management as a visionary proactive approach: Transition management is seen as a response to lagging, or mis-fit approaches for public engagement and strategic urban planning (chapters "Urban Planning and Transition Management: Rationalities, Instruments and Dialectics", "Empowering Actors in Transition Management in and for Cities", "Transition Management in Urban Neighbourhoods: The Case of Carnisse, Rotterdam, The Netherlands", "Transition Management for Local Sustainability: A Case Study from La Botija Protected Area, San Marcos de Colón, Honduras", and "Transition Management in Urban China: Co-creation of New Homes or a Monopoly Game?" in this volume). In these cases, transition management is applied as in response, or even as a 'reaction' to an emerging urban problem. This may come to undermine the strength of transition management in facilitating the co-creation of transformative visions as an outcome of the orientation phase. Visions can be powerful narrative and visual devices to attract action and to orient experimentation. They can play a role in unlocking perceptions of planners and citizens alike on what is possible and desirable for urban landscapes (Frantzeskaki and Tefrati 2016; think of Smart Cities visions cf. Vanolo 2014). Transition management can be an approach to open dialectic and governance innovation spaces for urban planning in a proactive manner, to seize and to create opportunities for more visionary urban planning.

Taking a closer look at the contributions of this book, we deduce that there are some additional avenues for theoretical and methodological improvement of transition management next to those recommended from our literature review. We summarize them here to be taken into consideration for future advancement of transition management. Some of these implications are applicable also for tools, methods and approaches for urban sustainability transitions' governance as well, not only for transition management. Additionally, these implications and suggestions for future research are also complemented by the literature review findings. There are two major points (Future directions #4 and #5) addressed by researchers who had applied transition management on future directions for development that our literature review brought to surface: the role of agency and the consideration of politics and power dynamics.

Future Direction #4: Consider the role of agency and the role of individuals in transition processes and in agenda realisation: On a more conceptual level, transition management lacks a deeper understanding of individuals and their agency. There is insufficient knowledge on the individuals engaging for example in transition experiments and what drives them, on different types of roles needed in transition processes, and on changes occurring at the level of the participating individual in transition management processes (Farla et al. 2012). In the same vein, Holscher et al. (chapter "Empowering Actors in Transition Management in and for Cities", this volume) introduce frameworks and concepts to understand actors and roles in transition (management) processes, they point out that yet "there is a demand for being more specific on the types of actors and roles (needed, desired) in a particular transition management context – and by 'whom' they are needed or desired."

Future Direction #5: Consider issues of conflict, politics, power, and democratic legitimacy of transition management processes and outcomes: One of the prevailing criticisms on transition management is that is ignoring or being naïve to politics (Avelino and Grin 2017). It is also criticised for simplifying agency dynamics and neglecting issues of conflict, democratic legitimacy and power (Voß et al. 2009; Shove and Walker 2007; Kern and Smith 2008; Smith and Stirling 2010; Hendriks 2009; Meadowcroft 2009a). Specifically, Kenis et al. (2016, p. 570) mention that "transition management fails to fully acknowledge power relations, radical pluralism and the possible constitutive role of conflict in society." In the same vein, Jhagroe (chapter "Urban Transition Management as a Democratic Practice? The Case of Rotterdam's Waterfront Regeneration", this volume) addresses that "transition management navigates between urban regimes and institutions on the one hand and alternative urbanisation visions and practices on the other hand" and proposes a democratic transition management version that "adds insights from critical urban research, drawing attention to the ways in which hierarchical power relations, uneven spatial developments and forms of resistance are part and parcel of urban transformations." It therefore remains to extend either the phases of transition management or enrich it methodologically in a way that can incorporate politics and power dynamics, and to address the political dimension of all outcomes of the transition management process itself (e.g. the vision, pathways, agenda).

Future Direction #6: Theoretically systematize the principles and methodologically the phases of transition management: From the literature review on transition management applications we found that generally there is no feedback loop that questions the fundamental assumptions underlying transition management despite a rich discourse and criticism due to underdeveloped methodological guidelines for the operational applications of transition management. New concepts are only introduced to a limited extent. From the contributions of this book, we also found that the same need for further systematization of the phases and associated steps is required. Holscher et al. (chapter "Empowering Actors in Transition Management in and for Cities" in this volume), presented a new framework for systematizing actor selection and actor mapping: the Multi-Actor Perspective. Maas et al. (chapter "Starting Up Transition Management: A Closer View on the Systems Analysis and How It Initiated Transformative Thinking in Ghent and Aberdeen Cities" in this volume), presented a systematized systems' approach for the first phase of transition management and how it was applied in two European cities.

Future Direction #7: Identify methods to connect every phase of transition management with on-going urban planning processes and practices beyond communication tools: Many of the contributions in this book pointed to the opportunities that transition management can bring to urban planning and governance. They all also point to the fact that there is a gap in how to establish a constructive and coordinated flow of knowledge and strategic information or ideas co-created in transition management processes to urban planning processes. A specific direction may be how to coordinate actions and innovations happening outside the negotiation space that transition management arenas/processes create and the innovations empowered through transition management. Coordination and cooperation in governance for urban sustainability is rather important for implementing sustainability agendas in cities (Smith and Wiek 2012). Future research needs to consider how transition management processes can also connect to place transformations (Ernst et al. 2016) and position them to the bigger systemic transformation through co-creation and co-production processes facilitated with the transition management methodology. Wolfram (chapter "Urban Planning and Transition Management: Rationalities, Instruments and Dialectics", this volume) points at this with very specific critical reflections: First he argues that "transition management in turn needs to better resonate with the existing *spatial* patterns of urban governance and planning practices" and then he suggests that "reconnecting the regulatory and visionary modes of urban planning through a differentiated foresight approach, and especially adding novel techniques for reflexivity and social learning that have remained largely absent in urban planning". Hölscher and Wittmayer (chapter "A German Experience: The Challenges of Mediating 'Ideal-Type' Transition Manage ment in Ludwigsburg", this volume) point at the same need to methodologically connect transition management processes and outcomes with urban planning processes so as to overcome the 'isolation' or even de-legitimization of the produced vision, pathways and experiments that can steer planning decisions to more socially accepted and sustainability oriented outcomes. Schwarz (chapter "A Transition Management Approach for Shrinking Cities in the United States", this volume) further points to how transition management can complement existing urban planning approaches for shrinking cities and proposes ways for how every phase of transition management can contribute to facilitating dialogues for change of current situations and opportunity seeking rather than incremental adaptation plans that are proposed with planning approaches.

Future Direction #8: Consider the spatial perspective when drafting visions, pathways and agendas with transition management in cities: Visions for transformative change that are to be formulated with the transition management process methodology require to go beyond 'systems' and include 'places'; or simply to be place relevant and specific. This is an added consideration discussed and proposed by Wittmayer et al. (chapter "Transition Management in Urban Neighbourhoods: The Case of Carnisse, Rotterdam, The Netherlands", this volume) when discussing how a neighborhood-level transition management process contributed to a place transition and created conditions for local transformative agency to be enacted and mobilized. At the same time, visions can bring in the place dimension by encountering images of future personalized experiences in places, like the personalized future lives of Ludwiga and Ludwiger presented by Hoelscher and Wittmayer (chapter "A German Experience: The Challenges of Mediating 'Ideal-Type' Tran sition Management in Ludwigsburg", this volume). In a similar argument, Lachman et al. (chapter "Context-Driven Transition Management as a Necessary Vehicle for Sustainable Urban Futures in Suriname", this volume) propose to make transition management more 'context-driven', specifically highlighting that: "specific characteristics from the context (which can be on a macro or micro level, depending on the scale and impact of the transition) need to be heeded in such a way that the functions of each aspect of transition management can come to full fruition on all execution levels". Additionally, a spatial perspective will further enrich the understanding on the politics of space (Healey 2006), and of the divides at territorial level (Rauws and de Roo 2011). This will require new methodological steps to integrate spatial perspectives into the transformative agenda building phase of transition management.

Future Direction #9: Consider how transition management steers into, or steers clear from contemporary urban politics: As a governance experimentation process that invites change agents (frontrunners as innovative thinkers and practitioners across sectors) to co-create visions, pathways of solutions and actions and experiments, transition management plays out with contemporary power dynamics and establishments in cities. Mai (chapter "Transition Management in Urban China: Co-creation of New Homes or a Monopoly Game?", this volume) discusses that it remains to be examined how transition management cab "ensure an equitable process and rebalance the power dynamics between the state and society given the impacts of system disequilibrium". From Mai's contribution in this book, we also see how state-led visions for urban futures can be powerful tools but need to be democratized. This is the mirroring opposite of what co-created visions with transition management require: a process of legitimization to complement their social value from the co-creation production. A deeper understanding of the strength and lastingness of networks, relations and partnerships established during the transition arena and how they can contribute to realizing the transition agenda, experiments and facilitate broader systemic transformations in cities is required that considers the evolution of roles of single and network actors over time (inspired by Holscher et al., chapter "Empowering Actors in Transition Management in and for Cities", this volume). Jhagroe (chapter "Urban Transition Management as a Dem ocratic Practice? The Case of Rotterdam's Waterfront Regeneration", this volume) similarly reflects on the legitimacy of transition management interventions as alternative democratic processes. Again, it becomes pertinent to examine the (characteristics of the) governance context in which transition management can operate in coordinated harmony with urban planning or even, regional and national planning and governance processes.

Future Direction #10: Strengthen co-creation and participatory nature of transition management in cities: It is prominently recognised by all contributing chapters of the book that transition management's strength is the participatory co-creation process on making transformative strategic elements: visions, pathways of actions and agendas that can be further operationalised into tactical experiments. In all chapters, this aspect of bringing co-creation, opening to participation with specific facilitation steps and phases with identified scope and outcomes make transition management stand out as a governance experimentation approach that fits in diverse contexts (see chapters "Empowering Actors in Transition Manage ment in and for Cities", "Starting Up Transition Management: A Closer View on the Systems Analysis and How It Initiated Transformative Thinking in Ghent and Aberdeen Cities", "Transition Management in Urban Neighbourhoods: The Case of Carnisse, Rotterdam, The Netherlands", "A German Experience: The Challenges of Mediating 'Ideal-Type' Transition Management in Ludwigsburg", and "Transition Management for Local Sustainability a Case Study from La Botija Protected Area, San Marcos De Colón, Honduras") and can bring new ideas and unveil new innovations for urban planning (see chapters "Translating Transitions Thinking and Transition Management into the City Planning World", "Transition Manage ment in Urban China: Co-creation of New Homes or a Monopoly Game?", "A Transition Management Approach for Shrinking Cities in the United States", and "Context-Driven Transition Management as a Necessary Vehicle for Sustainable Urban Futures in Suriname"). At the same time, transition management can develop into a governance approach that can bring together urban change agents from civil society and other sectors on equal grounds, to navigate between 'facilitation for producing public services' (Grotenbreg and van Buuren 2017) and 'driving innovation in cities' (Frantzeskaki et al. 2016). It is therefore important to consider how to further strengthen the co-creation element of transition management jointly with all the other conceptual and methodological advancements proposed above.

Outline of the Book

After providing an analysis of the literature review of the applications of transition management and a guidance for future research directions, we want to present you the outline of the book. We feel that knowing not only the applications of transition management that frame this book but also where the research of transition management is heading and requires advancement, you as readers will feel enticed to dive into the chapters of the book that showcase different applications of transition management.

The book includes four different parts: Part I introduces the urban context as a context for transition management interventions and sustainability transitions, the theoretical grounds of transitions' thinking and transition management with a critical reflection on its applications and limitations. Part II includes applications of transition management in different cities. In this part of the book, we gain an insight view from researchers that applied transition management, designed the process to fit the socio-political, cultural and ecological context of the city of application and analysed the results of the process using different lenses. This part includes transition management cases from Europe (Aberdeen/UK, Ghent/Belgium, Montreuil/France, Rotterdam/The Netherlands, Ludwigsburg/Germany) and Latin America (Honduras). Part III presents explorations for what transition management could mean in new contexts. This part includes a mix of practitioners' views and undertakings of what transition management can contribute to urban planning and researchers' critical reflections on the applicability of transition management as analytical and heuristic model for on-going urban planning processes and programs. This part includes cases from Sydney/Australia, Shenzhen/China, the Cleveland, Ohio and Flint, Michigan/United States, Surinam/Surinam, and Rotterdam/The Netherlands. Returning from our world tour of Transition Management, we will look at the thorny issues of power and agency, before drawing lessons for initiating and steering radical change at the urban level. We hope that you will enjoy this journey into the world of urban sustainability transitions and the practice of Transition Management in cities. Our aim is to spark your personal interest and enrich your (future) professional practice, while giving you a sense that the future is not just doom and gloom. Transformative change is both real and possible; we are not condemned to reenacting today's problems. Part IV of the book responds to that. Part IV includes a synthesis chapter on the impacts of transition management taking a capacities' framework into consideration for examining the governance outcomes, outputs and limitations of transition management applications in cities.

Last but not least, the book includes an appendix chapter with exercises, exam questions, group assignments and briefs on dialogue assignments that have been developed by the Transition Management Research team of the Dutch Research Institute For Transitions with Erasmus University Rotterdam, the Netherlands (www.drift.eur.nl) with more than 15 years of experience. The group assignments and exercises have been effectively used in lectures, trainings, courses and workshop sessions of transition management and urban sustainability transitions.

We hope that the richness of the very different case studies provides inspiration and practical insights to young planners on how to create sustainable urban futures in collaboration with citizens. This book is unique, because it is courageous in reflecting upon novel possibilities for reshaping the relationships between actors in the city, it offers food for thought and welcomes criticism. At the same time the book introduces additional lenses to understand the bigger picture that co-creation dynamics play on: the lens of power, examining empowerment and disempowerment dynamics, the lens of self-organisation, examining bottom-up innovation dynamics and the lens of governance, examining how agenda setting processes can link up and leak to broader governance processes. In this way, the broader picture of dynamics present in urban context is also mapped.

We hereby present what every part of the book includes and welcome readers of our book to dive in this unique set of case studies and lenses of transition management.

Part I: Introduction of Transition Thinking in Urban Context and Theoretical Grounds of Transition Management in and for Cities

Frantzeskaki, et al. – Understanding the urban context and its challenges. This chapter introduces a new understanding of the urban context and elaborates on its challenges, from a transitions' governance perspective. It positions cities in the landscape of change and persistent problems, looks at inherent and innovation characteristics of cities and relates to urban sustainability debate. We walk the reader from the stepping stones of theoretical foundations of urban change to what it means for practical actions in cities. In this way sets the scene on 'why cities' and which type of challenges are at hand for new governance approaches like transition management to be relevant and a fit approach to urban governance and planning, complementing strategic environmental planning approaches. As such, this chapter offers an insight on how multi-level governance of cities relates and requires participatory approaches for sustainability.

Frantzeskaki et al. – Introducing sustainability transitions' thinking in urban contexts. This chapter describes the basics of transitions thinking, its rationale and relation to sustainability transitions. We outline the fundamental conceptual models and frameworks (multi-phase model, multi-level perspective, multi-path, power and agency) and make a quick introduction to the relation of the sustainability transitions field to environmental governance.

Wittmayer et al. – Transition Management: Guiding Principles and Applications. This chapter introduces transition management as a way of influencing sustainability transitions, also reflecting on its strengths and limitations as a framework for strategic environmental planning in cities. The chapter includes a description of the guiding principles of transition management, their translation into a governance framework and the operationalization thereof into process guidelines. For each phase of the process, rationale and methods as well as outcomes are outlined.

Wolfram – Urban planning and transition management: Rationalities, instruments and dialectics. This chapter juxtaposes the transition management approach with the rationalities and instruments of urban planning. It introduces us to urban planning considering the historic evolution and current characteristics of mainstream urban planning in theory and practice, and then it elaborates on the key challenges and opportunities for engaging with transition management. The chapter takes a reflective but very well theoretically grounded view on what transition management offers to urban planning, and what are the merits and caveats of its application in cities. The discussion underlines the added value but also incompatibility of both approaches, and thus the need to create synergies and mutually compensate for critical deficits to effectively perform urban sustainability transitions.

Part II: Applications of Transition Management in Cities

Holscher et al. – Empowering actors in transition management in and for cities. The multi-actor nature of urban sustainability transitions challenges existing social fabrics and local governance settings, and raises questions about who takes decisions, with which agenda and to what end. A closer look is required for understanding who the actors involved in transition management processes are, what their roles and shifting roles are due to their engagement, what the responsibilities and relations of those actors and their affiliated networks entail. The chapter presents the notions of empowerment and disempowerment in view of transition arena processes, with the aim to support actor selection and analysis of their roles and the evolution of those thereof. The methodological propositions and reflections are further enriched with insights from empirical experiences with transition management in the cities of Ghent, in Belgium and Montreuil, in France.

Maas et al. – Starting up transition management: A closer look on the system analysis and how it initiated transformative thinking in Ghent and Aberdeen cities. Maas et al. present methodological steps for conducting a systems analysis and reflect on how these steps were adapted during the transition management processes in Ghent, Belgium, and Aberdeen, UK. A systems analysis enriches the start of transition management by stimulating holistic and integrative thinking about a cities' problem domains and their interconnections, providing a common information base and pin-pointing areas and actors in the system both contributing to problems of unsustainability and solutions. In Ghent and Aberdeen, the systems analysis resulted in an overview of interconnected challenges and opportunities in both cities regarding their energy use, CO_2 emissions reductions and issues of sustainability and liveability. This revealed institutional barriers to sustainable development, highlighted opportunities for co-benefits and innovative actors and initiatives and broadened problem perceptions and integrated different viewpoints towards mutual understandings. While a systems analysis is thus critical for thinking about envisioned sustainability transitions in the later process phases, the authors emphasise that the methodology is not a cookbook and needs to be flexibly adapted according to the needs of a specific context and the involved actors.

Wittmayer et al. – Sustainability transitions in urban neighbourhoods: The case of Carnisse, Rotterdam, The Netherlands. The case presented by Wittmayer et al. marks a socially disadvantaged community in the South of Rotterdam, the Netherlands: Carnisse. Transition management was applied in Carnisse in the context of decreasing public service provisions, closing of public facilities, high unemployment and crumbling social cohesion to facilitate the self-organisation of inhabitants to address persistent sustainability problems. Wittmayer et al. illustrate how the process implementation has been influenced by political, institutional and emotional factors related to the place-specific characteristics of urban neighbourhood transitions. While urban neighbourhoods mark opportunities for self-organisation – through for example emotional connectedness – the political playing field can provide support, or restrain it. The transition management application in Carnisse shows several advantages of applying transition management on the scale of an urban community. It provided a new perspective for the development of that community, increased social capital, supported reflexivity on roles and own behaviours and revealed opportunities for self-organisation.

Hölscher and Wittmayer – A German experience: the challenges of mediating 'ideal-type' transition management in Ludwigsburg. Hölscher and Wittmayer review the transition management process phases and results in Ludwigsburg, Germany, from the perspective of how the 'ideal-type' process methodology gets mediated with respect to context-specific characteristics. Ludwigsburg represents a context that, before the transition management intervention, had already well-developed sustainability strategies and programmes and implemented a number of participatory processes to collect citizen input for strategic planning. The authors highlight several recurring challenges that occur from the need to differentiate transition management and position its added value to addressing the problems of a city. In Ludwigsburg, there was no perceived need for searching for alternative approaches to participation. In combination with a limited space to engage with the transitions thinking approach underpinning transition management this contributed to a framing of the process as resembling the other participatory efforts of the city. This led to a focus on the process on collecting new input to the city's strategies rather than (also) inspiring a co-creative process of jointly thinking and acting for a sustainable future of the city.

Silvestri and Frantzeskaki – Transition Management for local sustainability- A case study from La Botija protected area, San Marcos de Colón, Honduras. This chapter presents two methods that we employed to facilitate the exploration for environmental stewardship in the La Botija area: Transition management was the transdisciplinary methodology that connected a co-created sustainability vision with the co-development and mobilization of local action. The ecosystem services framework was employed to systematize knowledge about resource use and ecosystem degradation and identify links between malpractices and ecosystems' deterioration. The transition management process activated participants to realise the co-created transition agenda by forging new partnerships, enabling self-organisation in community groups and empowering existing community groups. A highlighted learning outcome is the recognised need to develop own knowledge and community capacity to overcome local challenges deviating from the dependence on external aid.

Part III: What Transition Management Could Offer in Different Contexts

Walsh – Translating transitions thinking and transition management into the city planning world. This chapter needs to be read jointly with chapter "Urban Planning and Transition Management: Rationalities, Instruments and Dialectics" – Wolfram, this volume, so as to contrast the theoretical take on transition management versus strategic urban planning with the in-practice integration of transition management with urban planning in Australia that chapter "Translating Transitions Thinking and Transition Management into the City Planning World" offers. What is unique in this chapter is the practitioner's view and experience in how to translate the principles of transition management and transitions' thinking into more operational 'forms' of planning. Specifically, this chapter raises the question of whether, and how, transition management ideas might more directly enter and influence this "turbulent" reform space. Sydney provides the case-setting for the introduction of transition management, and related conceptions, to a recent major planning system reform project for that city. Working from a theoretical meshing of planning, relational-institutional and transitions studies, a new process model for planning system reform projects is introduced: the Planning System Transition (PST) framework. This framework's development and testing, through empirical research, is explained.

Mai – Transition management in urban China: Co-creation of new homes or a monopoly game? This chapter is an interpretative and analytical take on what transition management applied as a heuristic can offer to urban planning of metropolitan coastal cities in China, including Shenzhen, Guangzhou, Ningbo, Beijing, and Tianjini. This chapter places empirical applications and review of established Transition Management principles in the metropolitan areas in coastal China. Based on content analysis and semi-structured interviews, an overview of state envisioning processes of urban sustainability is first presented as a conflicting context for the disequilibria observed in the environmental governance system across the metropolitan areas. The interactions of observed disequilibria grounded locally potentially weaken the state-led envisioning processes. With the presence of challenges in state-led transitions, the chapter proceeds to case studies in Shenzhen, the city that is practicing the state-led sustainability visions, to analyse how local dynamics of resource and organisational landscape are integrated in the transition process.

Schwartz – A transition management approach for shrinking cities in the United States. This chapter applies transition management approach as a heuristic to identify areas of possible transformative change and opportunities for new types of governance using the transition management tenets and phases. It comes with the motivation that transition management can be a valuable approach to shrinking cities in the United States. Specifically, the principles of Transition Management may prove valuable to planning and policymaking efforts in the US, particularly in the shrinking cities of the Rustbelt region. In these cities, population loss and deindustrialization pose daunting challenges, but also offer robust and multi-faceted opportunities to pursue more sustainable urban futures. Established planning practices in the US are primarily geared toward managing growth. These practices are often ineffective in cities that are not growing. New, more flexible approaches to planning and community engagement are needed. The principles of Transition Management may prove especially relevant for shrinking cities in the US because the overarching idea that transition management is not about adapting or correcting current conditions, but rather a focus on transformative change.

Lachman et al. – Context-driven transition management as a necessary vehicle for sustainable urban futures in Suriname. This chapter presents how transition management can be applied to inform sustainable urban development in a fast urbanizing city, Suriname while considering climate change and housing challenges simultaneously. The analysis of the urbanization context and its challenges and the insight view of the existing planning processes posit this chapter as a special contribution to our book. Readers should consider this chapter as an analysts' view on possibilities for methodological advancement of transition management to fit to a very challenging context. Authors summarise their contribution as follows: The capital of Suriname has experienced enormous growth over the past decennia. The decline of industries in other urban settlement have resulted in migration to the capital along with population growth and ongoing country immigration. With the absence of sound urban and spatial planning, the result is urban sprawl where increasing population and traffic density create ongoing specialization of services in the capital with continuously decreasing action radius; there is a tendency where people opt to live further from the city core but which makes the utilization of services in the city's core cost-prohibitive. The answer to this is the creation of new urban settlements which build on new industries with increasing returns. However, this implies a break from existing paradigms with respect to strategizing, policy, planning, tourism, logistics, urbanism, etc. This requires a transition for which management has been broken down in several steps in literature. However, transitions tend to differ along geographical parameters; for instance, regimes tend to be relatively stronger in developing countries. Thus, transition management needs to fit the context which requires assessment tools such as Quinn's culture model and the STEEPL and OPTIMISM frameworks to anticipate pitfalls and opportunities during the transition.

Jhagroe – Urban Transition Management as a democratic practice? The case of Rotterdam's waterfront regeneration. This chapter examines the democratic aspects and challenges associated with urban transition management. It does so at a conceptual level and critically reflecting on the application of transition management in the context Rotterdam's sustainable waterfront regeneration. As such, it aims to shed a more reflexive light on the increasing academic and public interest in transition management.

Part IV: Reflecting on the Impacts of Transition Management in Cities

Holscher – So what? Transition management as a transformative approach to support governance capacities of cities. This chapter synthesises the insights from the various book contributions to critically reflect on what transition management offers when applied in and for cities. In this book, transition management has been suggested as a transformative governance approach in and for cities. Transition management seeks to provide impulses for systemic change by creating spaces for developing innovative ideas and practices and for empowering actors to develop transition experiments. It also holds the potential to fundamentally challenge existing planning and governance systems by aligning a diversity of actors, networks and initiatives under a shared and long-term sustainability vision and questioning existing roles and responsibilities of actors. We explore how transition management thus contributes to building *transformative capacity for systemic change* and *orchestrating capacity for the coordination of self-organisation.* On the basis of the book contributions we discuss the potentials, opportunities and challenges of transition management as a prescriptive and transformative governance approach. We conclude by identifying implications for transition management to strengthen governance capacities for navigating sustainability transitions in cities.

What This Book Is About and How to Read It

This book illustrates how to design and analyse participatory co-creation processes of transformative agendas in cities with real-life cases from Europe (Aberdeen, Ghent, Ludwigsburg, Rotterdam), and Latin America (Honduras). It is the first book that includes international case studies on transition management. In total, the book is not only a very practical guide for understanding (the rationale of) transition management in urban contexts, but also for designing and implementing transition management processes. The aim is to be a primer on the fundamental theoretical propositions as well as on the applications of transition management in different urban contexts to show the internal dynamics of a transformative agenda co-creation process with policy and social urban entrepreneurs. With covering both theoretical as well as empirical grounds with an inside-view and analysis of the negotiation and knowledge sharing processes, the novice planners and urban professionals will get familiar with the methodology on transformative agenda setting that employs tacit and scientific knowledge while fostering transformative capacity for sustainable cities.

It is the first book that comprehensively yet concisely introduces the transition management approach in urban contexts and includes worldwide case studies. The book covers the fundamental theoretical grounds of transformative science for policy and planning (the sustainability transitions approach in its participatory form that is transition management at the current state of the art of the sustainability transitions' studies) and shows the application of the transition management in real-life cases. What is unique is that the author teams of each chapter have been action researchers and analysts of the transition management cases and in this way the chapters have an insider's knowledge take on the application and the dynamics in place. This does not only cover the theoretical aspects of transition management and how it compares to strategic urban planning, but in particular the link of theory to practice. Practical case studies will present applications of transition management in very different localities, analyse the results, derive lessons learned and present opportunities and challenges. Based on the experience of DRIFT in the implementation of transition management, the book considers novelties that have been integrated to the approach over the past years. Readers will be surprised from what a complexity can be drawn out of urban sustainability governance, but, at the same time, how solutions are hidden into simplicity.

The book is particularly targeted towards achieving learning outcomes that prepare the next generation of urban planners and professionals to especially acquire a process-content thinking and to reflect on how to navigate societal complexity while mobilising innovative and transformative societal potential towards action for sustainability. This means that the book aims to inform educational pursuits and build skills, include exercises, systematize vocabulary and methods implicated in urban governance and transition management, give a glimpse into other cultures and places and expand horizons, build the self-esteem of urban planners and professionals, improve creativity in finding real-time solutions, facilitate exchange of best practices, and decrease mistakes to be done in the future. The reader will benefit from manifold practical examples, which illustrate the process application. After reading this book, the reader will have come in close contact with a variety of possible aspects of transition management in urban context and will be in a position to contribute in a constructive way and to offer suggestions that might shape and actively contribute to sustainable and livable urban futures.

Acknowledgements This book has been made possible with the research and funding support of two European research projects: EU FP7 funded IMPRESSIONS (www.impressions-project.eu) and Erasmus Mundus funded SUSTAIN (www.sustainedu.com). The majority of the chapters of Part I and Part II have been used as educational material for short-courses, an e-learning course, master courses for urban professionals realized in 2014–2017 by the Dutch Research Institute For Transitions, Erasmus University Rotterdam, The Netherlands, and lectures for transition management in cities given in Central European University, Budapest/Hungary (2014–2017), LUCSUS Sustainability Master Program, Lund University, Sweden (2014–2015), Environmental Management Master Program in Freiburg University, Germany (2015) and in Climate Governance Lecture Series with the Institute of Housing Studies (IHS), Erasmus University Rotterdam, The Netherlands (2014–2016). The comments, feedback and reflections after their 'use' as educational material resulted in the revised and improved versions that you find in this book. For this, we are thankful to all students and professionals as well as to all fellow lecturers in these short-courses, e-learning module, master courses and master classes for their critical and constructive feedbacks. Additionally, the book editors want to acknowledge the following research projects that inspired them, informed the conception of the book and the chapters they are involved in as research foundations of the book: MUSIC, InContext, GUST, RESILIENT EUROPE, KSI System Transitions and Melbourne's Transition to Water Sensitive Cities pilot project.

References and Further Readings

Aagaard Thuesen A (2013) Experiencing multi-level meta-governance. Local Gov Stud 39 (4):600–623. https://doi.org/10.1080/03003930.2012.755463

Abbott KW, Genschel P, Snidal D, Zangl B (eds) (2015) International organizations as orchestrators. Cambridge University Press, Cambridge

Adger WN (2003) Social capital, collective action and adaptation to climate change. Econ Geogr 79:387–404

Adger WN, Jordan A (2009) Sustainability: exploring the processes and outcomes of governance. In: Adger WN, Jordan A (eds) Governing Sustainability. Cambridge University Press, Cambridge

Adger WN, Vincent K (2005) Uncertainty in adaptive capacity, external geophysics. Clim Environ 337(4):399–410. https://doi.org/10.1016/j.crte.2004.11.004

Adger WN, Brooks N, Bentham G, Agnew M, Eriksen S (2004) New indicators of vulnerability and adaptive capacity, 122, Tyndall Centre for Climate Change Research Norwich, https://ipcc-wg2.gov/AR5-tools/author/njlite/data/7785.pdf

Adger WN, Arnell NW, Tompkins EL (2005) Successful adaptation to climate change across scales. Glob Environ Chang – Hum Policy Dimens 15:77–86

Adger WN, Dessai S, Goulden M, Hulme M, Lorenzoni I, Nelson DR, Naess LO, Wolf J, Wreford A (2009) Are there social limits to adaptation to climate change? Clim Chang 93:335–354. https://doi.org/10.1007/s10584-008-9520-z

Alfasi N, Portugali J (2004) Planning just-in-time versus planning just-in-case. Cities 21(1):29–39

Alkemade F, Frenken K, Hekkert MP, Schwoon M (2009) A complex systems methodology to transition management. J Evol Econ 19(4):527–543

Anderies JM, Folke C, Walker B, Ostrom E (2013) Aligning key concepts for global change policy: robustness, resilience, and sustainability. Ecol Soc 18(2):8. https://doi.org/10.5751/ES-05178-180208

Anderson K, Bows A (2008) Reframing the climate change challenge in light of post-2000 emission trends. Philos Trans A 366(1882). https://doi.org/10.1098/rsta.2008.0138

Armitage D (2005) Adaptive capacity and community-based natural resource management. Environ Manag 35(6):703–715

Armitage D, Plummer R (2010) Adapting and transforming: governance for navigating change, chapter 14. In: Armitage D, Plummer R (eds) Adaptive capacity and environmental governance. Springer, Heidelberg/Dordrecht/London/New York, pp 287–302

Armitage D, Marschke M, Plummer R (2008) Adaptive co-management and the paradox of learning. Glob Environ Chang 18:86–98
Avelino F (2009) Empowerment and the challenge of applying transition management to ongoing projects. Policy Sci 42(4):369–390
Avelino F (2011) Power in transition. Empowering discourses on sustainability transitions. PhD thesis, Erasmus Universiteit, Rotterdam
Avelino F, Grin J (2017) Beyond deconstruction. A reconstructive perspective on sustainability transition governance. Environ Innov Soc Trans 22:15–25
Avelino F, Wittmayer JM (2015) Shifting power relations in sustainability transitions: a multi-actor perspective. J Environ Policy Plan 18:628–649
Aylett A (2015) Institutionalizing the urban governance of climate change adaptation: results of an international survey. Urban Clim 14(1):4–16
Baba M (2013) Fukushima accident: what happened? Radiat Meas 55:17–21
Bahadur A, Tanner T (2014) Transformational resilience thinking: putting people, power and politics at the heart of urban climate resilience. Environ Urban 26:200. https://doi.org/10.1177/0956247814522154
Baldwin R, Teulings C (eds) (2014) Secular stagnation: facts, causes and cures. CEPR Press
Barnett J, O'Neill S (2010) Maladaptation. Glob Environ Chang 20(2):211–213
Barreteau O, Bots PWG, Daniell KA (2010) A framework for clarifying "participation" in participatory research to prevent its rejection for the wrong reasons. Ecol Soc 15(2):1
Beers PJ, Veldkamp A, Hermans F, Van Apeldoorn D, Vervoort JM, Kok K (2010) Future sustainability and images. Futures 42(7):723–732
Beisheim M, Simon N (2015) Meta-governance of partnerships for sustainable development. Actors' perspectives on how the UN could improve partnerships' governance services in areas of limited statehood. SFB-governance working paper series no. 68, Aug 2015. https://www.google.com/url?sa=t&rct=j&q=&esrc=s&source=web&cd=1&cad=rja&uact=8&ved=0ahUKEwi8x67Cmt_JAhVF0hoKHVxbCmAQFggcMAA&url=https%3A%2F%2Fwww.swp-berlin.org%2Ffileadmin%2Fcontents%2Fproducts%2Ffachpublikationen%2FBeisheim_etal2015_Meta-Governance_of_Partnerships_for_Sustainable_DevelopmentWP68.pdf&usg=AFQjCNEiy5h-r6Hn9NvzLsewgZKx4ursGg&bvm=bv.110151844,d.amc
Berkes F (1999) Sacred ecology: traditional ecological knowledge and management systems. Taylor & Francis, Philadelphia/London
Berkes F (2009) Evolution of co-management: role of knowledge generation, bridging organizations and social learning. J Environ Manag 90(5):1692–1702
Berkes F, Folke C (eds) (1998) Linking social and ecological systems: management practices and social mechanisms for building resilience. Cambridge University Press, Cambridge
Berkes F, Colding J, Folke C (eds) (2003) Navigating social-ecological systems: building resilience for complexity and change. Cambridge University Press, Cambridge
Berkhout F, Smith A, Stirling A (2004) Socio-technological regimes and transition contexts. In: Elzen B, Geels FW, Green K (eds) System innovation and the transition to sustainability: theory, evidence and policy. Edward Elgar, Cheltenham, pp 48–75
Berry PM, Brown S, Chen S, Kontogianni A, Rowlands O, Simpson G, Skourtos M (2015) Cross-sectoral interactions of mitigation and adaptation measures. Clim Chang 128:381–393. https://doi.org/10.1007/s10584-014-1214-0
Betsill MM, Bulkeley H (2006) Cities and the multilevel governance of global climate change. Glob Gov 12:141–159
Bettini YH (2013) Adapting institutions: processes and instruments behind urban water transitions. PhD thesis, School of Geography and Environmental Science, Monash University Melbourne
Bettini Y, Brown R, de Haan FJ (2015) Exploring institutional adaptive capacity in practice: examining water governance adaptation in Australia. Ecol Soc 20(1):47. https://doi.org/10.5751/ES-07291-200147
Bevir M (2009) Key concepts in governance. Sage, London

Bevir M (ed) (2011) The sage handbook of governance. Sage, London

Biggs R, Westley FR, Carpenter SR (2010) Navigating the back loop: fostering social innovation and transformation in ecosystem management. Ecol Soc 15(2):9

Bishop P, Hines A, Collins T (2007) The current state of scenario development: an overview of techniques. Foresight 9(1):5–25

Blackstock KL, Waylen KA, Dunglinson J, Marshall KM (2012) Linking process to outcomes – internal and external criteria for a stakeholder involvement in river basin management planning. Ecol Econ 77:113–122

Blong RJ (2013) Volcanic hazards: a sourcebook on the effects of eruptions. Elsevier, Amsterdam

Boonstra WJ, de Boer FW (2014) The historical dynamics of social–ecological traps. Ambio 43:260–274. https://doi.org/10.1007/s13280-013-0419-1

Bos JJ, Brown RR (2012) Governance experimentation and factors of success in socio-technical transitions in the urban water sector. Technol Forecast Soc Chang 79(7):1340–1353

Bos JJ, Brown RR (2013) Realising sustainable urban water management: can social theory help? Water Sci Technol 67(1):109–116

Bos JJ, Brown RR, Farrelly M, de Haan FJ (2013a) Enabling sustainable urban water management through governance experimentation. Water Sci Technol 67(8):1708–1717

Bos JJ, Brown RR, Farrelly M (2013b) A design framework for creating social learning situations. Glob Environ Chang 23(2):398–412

Bosman R, Rotmans J (2016) Transition governance towards a bioeconomy: a comparison of Finland and the Netherlands. Sustainability 8:1017. https://doi.org/10.3390/su8101017

Brones FA, De Carvahlo MM, de Senzi Zancul E (2017) Reviews, action and learning on change management for eco-design transition. J Clean Prod 142:8–22

Brown K, Westaway E (2011) Agency, capacity, and resilience to environmental change: lessons from human development, well-being, and disasters. Annu Rev Environ Resour 2011 (36):321–342. https://doi.org/10.1146/annurev-environ-052610-092905

Brown RR, Farrelly MA, Loorbach D (2013) Actors working the institutions in sustainability transitions: the case of Melbourne's stormwater management. Glob Environ Chang 23 (4):701–718

Bryson JM (2004) What to do when stakeholders matter. Publ Manag Rev 6(1):21–53

Bulkeley H (2010) Cities and the governing of climate change. Ann Rev Environ Res 12:141–159

Bulkeley H, Betsill M (2010) Rethinking sustainable cities: multilevel governance and the 'urban' politics of climate change. Environ Polit 14(1):42–63

Bulkeley H, Castan Broto V (2013) Government by experiment? Global cities and the governing of climate change. Trans Inst Br Geogr 38:361–375. https://doi.org/10.1111/j.1475-5661.2012.00535.x

Burch S (2010) Transforming barriers into enables of action on climate change: insights from three municipal case studies in British Columbia, Canada. Glob Environ Chang 20:287–297

Burch S, Harris S (2014) Understanding climate change: science, policy, and practice. Paperback

Burch S, Robinson J (2007) A framework for explaining the links between capacity and action in response to global climate change. Clim Policy 7:304–316

Burch S, Shaw A, Dale A, Robinson J (2014) Triggering transformative change: a development path approach to climate change response in communities. Clim Policy 14(4):467–487. https://doi.org/10.1080/14693062.2014.876342

Butzengeiger-Geyer S, Michaelowa M, Köhler M, Stadelmann M (2011) Market mechanisms for adaptation to climate change – lessons from mitigation and a pathway to implementation. Center for Comparative and International Studies (CIS), IPZ University of Zurich. Working paper http://www.cis.ethz.ch/publications/publications/WP_71.pdf

Cairns G, Ahmed I, Mullet J, Wright G (2013) Scenario method and stakeholder engagement: critical reflections on a climate change scenarios case study. Technol Forecast Soc Chang 80:1–10

Campos IS, Alves FM, Dinis J, Truninger M, Vizinho A, Penha-Lopes G (2016) Climate adaptation, transitions, and socially innovative action-research approaches. Ecol Soc 21 (1):13. http://dx.doi.org/10.5751/ES-08059-210113

Capano G, Howlett M, Ramesh M (2015) Bringing governments back in: governance and governing in comparative policy analysis. J Comp Policy Anal: Res Pract 17(4):311–321. https://doi.org/10.1080/13876988.2015.1031977

Caron-Flinterman F, Broerse JEW, Bunders JFG (2007) Patient partnership in decision-making on biomedical research: changing the network. Sci Technol Hum Values 32(3):339–368

Carpenter SR, Brock WA (2008) Adaptive capacity and traps. Ecol Soc 13(2):40. URL:http://www.ecologyandsociety.org/vol13/iss2/art40/

Carpenter SR, Walker M, Anderies JM, Abel N (2001) From metaphor to measurement: resilience of what to what? Ecosystems 4:765–781

Chang RD, Zuo J, Zhao ZY, Soebarto V, Zillante G, Gan XL (2017) Approaches for transitions towards sustainable development: status quo and challenges. Sustain Dev. https://doi.org/10.1002/sd.1661

Chapin SF III, Folke C, Konfinas G (2009) Principles of ecosystem stewardship: resilience-based natural resource management in a changing world. Springer, New York

Chapin SF III, Carpenter SR, Kofinas GP, Folke C, Abel N, Clark WC, Olsson P, Stafford Smith DM, Walker B, Young OR, Berkes F, Biggs R, Grove JM, Naylor RL, Pinkerton E, Steffen W, Swanson FJ (2010) Ecosystem stewardship: sustainability strategies for a rapidly changing planet. Trends Ecol Evol 25(4):241–249

Christodouleas JP, Forrest RD, Ainsley CG, Tochner Z, Hahn SM, Glatstein E (2011) Short-term and long-term health risks of nuclear-power-plant accidents. N Engl J Med 364:2334–2341. https://doi.org/10.1056/NEJMra1103676

Coenen L, Lopez FJD (2010) Comparing systems approaches to innovation and technological change for sustainable and competitive economies: an explorative study into conceptual commonalities, differences and complementarities. J Clean Prod 18:1149–1160

Cramer J (2013) Material efficiency: from top-down steering to tailor-made governance. Philos Trans R Soc A Math Phys Eng Sci 371(1986):20120496

Cumming GS, Olsson P, Chapin FS III, Holling CS (2012) Resilience, experimentation, and scale mismatches in social-ecological landscapes. Landsc Ecol. https://doi.org/10.1007/s10980-012-9725-4

Cutter SL, Barnes L, Berry M, Burton C, Evans E, Tate E, Webb J (2008) A place-based model for understanding community resilience to natural disasters. Glob Environ Chang 18(4):598–606. https://doi.org/10.1016/j.gloenvcha.2008.07.013

De Bruijne MLC, van de Rieet OAWT, de Haan ARC, Koppenjan JFM (2010) Dealing with dilemma's: how can experiments contribute to a more sustainable mobility system? Eur J Transp Infrastruct Res 10(3):274–289

De Gooyert V, Rouwette E, van Kranenburg H, Freeman E, van Breen H (2016) Sustainability transition dynamics: towards overcoming policy resistance. Technol Forecast Soc Chang 111:135–145

De Graaf R, van der Brugge R (2010) Transforming water infrastructure by linking water management and urban renewal in Rotterdam. Technol Forecast Soc Chang 77(8):1282–1291

De Grey A (1999) The mitochondrial free radical theory of aging. R.G. Landes Company, Texas, pp 1–199

De Grey A (2003) The foreseeability of real anti-aging medicine: focusing the debate. Exp Gerontol 38:927–934

De Haan J (2006) How emergence arise. Ecol Complex 3(4):293–301

De Haan J, Rotmans J (2011) Patterns in transitions: understanding complex chains of change. Technol Forecast Soc Chang 78(1):90–102

De la Cruz-Reyna S, Tilling RI (2008) Scientific and public responses to the ongoing volcanic crisis at Popocatépetl Volcano, Mexico: importance of an effective hazards-warning system. J Volcanol Geotherm Res 170:121–134

Doyle EEH, McClure J, Paton D, Johnston DM (2014) Uncertainty and decision making, volcanic crisis scenarios. Int J Disaster Risk Reduction 10:75–101

Dunford R, Harrison PA, Jäger J, Rounsevell MDA, Tinch R (2013) Vulnerability hotspot assessment. Report on the assessment of vulnerability across Europe and the identification of vulnerability hotspots. CLIMSAVE Deliverable 5.2. http://www.climsave.eu/climsave/doc/Report_on_assessing_vulnerability_hotspots.pdf

Eames M, Egmose J (2011) Community foresight for urban sustainability: insights from the Citizens Science for Sustainability (SuScit) project. Technol Forecast Soc Chang 78 (5):769–784

Eichengreen B (2015) Secular stagnation: the long view. NBER Working Paper, No. 20836

Elzen B, Wieczorek A (2005) Transitions towards sustainability through system innovation. Technol Forecast Soc Chang 72(6):651–661

Elzen B, Geels FW, Hofman PS, Green K (2004). 11. Socio-technical scenarios as a tool for transition policy: an example from the traffic and transport domain. System innovation and the transition to sustainability: theory, evidence and policy, 251

Elzen B, Geels FW, Leeuwis CS, Van Mierlo B (2011) Normative contestation in transitions 'in the making': animal welfare concerns and system innovation in pig husbandry (1970–2008). Res Policy 40:263–275

Emirbayer M, Mische A (1998) What is agency? Am J Sociol 103(4):962–1023

Enfors E (2013) Social-ecological traps and transformations in dryland agro-ecosystems: using water system innovations to change the trajectory of development. Glob Environ Chang 23:51–60. https://doi.org/10.1016/j.gloenvcha.2012.10.007

Engle NL (2011) Adaptive capacity and its assessment. Glob Environ Chang 21:647–656

Ensor J, Berger R (2009) Conclusion: community-based adaptation in practice, understanding climate change adaptation: lessons from community-based approaches. Practical Action Publishing, Rugby, chapter 10

Ernst L, de Graaf-Van Dinther RE, Peek GJ, Loorbach DA (2016) Sustainable urban transformation and sustainability transitions: conceptual framework and case study. J Clean Prod 112:2988–2999

Escobar Rangel L, Leveque F (2014) How Fukushima Dai-ichi core meltdown changed the probability of nuclear accidents. Saf Sci 64:90–98

EU Ad Hoc Expert Working Group on Biodiversity and Climate Change (2009) Working with nature: towards a strategy on climate change, ecosystem services and biodiversity. Discussion Paper. http://ec.europa.eu/environment/nature/pdf/discussion_paper_climate_change.pdf

Fankhauser S, Smith JB, Tol RSJ (1999) Weathering climate change: some simple rules to guide adaptation decisions. Ecol Econ 30:67–78

Farla J, Markard J, Raven R, Coenen L (2012) Sustainability transitions in the making: a closer look at actors, strategies and resources. Technol Forecast Soc Chang 79(6):991–998. https://doi.org/10.1016/j.techfore.2012.02.001

Ferguson B, Frantzeskaki N, Brown R (2013a) A strategic program for transitioning to a water sensitive city. Landsc Urban Plan 117:32–45

Ferguson B, Brown R, Frantzeskaki N, de Haan FJ, Deletic A (2013b) The enabling institutional context for integrated water management: lessons from Melbourne. Water Res 47:7300–7314

Findeisen W, Quade ES (1985) The methodology of systems analysis, chapter 4. In: Miser HJ, Quade ES (eds) Handbook of systems analysis: overview of uses, procedures, applications, and practice. Elsevier Science Publishing Co., New York

Fischer-Kowalski M, Rotmans J (2009) Conceptualizing, observing, and influencing social-ecological transitions. Ecol Soc 14(2):3. http://www.ecologyandsociety.org/vol14/iss2/art3/

Folke C (2006) Resilience: the emergence of a perspective for social-ecological system analyses. Glob Environ Chang 16(3):253–267

Folke C, Carpenter S, Elmqvist T, Gunderson L, Holling CS, Walker B (2002) Resilience and sustainable development: building adaptive capacity in a world of transformations. AMBIO J Hum Environ 31(5):437–440. https://doi.org/10.1579/0044-7447-31.5.437

Folke C, Colding J, Berkes F (2003) Synthesis: building resilience and adaptive capacity in social-ecological systems. In: Berkes F, Colding J, Folke C (eds) Navigating social-ecological systems: building resilience for complexity and change. Cambridge University Press, Cambridge, pp 352–387

Folke C, Hahn T, Olsson P, Norberg J (2005) Adaptive governance of social-ecological systems. Annu Rev Environ Resour 30(1):441–473

Folke C, Carpenter SR, Walker B, Scheffer M, Chapin T, Rockström J (2010) Resilience thinking: integrating resilience, adaptability and transformability. Ecol Soc 15(4):20. DOI: http://www.ecologyandsociety.org/vol15/iss4/art20/

Foxon T, Reed MS, Stringer LC (2009) Governing long-term social-ecological change: what can the adaptive management and transition management approaches learn from each other? Environ Policy Gov 19(1):3–20

Foxon TJ, Hammond G, Pearson PJG (2010) Developing transition pathways for a low carbon electricity system in the UK. Technol Forecast Soc Chang 77(8):1203–1213

Frantzeskaki N, Loorbach D (2010) Towards governing infrastructure transitions: reinforcing lock-in or facilitating change? Technol Forecast Soc Chang 77:1292–1301

Frantzeskaki N, Shiroyama H (2016) Sketching future research directions for transition management applications in cities. In: Loorbach D, Wittmayer JM, Shiroyama H, Fujino J, Mizuguchi S (eds) Governance of urban sustainability transitions. European and Asian experiences. Springer, Tokyo/Heidelberg/New York/Dordrecht/London

Frantzeskaki N, Tefrati N (2016) A transformative vision unlocks the innovative potential of Aberdeen City, UK. In: Loorbach D, Wittmayer JM, Shiroyama H, Fujino J, Mizuguchi S (eds) Governance of urban sustainability transitions. European and Asian experiences. Springer, Tokyo/Heidelberg/New York/Dordrecht/London, pp 49–68

Frantzeskaki N, Tillie N (2014) The dynamics of urban ecosystem governance in Rotterdam, The Netherlands. Ambio 43:542–555

Frantzeskaki N, Walker W (2013) Concepts and methods of policy analysis, 261–272. In: Thissen WAH, Warren W (eds) Public policy analysis: new developments, international series in operations research and management science. Springer, Berlin. ISBN:1461446015

Frantzeskaki N, van Daalen E, Slinger JH (2008) Detecting and directing socioecological transitions. A transitions management approach. International Society of Ecological Economics (ISEE), 8th international conference, Nairobi: applying ecological economics for social and environmental sustainability pp 7–11

Frantzeskaki N, Slinger JH, Vreugdenhil H, van Daalen E (2010) Social-ecological systems governance: from paradigm to management approach. Nat Cult 5(1):84–98

Frantzeskaki N, Loorbach D, Meadowcroft J (2012) Governing societal transitions to sustainability. Int J Sustain Dev 15(1):19–36

Frantzeskaki N, Wittmayer J, Loorbach D (2014a) The role of partnerships in 'realizing' urban sustainability in Rotterdam's City Ports Area, the Netherlands. J Clean Prod 65:406–417. https://doi.org/10.1016/j.jclepro.2013.09.023

Frantzeskaki N, Bach M, Hölscher K, Wittmayer JM (2014b) Applications of transition management: taking stock of 13 years of transition management and lessons for future research, 5th International Sustainability Transitions (IST) conference, 27–29 Aug 2014, Utrecht, The Netherlands

Frantzeskaki N, Kabisch N, McPhearson T (2016a) Advancing urban environmental governance: understanding theories, practices and processes shaping urban sustainability and resilience. Environ Sci Policy 62:1–6

Frantzeskaki N, Dumitru A, Anguelovski I, Avelino F, Bach M, Best B, Binder C, Barnes J, Carrus J, Egermann M, Haxeltine A, Moore ML, Mira RG, Loorbach D, Uzzell D, Omman I, Olsson P, Silvestri G, Stedman R, Wittmayer J, Durrant R, Rauschmeyer F (2016b) Elucidating the changing roles of civil society in urban sustainability transitions. Curr Opin Environ Sustain 22:41–50

Frantzeskaki N, Castan-Broto V, Coenen L, Loorbach D (eds) (2017) Urban sustainability transitions. Routledge, New York. ISBN 978-0-415-78418-4
Gallopín GC (2006) Linkages between vulnerability, resilience, and adaptive capacity. Glob Environ Chang 16:293–303
Garud R, Hardy H, Maguire S (2007) Institutional entrepreneurship as embedded agency: an introduction to the special issue. Organ Stud 28(7):957–969
Geels FW (2005) The dynamics of transitions in socio-technical systems: a multi-level analysis of the transition pathway from horse-drawn carriages to automobiles (1860–1930). Tech Anal Strat Manag 17(4):445–476
Geels FW, Schot J (2007) Typology of sociotechnical transition pathways. Res Policy 36:399–417
Genus A, Coles A-M (2008) Rethinking the multi-level perspective of technological transitions. Res Policy 37(9):1436–1445
Giddens A (1984) The constitution of society: outline of the theory of structuration. University of California Press, Berkeley/Los Angeles
Gidley JM, Fien J, Smith J-A, Thomsen DC, Smith TF (2009) Participatory futures methods: towards adaptability and resilience in climate-vulnerable communities. Environ Policy Gov 19:427–440
Godet M (2000) The art of scenarios and strategic planning: tools and pitfalls. Technol Forecast Soc Chang 65(1):3–22
Göpel M (2014) Navigating a new agenda. Questions and answers on paradigm shifts & transformational change. Working Paper. Wuppertal Institute for Climate, Environment and Energy GmbH, Berlin. https://www.google.com/url?sa=t&rct=j&q=&esrc=s&source=web&cd=1&cad=rja&uact=8&ved=0ahUKEwizxODb-mt_JAhXEtRoKHSbsBeAQFggcMAA&url=https%3A%2F%2Fwww.giz.de%2Ffachexpertise%2Fdownloads%2Fgiz2014-en-climate-finance-navigating-new-agenda.pdf&usg=AFQjCNHNPmmYQzxwwuXBIZTiPmA8lUF9cQ&bvm=bv.110151844,d.amc
Gordon R (2012) Is US economic growth over? Faltering innovation confronts the six headwinds. NBER Working Paper No. 18315
Gordon A (2015) Implementing backcasting for conservation: determining multiple policy pathways for retaining future targets of endangered woodlands in Sydney, Australia. Biol Conserv 181:182–189
Gössling S, Hall CM, Ekström F, Engeset AB, Aall C (2012) Transition management: a tool for implementing sustainable tourism scenarios? J Sustain Tour 20(6):899–916
Grin J (2012) The politics of transition governance in Dutch agriculture. Conceptual understanding and implications for transition management. Int J Sustain Dev 15(1–2):72–89
Grin J, Rotmans, Schot J (2010) Introduction: from persistent problems to system innovations and transitions. In: Grin J, Rotmans J, Schot J (eds) Transitions to sustainable development. New directions in the study of long term transformative change. Routledge, New York/London, pp 1–8
Grosskurth J, Rotmans J (2005) The SCENE model: getting a grip on sustainable development in policymaking. Environ Dev Sustain 7:135–151
Grotenbreg S, van Buuren A (2017) Facilitation as a governance strategy: unravelling governments' facilitation frames. Sustainability 9:160. https://doi.org/10.3390/su9010160
Gunderson LH, Carpenter SR, Folke C, Olsson P, Peterson G (2006) Water RATs (Resilience, Adaptability, and Transformability) in lake and wetland social-ecological systems. Ecol Soc 11 (1):16. URL:http://www.ecologyandsociety.org/vol11/iss1/art16/
Gupta J, Termeer C, Klostermann J, Meijerink S, van den Brink M, Jong P, Nooteboom S, Bergsma E (2010) The adaptive capacity wheel: a method to assess the inherent characteristics of institutions to enable the adaptive capacity of society. Environ Sci Policy 13:459–471
Hansen A (1938) Economic progress and declining population growth. Am Econ Rev 29:1–15
Harman D (1992) Free radical theory of aging. Mutat Res 275:257–266

Haslauer E, Biberacher M, Blaschke T (2015) A spatially explicit backcasting approach for sustainable land-use planning. J Environ Plan Manag. https://doi.org/10.1080/09640568.2015.1044652
Healey P (2006) Transforming governance: challenges of institutional adaptation and a new politics of space. Eur Plan Stud 14(3):299–320. https://doi.org/10.1080/09654310500420792
Heiskanen E, Kivisaari S, Lovio R, Mickwitz P (2009) Designed to travel? Transition management encounters environmental and innovation policy histories in Finland. Policy Sci 42(4):409–427
Hekkert MP, Suurs RAA, Negro SO, Kuhlmann S, Smits REHM (2007) Functions of innovation systems: a new approach for analysing technological change. Technol Forecast Soc Chang 74 (4):413–432
Helling A (1998) Collaborative visioning: proceed with caution!: results from evaluating Atlanta's vision 2020 project. J Am Plan Assoc 64(3):335–349
Hendriks C (2009) Policy design without democracy? Making democratic sense of transition management. Policy Sci 42(4):341–368
Hendriks CM, Grin J (2007) Contextualising reflexive governance: the politics of Dutch transitions to sustainability. J Environ Policy Plan 9(3/4):333–350
Henstra D (2015) The tools of climate adaptation policy: analyzing instruments and instrument selection. Clim Policy 03. https://doi.org/10.1080/14693062.2015.1015946
Hermans L (2010) Actor analysis. In: Enserink B, Hermans L, Kwakkel J, Thissen W, Koppenjan J, Bots P (eds) Policy analysis of multi-actor systems. Boom Lemma Uitgevers, The Hague, pp 79–108
Hermans LM, Thissen WAH (2009) Actor analysis methods and their use for public policy analysts. Eur J Oper Res 196:808–818
Hill M, Engle N (2013) Adaptive capacity: tensions across scales. Environ Policy Gov 23:177–192
Hisschemöller M, Hoppe R (1995) Coping with intractable controversies: the case for problem structuring in policy design and analysis. Knowl Technol Policy 8(4):40–60
Hodson M, Marvin S (2010) Can cities shape socio-technical transitions and how would we know if they were? Res Policy 39:477–485
Holdren JP (2006) The energy innovation imperative – addressing oil dependence, climate change, and other 21st century energy challenges. Innovations 1(2):3–23
Holland JH (1995) Hidden order: how adaptation builds complexity. Helix Books/Perseus Books, Cambridge
Holling CS (ed) (1978) Adaptive environmental assessment and management. Wiley, Chichester
Holling CS, Gunderson L, Peterson G (2002a) Sustainability and panarchies. In: Gunderson LH, Holling CS (eds) Panarchy: understanding transformations in human and natural systems. Island, Washington, DC, pp 63–102
Holling CS, Gunderson L, Ludwig L (2002b) In quest of a theory of adaptive change. In: Gunderson LH, Holling CS (eds) Panarchy: understanding transformations in human and natural systems. Island, Washington, DC, pp 3–24
Hölscher K, Frantzeskaki N, Loorbach D (2014) Two sides of the same coin? Exploring resilience and transitions thinking for climate change governance, Deltas in times of climate change conference II, 24–26 Sept 2014, Rotterdam, The Netherlands
Hölscher K, Frantzeskaki N, Loorbach D (2015) Orchestrating adaptation, mitigation and transformation in urban areas. Agency's capacities for climate governance in Rotterdam, 6th International Sustainability Transitions (IST) conference, 25–28 Aug 2015, Brighton, UK
Hölscher K, Wittmayer JM, Avelino F, Giezen M (2017). Opening up the transition arena: An analysis of (dis)empowerment of civil society actors in transition management in cities. Technological forecasting and social change. (Online first). https://doi.org/10.1016/j.techfore.2017.05.004
Holtz G, Brugnach M, Pahl-Wostl C (2008) Specifying "regime" – a framework for defining and describing regimes in transition research. Technol Forecast Soc Chang 75(5):623–643
Hoppe R (2011) The governance of problems, puzzling, powering and participation. Policy Press, University of Bristol, Bristol

Hordijk M, Sara LM, Sutherland C (2014) Resilience, transition or transformation? A comparative analysis of changing water governance systems in four southern cities. Environ Urban 26:130. https://doi.org/10.1177/0956247813519044

House of Lords – Select Committee on Intergovernmental Organisations (2008) Diseases know no frontiers: how effective are intergovernmental organisations in controlling their spread? The Stationery Office, London, pp 1–79. http://www.publications.parliament.uk/pa/ld200708/ldselect/ldintergov/143/143.pdf

Howlett M (2014) Why are policy innovations rare and so often negative? Blame avoidance and problem denial in climate change policy-making. Glob Environ Chang 29:395–403

Huétink F (2010) Initial infrastructure development strategies for the transition to sustainable mobility. Technol Forecast Soc Chang 77(8):1270–1281

Hughes N (2013) Towards improving the relevance of scenarios for public policy questions: a proposed methodological framework for policy relevant low carbon scenarios. Technol Forecast Soc Chang 80:687–698

Hurlbert M, Mcnutt K, Rayner J (2011) Pathways to power: policy transitions and the reappearance of the nuclear power option in Saskatchewan. Energ Policy 39(6):3182–3190

Hyde WT, Crowley TJ (2000) Probability of future climatically significant volcanic eruptions. J Clim 13:1445–1450

IEA (2014) World energy investment outlook. International Energy Agency, Paris. https://www.iea.org/publications/freepublications/publication/WEIO2014.pdf

Innes JE, Booher DE (1999) Consensus building and complex adaptive systems. J Am Plan Assoc 65(4):412–423

IPCC (2014) Climate change 2014: impacts, adaptation and vulnerability. IPCC working group II contribution to AR5. Summary for policymakers. Cambridge University Press, Cambridge/New York

Jaeger J (2009) The governance of science for sustainability, as chapter 7. In: Adger WN, Jordan A (eds) Governing sustainability. Cambridge University Press, Cambridge, pp 142–158

Jessop B (1998) The rise of governance and the risks of failure: the case of economic development. Int Soc Sci J 50(155):29–45

Jessop B (2002) Governance and metagovernance: on reflexivity, requisite variety, and requisite irony. Lancaster University, Department of Sociology, Lancaster. http://www.comp.lancs.ac.uk/sociology/papers/Jessop-Governance-and-Metagovernance.pdf

Jessop B (2004) Multilevel governance and multi-level metagovernance. In: Backe I, Flinders M (eds) Multilevel governance. Oxford University Press, Oxford

Jessop B (2009) From governance to governance failure and from multi-level governance to multiscalar meta-governance. In: Arts B, Lagendijk A, van Houtum H (eds) The disoriented state: shifts in governmentality, territoriality and governance. Springer, Berlin, pp 79–100

Jessop B (2011) Metagovernance. In: Bevir M (ed) The Sage handbook of governance. Sage, London, pp 106–123

Jhagroe S, Frantzeskaki N (2012) A state of exceptional democracy: unravelling the politics of crisis in Dutch infrastructure governance. Paper presented at the 7th international conference on interpretive policy analysis understanding the drama of democracy: policy work, power and transformation. 5–7 Jul 2012, Tilburg, The Netherlands

Jhagroe S, Loorbach D (2015) See no evil, hear no evil: the democratic potential of transition management. Environ Innov Soc Trans 8: https://doi.org/10.1016/j.eist.2014.07.001

Johansen F, van den Bosch S (2017) The scaling-up of neighborhood care: from experiment towards a transformative movement in healthcare. Futures 89:60–73. https://doi.org/10.1016/j.futures.2017.04.004

Johansson TB, Nakicenovic N, Patwardhan A, Gomez-Echeverri L (2012) Global Energy Assessment (GEA) – toward a sustainable future. Cambridge University Press, Cambridge

Kates RW, Travis WR, Wilbanks TJ (2012) Transformational adaptation when incremental adaptations to climate change are insufficient. PNAS 109(19):7156–7161

Keeney RL (1996a) Value-focused thinking: identifying decision opportunities and creating alternatives. Eur J Oper Res 92:537–549
Keeney RL (1996b) Value-focused thinking: a path to creative decisionmaking, chapter 3. Harvard University Press, Cambridge, MA
Kemp R, Rotmans J (2009) Transitioning policy: co-production of a new strategic framework for energy innovation policy in the Netherlands. Policy Sci 42(4):303–322
Kemp R, Parto S, Gibson R (2005) Governance for sustainable development: moving from theory to practice. Int J Sustain Dev 8(1–2):12–30
Kemp R, Loorbach D, Rotmans J (2007a) Transition management as a model for managing processes of co-evolution towards sustainable development. Int J Sust Dev World 14(1):78–91
Kemp R, Rotmans J, Loorbach D (2007b) Assessing the Dutch energy transition policy: how does it deal with dilemmas of managing transitions? J Environ Policy Plan 9(3–4):315–331
Kenis A, Bono F, Mathijs E (2016) Unravelling the (post-)political in transition management: interrogating pathways towards sustainable change. J Environ Policy Plan 18(5):568–584. https://doi.org/10.1080/1523908X.2016.1141672
Kern F, Howlett M (2009) Implementing transition management as policy reforms: a case study of the Dutch energy sector. Policy Sci 42:391–408
Kern F, Smith A (2008) Restructuring energy systems for sustainability? Energy transition policy in the Netherlands. Energ Policy 36(11):4093–4103
Klijn E, Edelenbos J (2007) Metagovernance as network management. In: Sørensen E, Torfing J (eds) Theories of democratic network governance. Palgrave Macmillan, New York, pp 199–214
Knill C, Lehmkuhl D (2002) Private actors and the state: internationalization and changing patterns of governance. Governance 15(1):41–63
Kok K, Christensen JH (2014) IMPRESSIONS scenario methodology, IMPRESSIONS living document
Kok K, van Vliet M, Bärlund I, Dubel A, Sendzimir J (2011) Combining participative backcasting and exploratory scenario development: experiences from the SCENES project. Technol Forecast Soc Chang 78:835–851
Kooiman J (1993) Modern governance: new government-society interactions. Sage, London
Kooiman J (2003) Governing as governance. Sage, London
Kooiman J, Jentoft S (2009) Meta-Governance: values, norms and principles, and the making of hard choices. Public Adm 87(4):818–836
Köves A, Király G, Pataki G, Balázs B (2013) Backcasting for sustainable employment: a Hungarian experience. Sustainability 5(7):2991–3005
Kuch P, Gigli S (2007) Economic approaches to climate change adaptation. GTZ, Eschborn
Lachman DA (2013) A survey and review of approaches to study transitions. Energ Policy 58 (07):269–276
Lebel L, Anderies JM, Campbell B, Folke C, Hatfield-Dodds S, Hughes TP, Wilson J (2006) Governance and the capacity to manage resilience in regional social-ecological systems. Ecol Soc 11(1):19. http://www.ecologyandsociety.org/vol11/iss1/art19/
Lebel L, Foran T, Garden P, Manuta BJ (2009) Adaptation to climate change and social justice: challenges for flood and disaster management in Thailand. In: Ludwig F, Kabiat P, van Schaik H, van der Valk M (eds) Climate change adaptation in the water sector. Earthscan, London, pp 125–141
Lebel L, Grothmann T, Siebenhüner B (2010) The role of social learning in adaptiveness: insights from water management. Int Environ Agreements 10:333–353
Lelieveld J, Kunkel D, Lawrence MG (2012) Global risk of radioactive fallout after major nuclear reactor accidents. Atmos Chem Phys 12(9):4245–4258
Lipton J, Arnold D, Nigl F, Lopez N, Cohen D, Noren N, Lipson H (2010) Multi-material food printing with complex internal structure suitable for conventional post-processing. 21st, international solid freeform fabrication symposium, Austin, USA
Little LK (2008) Plague and the end of antiquity. Paperback

Lockwood M, Raymond CM, Oczkowski E, Morrison M (2015) Measuring the dimensions of adaptive capacity: a psychometric approach. Ecol Soc 20(1):37. https://doi.org/10.5751/ES-07203-200137
Loorbach D (2007) Transition management. New mode of governance for sustainable development. PhD thesis, Erasmus Universiteit Rotterdam
Loorbach D (2010) Transition management for sustainable development: a prescriptive, complexity-based governance framework. Gov: Int J Policy Adm Inst 23(1):161–183
Loorbach D (2014) To transition! Governance panarchy in the new transformation. Inaugural lecture. Erasmus Universiteit Rotterdam, Rotterdam
Loorbach D, Frantzeskaki N (2012) Why taking complexity seriously implies a paradigm shift for policy studies, chapter 15. In: Gerrits L, Marks P (eds) COMPACT I: public administration in complexity. pp 327–345
Loorbach D, Rotmans J (2010) The practice of transition management: examples and lessons from four distinct cases. Futures 42(3):237–246
Loorbach D, Wijsman K (2013) Business transition management: exploring a new role for business in sustainability transitions. J Clean Prod 45:20–28
Loorbach D, Bakel J, Whiteman G, Rotmans J (2010) Business strategies for transitions towards sustainable systems. Bus Strateg Environ 19(2):133–146
Loorbach D, Frantzeskaki N, Thissen WH (2011) A transition research perspective on governance for sustainability. In: Jaeger CC, Tàbara JD, Jaeger J (eds) European research on sustainable development. Volume 1: transformative science approaches for sustainability. Springer, Berlin/Heidelberg, pp 73–90
Loorbach D, Frantzeskaki N, Huffenreuter LR (2015) Transition management: taking stock from governance experimentation. J Corp Citizenship 58:48–66
Lopes AM, Fam D, Williams J (2012) Designing sustainable sanitation: involving design in innovative, transdisciplinary research. Des Stud 33(3):298–317
Luers AL, Sklar LS (2013) The difficult, the dangerous, and the catastrophic: managing the spectrum of climate change risks. Earths Future 2:114–118. https://doi.org/10.1002/2013EF000192
Luiten W, van Sandick H (2007) Experiments for transitions: an interactive approach to setting up breakthrough experiments. Int J Innov Sustain Dev 2(2):215–229
Marquardt J, Steinbacher K, Schreuers M (2016) Driving force or forced transition? The role of development cooperation in promoting energy transitions in the Philippines and Morocco. J Clean Prod 128:22–33
Marshall NA, Park SE, Adger WN, Brown K, Howden SM (2012) Transformational capacity and the influence of place and identity. Environ Res Lett 7:034022
Marzocchi W, Sandri L, Gasparini P, Newhall C, Boschi E (2004) Quantifying probabilities of volcanic events: the example of volcanic hazard at Mount Vesuvius. J Geophys Res 109:1–18
Matyas D, Pelling M (2014) Positioning resilience for 2015: the role of resistance, incremental adjustment and transformation in disaster risk management policy. Disasters 39(S1):S1–S18. https://doi.org/10.1111/disa.12107
Meadowcroft J (2009a) What about the politics? Sustainable development, transition management, and long term energy transitions. Policy Sci 42(4):323–340
Meadowcroft J (2009b) Climate change governance. Background paper to the 2010 world development report. Policy research working paper 4941. The World Bank
Meadows HD (1996) Envisioning a sustainable world. In: Costanza R, Segura O, Martinez-Alier J (eds) Getting down to earth, practical applications of ecological economics. Island Press, Washington, DC
Meuleman L (2008) Public management and the metagovernance of hierarchies, networks and markets. Physica, Heidelberg
Meuleman L (2013) Cultural diversity and sustainability metagovernance. In: Meuleman L (ed) Transgovernance. Advancing sustainability governance. Springer, Heidelberg/New York/Dordrecht/London

MIAVITA team (2012) Handbook for volcanic risk management – prevention, crisis management, resilience. Orleans, France. http://miavita.brgm.fr/Documents/Handbook-VolcRiskMgt-lr.pdf
Miles G, Grainger RG, Highwood EJ (2004) The significance of volcanic eruption strength and frequency for climate. Q J R Meteorol Soc 130:2361–2376
Miller CA, O'Leary J, Graffy E, Stechel EB, Dirks G (2015) Narrative futures and the governance of energy transitions. Futures 70:65–74
Monaghan A (2009) Conceptual niche management of grassroots innovation for sustainability: the case of body disposal practices in the UK. Technol Forecast Soc Chang 76(8):1026–1043
Moore M-L, Westley F (2011) Surmountable chasms: networks and social innovation for resilient systems. Ecol Soc 16(1):5. http://www.ecologyandsociety.org/vol16/iss1/art5/
Moore M-L, Westley FR, Tjornbo O, Holroyd C (2011) The loop, the lens and the lesson: using resilience theory to examine public policy and social innovation. In: Nicholls A, Murdock A (eds) Social innovation: blurring boundaries to reconfigure markets. Palgrave Macmillan, Basingstoke, p 320
Muro M, Jeffrey P (2012) Time to talk? How the structure of dialog processes shapes stakeholder learning in participatory water resources management. Ecol Soc 17(1):3
Nelson DR, Adger WN, Brown K (2007) Adaptation to environmental change: contributions of a resilience framework. Annu Rev Environ Resour 32:395–419
Nevens F, Roorda C (2013) A climate of change: a transition approach for climate neutrality in the city of Ghent (Belgium). Sustain Cities Soc 10:112–121
Nevens F, Frantzeskaki N, Gorissen L, Loorbach D (2013) Urban transition labs: co-creating transformative action for sustainable cities. J Clean Prod 50:111–122
Newman PWG (1999) Sustainability and cities: extending the metabolism model. Landsc Urban Plan 44:219–226
Nill J, Kemp R (2009) Evolutionary approaches for sustainable innovation policies: from niche to paradigm? Res Policy 38(4):668–680
Nooteboom S (2007) Impact assessment procedures for sustainable development: a complexity theory perspective. Environ Impact Assess Rev 27(7):645–665
Nutley SM, Eslyrt I, Fsbird HTO (2007) Using evidence: how research can inform public services. The Polity Press, Bristol
O'Brien K (2012) Global environmental change II: from adaptation to deliberate transformation. Prog Hum Geogr 36(5):667–676
O'Brien FA, Meadows M (2013) Scenario orientation and use to support strategy development. Technol Forecast Soc Chang 80(4):643–656
Olah GA, Goeppert A, Prakash GS (2008) Chemical recycling of carbon dioxide to methanol and dimethyl ether: from greenhouse gas to renewable, environmentally carbon neutral fuels and synthetic hydrocarbons. J Organomet Chem 74:487–498
Olsson P, Galaz V (2012) Social-ecological innovation and transformation. In: Nicholls A, Murdock A (eds) Social innovation: blurring boundaries to reconfigure markets. Palgrave Macmillan, Basingstoke, pp 223–243
Olsson P, Folke C, Hahn T (2004) Social-ecological transformation for ecosystem management: the development of adaptive co-management of a wetland landscape in Southern Sweden. Ecol Soc 9(4):2
Olsson P, Gunderson LH, Carpenter SR, Ryan P, Lebel L, Folke C, Holling CS (2006) Shooting the rapids: navigating transitions to adaptive governance of social-ecological systems. Ecol Soc 11(1):18. DOI: http://www.ecologyandsociety.org/vol11/iss1/art18
Olsson P, Galaz V, Boonstra WJ (2014) Sustainability transformations: a resilience perspective. Ecol Soc 19(4):1
Olsson L, Jerneck A, Thoren H, Persson, O'Byrne D (2015) Why resilience is unappealing to social science: theoretical and empirical investigations of the scientific use of resilience. Sci Adv 2015:1

Pahl-Wostl C, Becker G, Knieper C, Sendzimir J (2013) How multilevel societal learning processes facilitate transformative change: a comparative case study analysis on flood management. Ecol Soc 18(4):58. https://doi.org/10.5751/ES-05779-180458

Park SE, Marshall NA, Jakku E, Dowd AM, Howden SM, Mendham E, Fleming A (2012) Informing adaptation responses to climate change through theories of transformation. Glob Environ Chang 22(1):115–126. https://doi.org/10.1016/j.gloenvcha.2011.10.003

Parson EA (2003) Protecting the ozone layer: science and strategy. Oxford University Press, Oxford

Patton CV, Sawicki DS (1986) Basic methods of policy analysis and planning. Prentice-Hall, Englewood Cliffs

Pelling M (2011) Adaptation to climate change: from resilience to transformation. Routledge, London

Pelling M, Manuel-Navarrete D (2011) From resilience to transformation: the adaptive cycle in two Mexican urban centers. Ecol Soc 16(2):11. http://www.ecologyandsociety.org/vol16/iss2/art11/

Pelling M, High C, Dearing J, Smith D (2008) Shadow spaces for social learning: a relational understanding of adaptive capacity to climate change within organisations. Environ Plan A 40:867–884

Pesch U (2015) Tracing discursive space: agency and change in sustainability transitions. Technol Forecast Soc Chang 90:379–388

Phdungsilp A (2011) Future studies' backcasting method used for strategic sustainable city planning. Futures 43:707–714

Pichert D, Katsikopoulos KV (2008) Green defaults: information presentation and pro-environmental behavior. J Environ Psychol 28:63–73

Plummer R (2013) Can adaptive comanagement help to address the challenges of climate change adaptation? Ecol Soc 18(4):2

Plummer R, Armitage DR (2010) Integrating perspectives on adaptive capacity and environmental governance. In: Armitage D, Plummer R (eds) Adaptive capacity and environmental governance. Springer, Berlin, pp 1–19

Popper R (2011) Metodología de la Prospectiva. Manual de Prospectiva Tecnológica. Latin American Faculty of Social Science (FLACSO), Mexico

Poustie M, Frantzeskaki N, Brown R (2016) A transition scenario for leapfrogging to a sustainable urban water future in Port Vila, Vanuatu. Technol Forecast Soc Chang 105:129–139. https://doi.org/10.1016/j.techfore.2015.12.008

Quay R (2010) Anticipatory governance. A tool for climate change adaptation. J Am Plan Assoc 76(4):496–511

Quist J, Thissen W, Vergragt PJ (2011) The impact and spin-off of participatory backcasting: from vision to niche. Technol Forecast Soc Chang 78:883–897

Rauws WS, de Roo G (2011) Exploring transitions in the peri-urban area. Plan Theory Pract 12 (2):269–284. https://doi.org/10.1080/14649357.2011.581025

Raven R, van den Bosch S, Weterings R (2010) Transitions and strategic niche management: towards a competence kit for practitioners. Int J Technol Manag 51(1):57–74

Raworth K (2012) A safe and just space for humanity. Can we live within the doughnut? Oxfam Discussion Paper, https://www.oxfam.org/sites/www.oxfam.org/files/dp-a-safe-and-just-space-for-humanity-130212-en.pdf

Redman CL (2014) Should sustainability and resilience be combined or remain distinct pursuits? Ecol Soc 19(2):37. https://doi.org/10.5751/ES-06390-190237

REN21 (2013) Renewables global futures report, UNEP, Paris CEDEX, France. http://www.ren21.net/Portals/0/documents/activities/gfr/REN21_GFR_2013.pdf

REN21 (2014) Global status report. UNEP, Paris. http://www.ren21.net/Portals/0/documents/Resources/GSR/2014/GSR2014_full%20report_low%20res.pdf

Rijke J, Farrelly, Brown R, Zevenbergen C (2013) Configuring transformative governance to enhance resilient urban water systems. Environ Sci Policy 25(2013):62–72

Ringland G (2010) The role of scenarios in strategic foresight. Technol Forecast Soc Chang 77 (9):1493–1498
Robinson J, Burch S, Talwar S, O'Shea M, Walsh M (2011) Envisioning sustainability: recent progress in the use of participatory backcasting approaches for sustainability research. Technol Forecast Soc Chang 78:756–768
Rockström J, Steffen W, Noone K, Persson A, Chapin FS III, Lambin EF, Lenton TM, Scheffer M, Folke C, Schellnhuber HJ, Nykvist B, de Wit CA, Hughes T, van der Leeuw S, Rodhe H, Sörlin S, Snyder PK, Costanza R, Svedin U, Falkenmark M, Karlberg L, Corell RW, Fabry VJ, Hansen J, Walker B, Liverman D, Richardson K, Crutzen P, Foley JA (2009) A safe operating space for humanity. Nature 461:472–475
Rogers T, Bazerman MH (2008) Future lock-in: future implementation increases selection of "should" choices. Organ Behav Hum Decis Process 106(1):1–20
Roorda C, Wittmayer J, Henneman P, van Steenbergen F, Frantzeskaki N (2014) Transition management in the urban context: guidance manual. Erasmus University Rotterdam, DRIFT
Rotmans J, Loorbach D (2009) Complexity and transition management. J Ind Ecol 13(2):184–196
Rotmans J, Loorbach D (2010) Towards a better understanding of transitions and their governance: a systemic and reflexive approach. In: Grin J, Rotmans J, Schot J (eds) Transitions to sustainable development. New directions in the study of long term transformative change. Routledge, New York/London, pp 105–220
Rotmans J, Kemp R, van Asselt M (2001) Emerald article: more evolution than revolution: transition management in public policy. Foresight 3(1):15–31
Rowe G, Frewer LJ (2000) Public participation methods: a framework for evaluation. Sci Technol Hum Values 25(1):3–29
Sabina P, Stanghellini L (2010) Stakeholder involvement in water management: the role of the stakeholder analysis within participatory processes. Water Policy 10:675–694
Sage AP, Armstrong JE (2000) Introduction to systems engineering. Wiley, New York
Sandström A, Rova C (2010) Adaptive co-management networks: a comparative analysis of two fishery conservation areas in Sweden. Ecol Soc 15(3):14
Santos R, Antunes P, Baptista G, Mateus P, Madruga L (2006) Stakeholder participation in the design of environmental policy mixes. Ecol Econ 60:100–110
Saritas O, Smith JE (2011) The big picture – trends, drivers, wildcards, weak signals. Futures 43:292–312
Scarse I, Smith A (2009) The (non-)politics of managing low carbon socio-technical transitions. Environ Polit 18(5):707–726
Schapke N, Omann I, Wittmayer JM, van Steenbergen F, Mock M (2017) Linking transitions to sustainability: a study of the societal effects of transition management. Sustainability 9:737. https://doi.org/10.3390/su9050737
Scharpf FW (1994) Games real actors could play. Positive and negative coordination in embedded negotiations. J Theor Polit 6(1):27–53
Schreurs MA (2008) From the bottom up: local and subnational climate change politics. J Environ Dev 17:343–355
Schuitmaker TJ (2012) Identifying and unravelling persistent problems. Technol Forecast Soc Chang 79:1021–1031
Schultz PW, Nolan J, Cialdini RB, Goldstein NJ, Griskevicius V (2007) The constructive, destructive, and reconstructive power of social norms. Psychol Sci 18:429–434
Scott AJ (2001) Globalization and the rise of city-regions. Eur Plan Stud 9(7):813–826. https://doi.org/10.1080/09654310120079788
Scourfield P (2015) Implementing co-production in adult social care: an example of meta-governance failure? Soc Policy Soc 14(4):541–554. https://doi.org/10.1017/S1474746414000438
Scuttari A, Volgger M, Pechlaner H (2016) Transition management towards sustainable mobility in Alpine destinations: realities and realpolitik in Italy's South Tyrol region. J Sustain Tour 24 (3):463–483. https://doi.org/10.1080/09669582.2015.1136634

Sengers F, Raven RPJM (2015) Towards a spatial perspective on niche development: the case of bus rapid transit. Environmental innovation and societal transitions. In press
Seyfang G, Smith A (2007) Grassroots innovations for sustainable development: towards a new research and policy agenda. Environ Polit 16(4):584–603
Sharma S, Ghoshal SK (2015) Hydrogen the future transportation fuel: from production to applications. Renew Sust Energ Rev 43:1151–1158
Shaw A, Burch S, Kristensen F, Robinson J, Dale A (2014) Accelerating the sustainability transition: exploring synergies between adaptation and mitigation in British Columbian communities. Glob Environ Chang 25:41–51
Sheppard SRJ, Shaw A, Flanders D, Burch S, Wiek A, Carmichael J, Robinson J, Cohen S (2011) Future visioning of local climate change: a framework for community engagement and planning with scenarios and visualization. Futures 43:400–412
Shipley R (2000) The origin and development of vision and visioning in planning. Int Plan Stud 5 (2):225–236
Shipley R, Newkirk R (1999) Vision and visioning in planning: what do these terms really mean? Environ Plan B: Plan Des 26(4):573–591
Shove E, Walker G (2007) CAUTION! Transitions ahead: politics, practice, and sustainable transition management. Environ Plan A 39(4):763–770
Smil V (2013) Should we eat meat? Evolution and consequences of modern carnivory. Wiley-Blackwell, Hoboken, pp 1–276
Smit B, Wandel J (2006) Adaptation, adaptive capacity and vulnerability. Glob Environ Chang 16 (3):282–292
Smith A, Raven R (2012) What is protective space? Reconsidering niches in transitions to sustainability. Res Policy 41(6):995. https://doi.org/10.1016/j.respol.2011.12.012
Smith A, Stirling A (2010) The politics of social-ecological resilience and sustainable socio-technical transitions. Ecol Soc 15(1):11. http://www.ecologyandsociety.org/vol15/iss1/art11/
Smith R, Wiek A (2012) Achievements and opportunities in initiating governance for urban sustainability. Environ Plan C: Gov Policy 30:429–447
Sondeijker S, Geurts J, Rotmans J, Tukker A (2006) Imagining sustainability: the added value of transition scenarios in transition management. Foresight 8(5):15–30
Sørensen E (2006) Metagovernance. The changing role of politicians in processes of democratic governance. Public Adm 36(1):98–114. https://doi.org/10.1177/0275074005282584
Sørensen E (2014) The metagovernance of public innovation in governance networks. Paper presented at Policy and Politics conference, Bristol, 16–17 Sept 2014
Sørensen E, Torfing J (2009) Making governance networks effective and democratic through metagovernance. Public Adm 87(2):234–258
Späth P, Rohracher H (2010) 'Energy regions': the transformative power of regional discourses on socio-technical futures. Res Policy 39(4):449–458
Stavins R (2003) Market-based environmental policies: what can we learn from US experience and related research? Resources for the Future, Washington, DC
Stehfest E, Bouwman L, van Vuuren D, P., den Elzen MGJ, Eickhout B, Kabat P (2009) Climate benefits of changing diet. Clim Chang 102:95–83
Steinmueller K (2008) Wild cards – preparing for the unpredictable, conference paper for thinking about the future, strategic anticipation and RAHS by Risk Assessment and Horizon Scanning (RAHS). National Security Coordination Secretariat, Singapore
Stephens J, Graham A (2010) Toward an empirical research agenda for sustainability in higher education: exploring the transition management framework. J Clean Prod 18(7):611–618
Sterman JD (2000) Business dynamics, systems thinking, and modeling for a complex world. McGraw Hill, New York, pp 137–156
Steurer R (2013) Disentangling governance: a synoptic view of regulation by government, business and civil society. Policy Sci 46:387–410. https://doi.org/10.1007/s11077-013-9177-y
Stirling A (2008) "Opening up" and "closing down": power, participation, and pluralism in the social appraisal of technology. Sci Technol Hum Values 33(2):262–294

Stock GB (2004) The pitfalls of planning for demographic change. Ann N Y Acad Sci 1019:546–551

Stoker G (1998) Governance as theory: five propositions. Int Soc Sci J 50(155):17–28

Stringer LC, Dougill AJ, Fraser E, Hubacek K, Prell C, Reed MS (2006) Unpacking "participation" in the adaptive management of social–ecological systems: a critical review. Ecol Soc 11 (2):39

Summers L (2014) US economic prospects: secular stagnation, hysteresis and the zero lower bound, speech delivered to the national association for business economics' economic policy conference. 24 Feb 2014

Tàbara JD (2011) Integrated climate governance. In: Carlo C, Jäger J, Tabara D (eds) European research on sustainable development. Volume 1: transformative science approaches for sustainability. Springer, Berlin/Heidelberg, pp 91–110

Tàbara JD, Pahl-Wostl C (2007) Sustainability learning in natural resource use and management. Ecol Soc 12(2):3

Tàbara JD, Dai X, Jia G, McEvoy D, Neufeldt H, Serra A, Werners S, West JJ (2010) The climate learning ladder. A pragmatic procedure to support climate adaptation. Environ Pol Gov 20:1–11. https://doi.org/10.1002/eet.530

Tàbara JD, Jäger J Roventini A, Frantzeskaki N, Hölscher K, Tinch R, Mangalagiu D, Harrison P (2015) Integrated climate governance framework for synthesis. Milestone M19 working document for deliverable D5.4, IMPRESSIONS project

Tainter J (1990) The collapse of complex societies. Cambridge University Press, Cambridge

Tanner T, Horn-Phathanothai L (2014) Climate change and development. Routledge, Abingdon/New York

The World Bank Group (2014) The economic impact of the 2014 ebola epidemic: short and medium term estimates for West Africa, report

Tinch R, Jäger J, Omann I, Harrison PA, Wesely J, Dunford R (2015) Applying a capitals framework to measuring coping and adaptive capacity in integrated assessment models. Clim Chang 128:323–337

Tompkins EM, Few R, Brown K (2008) Scenario-based stakeholder engagement: incorporating stakeholders preferences into coastal planning for climate change. J Environ Manag 88 (4):1580–1592

Tukker A, Butter M (2007) Governance of sustainable transitions: about the 4(0) ways to change the world. J Clean Prod 15(1):94–103

Turner BL II (2003) A framework for vulnerability analysis in sustainability science. Proc Natl Acad Sci U S A 100:8074–8079

U.S. Nuclear Regulatory Commission (1990) Severe accident risks: an assessment for five U.S. nuclear power plants. Final summary report

UNDP and UNICEF (2002) The human consequences of the chernobyl nuclear accident a strategy for recovery. Report Commissioned by UNDP and UNICEF with the support of UN-OCHA and WHO. https://www.iaea.org/sites/default/files/strategy_for_recovery.pdf

UNEP (2010) Assessing the environmental impacts of consumption and production: priority products and materials, a report of the working group on the environmental impacts of products and materials to the international panel for sustainable resource management. United Nations Environment Programme (UNEP), Paris, pp 1–110. http://www.greeningtheblue.org/sites/default/files/Assessing%20the%20environmental%20ipacts%20of%20consumption%20and%20production.pdf

UNEP (2011) Annual report, United Nations Environment Programme (UNEP), Nairobi, Kenya. p 1–112. http://www.unep.org/annualreport/2011/docs/UNEP_ANNUAL_REPORT_2011.pdf

UNEP (2012) 21 issues for the 21st century: result of the UNEP foresight process on emerging environmental issues. United Nations Environment Programme (UNEP), Nairobi, pp 1–56. http://www.unep.org/pdf/Foresight_Report-21_Issues_for_the_21st_Century.pdf

UNEP (2013) Annual report. United Nations Environment Programme (UNEP), Nairobi, Kenya. p 1–55. http://www.unep.org/annualreport/2013/docs/ar_low_res.pdf

UNSCEAR (2008) Sources and effects of ionizing radiation, report to the general assembly
US Department of Energy (2014) Advanced research projects agency – energy annual report for FY2013. Report to Congress Apr 2014, United States Department of Energy, Washington, DC
Vabo SI, Røiseland A (2012) Conceptualizing the tools of government in urban network governance. Int J Publ Adm 35(14):934–946. https://doi.org/10.1080/01900692.2012.691243
Van Buuren A, Loorbach D (2009) Policy innovation in isolation? Conditions for policy renewal by transition arenas and pilot projects. Publ Manag Rev 11(3):375–392
Van de Kerkhof M, Wieczorek W (2005) Learning and stakeholder participation in transition processes towards sustainability: methodological considerations. Technol Forecast Soc Chang 72(6):733–747
Van den Hove S (2000) Participatory approaches to environmental policy making: the European commission climate policy process as a case study. Ecol Econ 33:457–472
Van der Brugge R (2009) The case of Dutch water management. PhD dissertation, Erasmus University Rotterdam, The Netherlands
Van der Brugge R, van Raak R (2007) Facing the adaptive management challenge: insights from transition management. Ecol Soc 12(2):33
Van der Brugge R, Rotmans J, Loorbach D (2005) The transition in Dutch water management. Reg Environ Chang 5(4):164–176
Van der Heijden J (2016) Experimental governance for low-carbon buildings and cities: value and limits of local action networks. Cities 53:1–7
Van der Helm R (2009) The vision phenomenon: towards a theoretical underpinning of visions of the future and the process of envisioning. Futures 41:96–104
Van der Lei TE, Enserink B, Thissen WAH, Bekebrede G (2010) How to use a systems diagram to analyse and structure complex problems for policy issue papers. J Oper Res Soc 62:1391–1402
Van der Voorn T, Pahl-Wostl C, Quist J (2012) Combining backcasting and adaptive management for climate adaptation in coastal regions: a methodology and a South African case study. Futures 44:346–364
Van Eijndhoven J, Frantzeskaki N, Loorbach D (2013) Connecting long and short-term via envisioning in transition arenas, as chapter 9. In: Edelenbos J, Bressers N, Scholten P (eds) Connective capacity in water governance. Ashgate Publications, London, pp 172–190
Van Vliet M, Kok K (2015) Combining backcasting and exploratory scenarios to develop robust water strategies in face of uncertain futures. Mitig Adapt Strateg Glob Chang 20:43–74
Vanolo A (2014) Smartmentality: the smart city as disciplinary strategy. Urban Stud 51 (5):883–898
Vasiliadou E, Safarzynsja K (2010) Transitions: taking complexity seriously. Futures. Article in Press
Verbong G, Loorbach D (eds) (2012) Governing the energy transition, reality, illusion or necessity? Routledge, New York
Vergragt PJ (2011) Beyond politization of technology and sustainability: a plea for visioning. Found Sci:1–5
Volkery A, Ribeiro T (2009) Scenario planning in public policy: understanding use, impacts and the role of institutional context factors. Technol Forecast Soc Chang 76:1198–1207
Vollenbroek F (2002) Sustainable development and the challenge of innovation. J Clean Prod 10 (3):215–223
Von Korff Y, Daniell KA, Moellenkamp S, Bots P, Bijlsma RM (2012) Implementing participatory water management: recent advances in theory, practice, and evaluation. Ecol Soc 17(1):30
Voß J, Bornemann B (2011) The politics of reflexive governance: challenges for designing adaptive management and transition management. Ecol Soc 16(2):9
Voß J-P, Smith A, Grin J (2009) Designing long-term policy: rethinking transition management. Policy Sci 42(4):275–302
Vreugdenhil H, Taljaard S, Slinger JH (2012) Pilot projects and their diffusion: a case study of integrated coastal management in South Africa. Int J Sustain Dev 15(1–2):148–172

Wachsmuth D, Aldana-Cohen D, Angelo H (2016) Expand the frontiers of urban sustainability. Nature 539:391–393
Walker WE (2000) Policy analysis: a systematic approach to supporting policymaking in the public sector. J Multicrit Decis Anal 9:11–27
Walker BH, Meyers JA (2004) Thresholds in ecological and social-ecological systems: a developing database. Ecol Soc 9(2):3. http://www.ecologyandsociety.org/vol9/iss2/art3
Walker G, Shove E (2007) Ambivalence, sustainability and the governance of socio-technical transitions. J Environ Policy Plan 9(3–4):213–225
Walker BH, Holling CS, Carpenter SR, Kinzig A (2004) Resilience, adaptability and transformability in social-ecological systems. Ecol Soc 9(2):5. DOI: http://www.ecologyandsociety.org/vol9/iss2/art5
Walker BH, Gunderson LH, Kinzig AP, Folke C, Carpenter SR, Schultz L (2006) A handful of heuristics and some propositions for understanding resilience in social-ecological systems. Ecol Soc 9(2):5. http://www. ecologyandsociety.org/vol9/iss2/art5/
Walker BH, Abel N, Anderies JM, Ryan P (2009) Resilience, adaptability, and transformability in the Goulburn-Broken Catchment, Australia. Ecol Soc 14(1):12
Wang Q, Chen X (2012) Regulatory failures for nuclear safety – the bad example of Japan – implication for the rest of the world. Renew Sust Energ Rev 16:2610–2617
Wang Q, Chen X, Yi-chong X (2013) Accident like the Fukushima unlikely in a country with effective nuclear regulation: literature review and proposed guidelines. Renew Sust Energ Rev 17:126–146
Wang MO, Vorwald CE, Dreher ML, Mott EJ, Cheng M, Cinar A, Mehdizadeh H, Somo S, Dean D, Brey EM, Fisher JP (2015) Evaluating-3D-printed-biomaterials-as scaffolds-for-vascularized-bone-tissue-engineering. Adv Mater 27:138–144
Wangel J (2011) Exploring social structures and agency in backcasting studies for sustainable development. Technol Forecast Soc Chang 78:872–882
Wardekker JA, de Jong A, Knoop JM, van der Sluijs JP (2010) Operationalising a resilience approach to adapting an urban delta to uncertain climate changes. Technol Forecast Soc Chang 77:987–998
Warner K (2014) Multiple resilience pathways: climate resilient pathways to sustainable development (IPCC AR5 WG2 Ch.20), presentation at UNU-EHS, 19 May 2014
WBGU (German Advisory Council on Global Change) (2011) World in transition: a social contract for sustainability, flagship report
WEF (2012) Risk and responsibility in a hyperconnected world – pathways to global cyber resilience. World Economic Forum (WEF), Cologny, pp 1–49. http://www3.weforum.org/docs/WEF_IT_PathwaysToGlobalCyberResilience_Report_2012.pdf
WEF (2014) Global risks 2014 – ninth edition WEF (2014). World Economic Forum (WEF), Geneva, pp 1–60. http://www3.weforum.org/docs/WEF_GlobalRisks_Report_2014.pdf
Westley F, Olsson P, Folke C, Homer-Dixon T, Vredenburg H, Loorbach D, Thompson J, Nilsson M, Lambin E, Sendzimir J, Banerjee B, Galaz V, van der Leeuw S (2011) Tipping toward sustainability: emerging pathways of transformation. Ambio 40:762–780
Westley FR, Tjornbo O, Schultz L, Olsson P, Folke C, Crona B, Bodin Ö (2013) A theory of transformative agency in linked social-ecological systems. Ecol Soc 18(3):27. https://doi.org/10.5751/ES-05072-180327
WHO (2012) Global health risks – mortality and burden of disease attributable to selected major risks. World Health Organization (WHO), Geneva, pp 1–56. http://www.who.int/healthinfo/global_burden_disease/GlobalHealthRisks_report_full.pdf
WHO (2013) Pandemic influenza risk management – WHO interim guidance. World Health Organization (WHO), Geneva, pp 1–56. http://www.who.int/influenza/preparedness/pandemic/GIP_PandemicInfluenzaRiskManagementInterimGuidance_Jun2013.pdf
WHO (2014) Outbreak surveillance and response in humanitarian emergencies, WHO guidelines for EWARN implementation

Wiek A, Iwaniec D (2014) Quality criteria for visions and visioning in sustainability science. Sustain Sci 9:497–512
Wiek A, Binder C, Scholz RW (2006) Functions of scenarios in transition processes. Futures 38 (7):740–766
Wilson S, Pearson LJ, Kashima Y, Lusher D, Pearson C (2013) Separating adaptive maintenance (resilience) and transformative capacity of social-ecological systems. Ecol Soc 18(1):22. DOI: http://dx.doi.org/10.5751/ES-05100-180122
Winkler H, Baumert K, Blanchard O, Burch S, Robinson J (2007) What factors influence mitigative capacity? Energ Policy 35:692–703
Wise RM, Fazey I, Stafford Smith M, Park SE, Eakin HC, Archer van Garderen ERM, Campbell B (2014) Reconceptualising adaptation to climate change as part of pathways of change and response. Glob Environ Chang 28:326–336. https://doi.org/10.1016/j.gloenvcha.2013.12.002
Wittmayer J, Schäpke N (2014) Action, research and participation: roles of researchers in sustainability transitions. Sustain Sci 9:483–496
Wittmayer J, van Steenbergen F, Quist J, Loorbach D, Hoogland C (2011) InContext. The community arena: a co-creation tool for sustainable behaviour by local communities. Methodological guidelines. http://incontext-fp7.eu/sites/default/files/Methodological%20guidelines_final.pdf
Wittmayer J, Van Steenbergen F, Loorbach D., Mock M, Omann I, Kirner B (2014) Exploring the transformative potential of communities. In: Wittmayer J, Roorda C, Van Steenbergen F (eds) Governing urban sustainability transitions – inspiring examples, DRIFT, creative commons. pp 83–89
Wittmayer JM, van Steenbergen F, Rok A, Roorda C (2016) Governing sustainability: a dialogue between local agenda 21 and transition management. Local Environ 21(8):939–955. https://doi.org/10.1080/13549839.2015.1050658
Yin RK (1994) Case study research: design and methods. Sage, Thousand Oaks
Yohe GW (2001) Mitigative capacity – the mirrow image of adaptive capacity on the emission side. Clim Chang 49:247–262
Zonneveld W, Spaans M (2014) Meta-governance and developing integrated territorial strategies: the case study of MIRT territorial agendas in the Randstad (Netherlands). Plan Theory Pract 15 (4):543–562

Part I
Introduction of Transitions Thinking in Urban Context and Theoretical Grounds of Transition Management in and for Cities

Chapter 2: Frantzeskaki et al. – Understanding the urban context and its challenges. This chapter introduces a new understanding of the urban context and elaborates on its challenges, from a transitions' governance perspective. It positions cities in the landscape of change and persistent problems, looks at inherent and innovation characteristics of cities and relates to urban sustainability debate. We walk the reader from the stepping stones of theoretical foundations of urban change to what it means for practical actions in cities. In this way sets the scene on 'why cities' and which type of challenges are at hand for new governance approaches like transition management to be relevant and a fit approach to urban governance and planning, complementing strategic environmental planning approaches. As such, this chapter offers an insight on how multi-level governance of cities relates and requires participatory approaches for sustainability.

Chapter 3: Frantzeskaki et al. – Introducing sustainability transitions' thinking in urban contexts. This chapter describes the basics of transitions thinking, its rationale and relation to sustainability transitions. We outline the fundamental conceptual models and frameworks (multi-phase model, multi-level perspective, multi-path, power and agency) and make a quick introduction to the relation of the sustainability transitions field to environmental governance.

Chapter 4: Wittmayer et al. – Transition Management: Guiding Principles and Applications. This chapter introduces transition management as a way of influencing sustainability transitions, also reflecting on its strengths and limitations as a framework for strategic environmental planning in cities. The chapter includes a description of the guiding principles of transition management, their translation into a governance framework and the operationalization thereof into process guidelines. For each phase of the process, rationale and methods as well as outcomes are outlined.

Chapter 5: Wolfram – Urban planning and transition management: Rationalities, instruments and dialectics. This chapter juxtaposes the transition management approach with the rationalities and instruments of urban planning. It introduces us to urban planning considering the historic evolution and current characteristics of mainstream urban planning in theory and practice, and then it elaborates on the key challenges and opportunities for engaging with transition management. The chapter takes a reflective but very well theoretically grounded view on what transition management offers to urban planning, and what are the merits and caveats of its application in cities. The discussion underlines the added value but also incompatibility of both approaches, and thus the need to create synergies and mutually compensate for critical deficits to effectively perform urban sustainability transitions.

Understanding the Urban Context and Its Challenges

Niki Frantzeskaki, Matthew Bach, and Patience Mguni

Abstract This chapter introduces a particular understanding of the urban context against the background of on-going societal change and persistent problems that threaten to induce disastrous change. The urban context is presented in relation to urban challenges, scopes of urban sustainability and the governance of the urban. Understanding urban context and its dynamics is important for finding the right approaches to tap into the potentials of cities and reroute their development pathways towards sustainability and resilience. The governance of urban sustainability transitions requires new forms of strategic environmental planning processes that integrate diverse sources of knowledge and diverse perspectives. The new planning processes need also to set connections between urban challenges and global, national and regional developments. The challenge remains for urban practitioners and scientists alike to take up a process-content thinking and to reflect on how to navigate societal complexity while mobilizing transformative societal potential towards action for sustainability.

Introduction

Today's urban areas face a host of complex problems, from the larger issue of climate change, to the more localized challenges of urban regeneration and the management of critical infrastructures. With the size and importance of cities set to increase over the next century, getting to grips with these issues acquires a more urgent meaning. However, addressing them is by no means an easy process.

N. Frantzeskaki (✉) · M. Bach
Dutch Research Institute For Transitions (DRIFT), Faculty of Social Sciences, Erasmus University Rotterdam, Rotterdam, The Netherlands
e-mail: n.frantzeskaki@drift.eur.nl; bach@drift.eur.nl

P. Mguni
Institute of Geosciences and Natural Resource Management, Copenhagen University, Copenhagen, Denmark
e-mail: pamg@ign.ku.dk

N. Frantzeskaki et al. (eds.), *Co-creating Sustainable Urban Futures*, Future City 11, https://doi.org/10.1007/978-3-319-69273-9_2

Complex problems commonly resist policy interventions, ignore optimization efforts, and frustrate even the most well-intentioned policies.

Lets turn our attention to Rotterdam, a mid-sized city in the Netherlands famous for its enormous port. It has been facing just these problems. On the one hand, its harbor activities have been relocated to the periphery of the city, leaving vast empty and, often, neglected spaces at the heart of the urban area. Deciding how to make use of this space will durably alter the character of the city. On the other hand, spurred by the threat of rising sea levels from climate change, the city has embraced sustainability as a guiding principle for its development.

Yet simply redesigning areas and committing to targets falls short of what is needed. At issue are not only the nature of the problems, but also how we respond to them. As in many other local governments, rigid hierarchies and the compartmentalization of crosscutting topics has led to a degree of institutional sclerosis. This has prevented the envisioning of a desired future for and by the city dwellers, and has generally led to symptoms being taken for causes and causes going untreated. Multi-dimensional topics such as climate change and urban renewal have been chopped up and distributed throughout the administration, which itself is frequently isolated from other urban actors. It is often the case that for sustainability solutions to be discovered and/or even legitimized, several local government departments have to collaborate and to be in agreement. This has made it all but impossible to put forward effective strategies and inspiring visions capable of addressing the underlying problems.

To address these combined challenges, the city has therefore had to consider its desired future and ways to get there. In doing so, local governments have found that it is not only a matter of changing approaches of planning and city making, but also how city making is practiced – it has had to reinvent its role without overlooking its institutional constraints, and open-up to experimenting. For example, the local government of Rotterdam has entered into new partnerships based on collaborative governance approaches. These involve devolving authority to actors not traditionally in charge of urban developments, while at the same time remaining ultimately accountable to citizens.

As such, an integral part of complex change processes in urban areas is learning about one's role, new pathways for action, and the interrelated challenges of the city. Traditional governance paradigms are rather ill-suited to the task: *persistent problems need to be tackled at their roots by deeper, transformative change*. Doing so then requires new approaches that will bring about new means of intervention, new meanings and discourses. As we will see in the following chapters, transition management, an offshoot of the transitions approach, seeks to provide just this: a new approach to co-create transformative action to deal with and/or steer clear from persistent problems of unsustainability (Tidball et al. 2016).

By way of this short example, we hope to provide a taste of the complex and complicated nature of the problems facing urban areas, and the need for radical societal change. In this chapter, we will further outline our understanding of the urban context in relation to on-going societal change and persistent problems. This kind of understanding is crucial for tapping into the potentials of cities and rerouting their development pathways towards sustainability and resilience. We also discuss scopes of urban sustainability and ways forward for urban governance for sustainability. This chapter provides a conceptual basis before turning to urban applications of transition management through a series of global case studies. The case study chapters will both enliven and make more concrete what we mean by transition management in and for cities and how the methodology relates to urban challenges.

Change and Its Challenges

Heraklietos noted that there is nothing more permanent than change. Crises, as a particular form of change, have commonly been seen as profoundly altering societies; they are perceived as critical events, upending routines and perceptions. Recent years have seen their fair share, whether economic (e.g., the 2008 financial crisis) or environmental (e.g., climate change). Apart from prompting national governments and international organizations to take decisive action, crises may also alter people's perceptions and opinions about the way they live and act, and the choices they make (Jhagroe and Frantzeskaki 2015).

Change, however, is not limited to cataclysmic events: it is a continuous process in both social and environmental systems. Its causes and conditions, however, cannot be reduced to a handful of factors. The ways in which change is analyzed are equally numerous. According to historical analyses, the prevailing conditions for social change include war, the adoption of a new government or institutional system (e.g. feudalism, democracy) and, more recently, the introduction of technologies (Parsons 1977). In political analyses, the focus is on the role of ideology: ideas, paradigms, and leaders provoke change. Economists write of market forces and the supply-demand nexus as the drivers of growth and progress. In all, change is seen as an on-going and unavoidable phenomenon that occurs due to changing conditions.

However, perceiving social change as an emergent phenomenon overlooks the impact that targets or visions can have in driving it. Societies typically seek to improve their circumstances, and different visions lead this process – the democratic society, the welfare society, the knowledge society, the innovation society, and the sustainable society. Since 1987 with the publication of the Brundtland report (WCED 1987), sustainable development has been introduced as a pathway for our societies to achieve a twofold balance: between environment, economy, and society, and between the rights of current and future generations (see Box 1).

Sustainable development has been cast as an antidote to environmental degradation and climate change, and politicians have readily embraced it as a new approach to the current string of crises. By highlighting the need to balance between 'people' (society), 'planet' (environment), and 'profit' (economy), sustainable development has been seen as a way of rebalancing our disproportional focus on economic growth, while mitigating some of the negative externalities it has generated in the other two areas. However, being aware of a desirable direction may not be sufficient to bring about desired changes.

Box 1: Brundtland Definition of Sustainable Development

Sustainable development is development that meets the needs of the present without compromising the ability of future generations to meet their own needs. It contains within it two key concepts: the concept of 'needs', in particular the essential needs of the world's poor, to which overriding priority should be given; and the idea of limitations imposed by the state of technology and social organization on the environment's ability to meet present and future needs. (WCED 1987: 41)

At the same time, as globalized societies become increasingly networked and interlinked, they also become more complex. This leads to a new set of problems and limitations for governance and planning, which requires new ways of thinking and governing, as well as an understanding of their urgency and complexity. Understanding societal change therefore becomes of high interest for understanding how our societies are going to learn from past experiences and to which extent they can shape their future.

Sustainable Development

As noted above, the concept of sustainable development underpins much current thinking on the problems facing cities and urban areas. Since the late 1980s, many countries have committed to it, but are struggling with how to achieve it. Following the Brundtland report *Our Common Future* (WCED 1987), sustainable development came to be defined as redirection of social development in ways which combine prosperity, environmental protection, and social cohesion. At the same time, it puts forward the notion that today's development cannot come at the expense of future generations. In this way, sustainable development introduces a subjective element, as assessing future needs is inevitably speculative and the very concept of needs is born of cultural, ecological and economic factors that can be weighted in more than one way (Martens and Rotmans 2005; UN 1997).

At the international level, there is a consensus on the need for sustainable development, as well as on key areas requiring progress over the next decade: poverty, hunger, health, education, etc. (UN 2005). Yet, there is little agreement on strategies and solutions, which vary from country to country. Many, for example, have formed sustainability councils and developed sustainability indicators (Mulder 2006: 148–165). In this context, sustainable development has been represented as the rebalancing of economic, social and environmental agendas, primarily through the mainstreaming of environmental concerns into policy.

The recently agreed upon New Urban Agenda (UN-Habitat 2016) also takes an agreement on sustainable development as given and promotes an urban future that centers on just and inclusive urban development. As such, it explicates desirable directions to take but does not indicate the pathways of action to achieve and pursue them.

Conceptually, there are several core characteristics of sustainable development that can be isolated. First, it is intergenerational, implying the need to take a longer-term view (e.g., one to two generations or 25–50 years). Second, scale is central – sustainability occurs at different levels (local to global) and acquires contextualized meanings, which interact. Third, sustainability is at the intersection of multiple domains, due to its concern with aligning social, economic, and environmental values (Kates and Parris 2003). Sustainability is a guiding value proposition: posits values on social, ecological and economic aspirations together in an integrative manner, making their achievement inseparable. Sustainable development therefore provides a general frame of reference for conceptualizing our common desired future.

Persistent Problems and System Failures

Understanding change and thinking in terms of sustainable development, however helpful, cannot succeed without a deeper understanding of the problems at hand. Societies have evolved into complex structures with interrelated and interdependent subsystems (e.g., energy) and functions (e.g., heating). Trends at the macro level such as demographic growth, globalization, and the spread of information and communication technologies (e.g., mobile phones, the internet), have all had a hand in this avalanche of complexity. It can, moreover, be seen as unfolding on three levels: on that of society itself, on that of the problems it faces, and on that of how we deal with these problems (Loorbach 2010: 163–164) (Fig. 1).

Simply being able to conceptualize such systems is a challenge. Approaches such as complex adaptive systems theory have attempted to do just this (Holland 1995). Deriving insights from the study of ecosystems, these approaches have sought to capture the characteristics of complex systems. Studies of complex adaptive systems have found that they involve multiple levels and scales, non-linear processes, and a diversity of responses to any one intervention (Holland 1995). The application of these approaches to societal systems has primarily operated through two system configurations: socio-technical and socio-ecological systems. In this volume, we will focus on the former (see Box 2), as the latter is more focused the interplay between social and environmental elements.

The study of societies from a systems perspective reveals that these cannot be governed through conventional top-down methods, which we refer to here as rationalist problem solving (Voß and Kemp 2006). This type of problem solving relies on precise goals and the use of predictions to choose between alternative options, which are then carried out through hierarchical institutions. For example, a

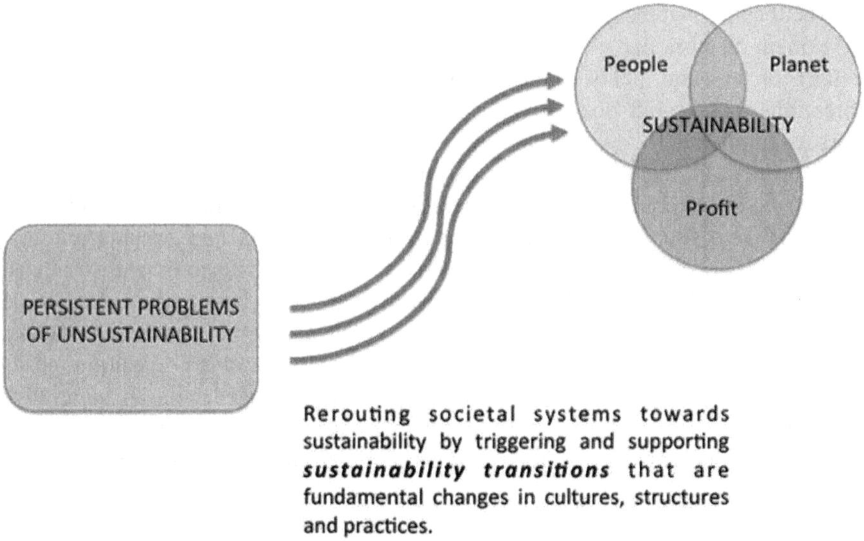

Fig. 1 A schematic of the basic concepts of sustainability transitions thinking

government facing high unemployment will seek to anticipate the outcomes of different measures (e.g., training programs, subsidies, etc.), and choose the one with the highest predicted effect. The new job creation program will then be implemented through different agencies, following a top-down pattern. In such an approach, the underlying principle is to isolate the problem, create a linear path to its solution, and devolve responsibilities for its resolution.

Box 2: Definition of Socio-technical Systems
Socio-technical: What we generally see as sectors can also be defined as socio-technical systems (e.g., energy). They contain actors (e.g., firms, individuals), institutions (e.g., behavioral norms, standards of practice), material artifacts (e.g., infrastructure), and knowledge (Markard et al. 2012). Viewing a sector as a socio-technical system helps to make the diversity and interdependency of its elements explicit.

Given that societal systems and their problems are complex, applying linear solutions yields unintended consequences, commonly understood as externalities. Indeed, by ignoring the messy nature of reality, rationalistic policies become blind to the interdependencies between the different societal subsystems, and to the potential consequences of actions targeting one subsystem on another one. These unintended consequences, however, often lead to more severe problems. For example, in response to demographic growth, governments sought to modernize agricultural systems (e.g. India's Green Revolution) by increasing the use of chemical fertilizers and pesticides. While these efforts initially led to a manifold

Table 1 Properties of persistent problems (Rotmans 2005)

Complex	Multiple causes and consequences
	Multi-sector, multi-scale
	Embedded in societal structures and institutions
Uncertain	No "ready-made" solutions
	Uncertainty cannot be reduced through knowledge acquisition
	Interventions primarily alter the problem perception, not the problem itself
Difficult to manage	Involvement of numerous autonomous actors with diverse interests at multiple scale levels
Hard to grasp	Resist interpretation
	Ill-structured
	Susceptible to power dynamics

increase in annual yields, they also polluted aquifers, destroyed soil fertility, and gave rise to a host of human health problems (e.g., birth defects, respiratory diseases).

These second-order problems are not only more numerous, but also more complex, interconnected and thus difficult to address, and are to a large degree the result of societal complexity. They can best be understood as **persistent problems**, which are rooted in the very structure of societal systems, making them difficult to manage. Persistent problems are all thorny, as they involve a wide range of actors from different sectors operating on multiple governance levels (local, national, supranational, global) and with divergent interests. Their name derives from their tendency to resist interventions that only address their symptoms – their impacts.

> Persistent problems are as a matter of fact, the superlative form of what Rittel and Webber (Rittel and Webber 1973) refer to as 'wicked problems'; their interrelationship to other societal problems and their entrenchment in our societal structures and institutions makes it impossible to analyze and solve them in isolation. Persistent problems could generally be considered to be symptoms of an unsustainable society. (Rotmans 2005: 8)

The core characteristics of persistent problems (Table 1) include the interdependence between actors, the need for action at multiple levels, the presence of feedback loops with negative impacts on the system, and the ill-structured nature of the problems themselves. This complexity requires new modes of governance, which is problematic: governance relies on institutions that themselves must be changed if a new mode of governance is to be enabled (Mayntz 2006).

Both persistent problems and the unintended consequences of rationalist problem solving be the result of system failures (Rotmans 2005). These are failures that have crept into our societal systems which, contrary to market failures, cannot be corrected by the market or current policies. Existing policies are necessary but not sufficient; much more is needed. It is this "much more" that we will explore throughout this volume.

The Urban Context

Understanding the nature of societal change and of the problems facing our societies is vital, though of little use without being placed into a specific context. As we will see, urban areas, and the cities within them, offer one of most fruitful grounds for both understanding persistent problems and experimenting with approaches to address them. We will therefore seek to gain insights into the urban context and its specific challenges.

Defining the Urban

> There are probably as many different ways of conceiving what a city is as there are cities. A simple definition therefore has its attractions. The simplest is that a city is a human settlement where strangers are likely to meet. Richard Sennett, The Fall of Public Man, 1977, p. 39

The first striking fact about this context is its sheer size – more than half of the world's population lives in urban areas and this number is set to continue increasing. There is no one definition of these concepts, though cities could broadly be seen as one of the main building blocks of urban areas. Khare et al. (2011) consider the city as "a conglomerate of people, dwellings and businesses with a politically defined boundary" (2011: 227). They also emphasize that a city is not just its administration, even if its governance is largely undertaken by local governments. Ernstson et al. (2010) go further in defining the city as "an agglomeration of contested spaces that generate a range of urban services, from transport, housing, and medical aid, to jobs and financial markets" (Ernstson et al. 2010: 531).

At the same time, there has been a shift towards separating cities from territory, and giving them a more relational reading as spatially connected and incredibly complex (Eames et al. 2013). Amin (2004) argues that cities are increasingly embedded in global networks and transnational flows of commodities, people, knowledge, and culture, making it particularly difficult to isolate one or the other: "in this emerging new order, spatial configurations and spatial boundaries are no longer necessarily or purposively territorial or scalar, since the social, economic, political and cultural inside and outside are constituted through the topologies of actor networks which are becoming increasingly dynamic and varied in spatial constitution" (Amin 2004: 33). Cities therefore become sites of imagination, constructed through interactions, and evading the prescription of immobile boundaries.

In more concrete terms, urban areas can be viewed through their endless appetite for (natural) resources and their ability to produce vast quantities of waste, which together can cause serious (environmental) impacts. These impacts have little regard for administrative boundaries, and often extend into neighbouring areas (e.g., rural areas supplying them with food), potentially threatening their functioning. Furthermore, due to globalization, the impacts from urban areas can spread globally, making it all the more difficult to erect conceptual or physical boundaries between them. Cities and the urban areas that emerge around them therefore become one of the key battlegrounds for a sustainable future, though due to the size and complexity of the problems, finding a starting point all too often remains elusive.

Viewing cities purely through their negative manifestations may, however, be counter-productive. Doing so occults their role and potential in resolving many of the larger and persistent problems that we face – e.g., climate change, energy security. It is all too easy to fall into either extreme: a blind faith in the "silver bullet" potential of cities to cure humanity of all its ills, or a Malthusian gloom that decries urban areas as the root of all evil. Needless to say, the reality lies somewhere in the middle, and the reason for this is that we – as researchers, civil society activists, business persons, local government officers, etc. – are all capable of playing a role in the direction taken. As Collier et al. (2013, p. 524) argue, "urban communities must be seen as the central stakeholders in transitioning objectives". Cities and their enclosing urban areas are neither inherently sustainable nor unsustainable; rather, specific policies, institutions or behaviours can be.

Urban Challenges

As noted above, cities, and more generally urban areas, hold a unique role in the nexus of sustainability and societal change. They are at once a source of problems – pollution, unsustainable patterns of resource consumption, poverty and inequality – and a setting for enacting new and high-impact solutions as they are the location of critical infrastructure networks and populations. They are increasingly complex entities, posing a challenge for governance – in general and for sustainability – and sites of experimentation for new forms of decision-making.

Those studying cities are often split between the apocalyptic – the 'doom and gloom' approach – and the evangelic – the 'silver bullet' mindset. Considering that the reality lies somewhere between these two poles, it is important to outline some of the challenges faced by urban areas, before we can turn our attention to their potential for radical change, for transitions.

In 2007, the United Nations Population Fund – a UN agency that promotes the right of all humans to enjoy a life of health and equal opportunity – released a landmark report on the potentials and pitfalls of urban growth (UNFPA State of the World Population 2007 – Unleashing the Potential of Urban Growth). Its publication came shortly before the world passed an invisible but momentous milestone: in 2008, for the first time in human history, over half of the world's population (ca. 3.3 billion people) inhabited urban areas. This number is only expected to increase, reaching the five billion mark by the year 2030.

With this large and ever-increasing concentration of populations in urban centers, the challenges faced by these also increase, not least since many of the new urban dwellers are and will continue to be poor. While this growth is not evenly spread – Africa and Asia account for the lion's share of it – the task of managing it is a truly global one. This may at first blush appear impossible, and urban areas can easily be cast in a negative light: they are, for instance, drivers of environmental degradation. At the same time, the overwhelming urbanization of the human population could very well be positive: "cities concentrate poverty, but they also represent the best hope of escaping it" (UNFPA 2007: 1).

This sense of the potential of urban areas to go from causing social, economic, and environmental problems, to being at the forefront of their mitigation is also highlighted in the IPCC's 5th report (IPCC 2013). On the one hand, urban areas concentrate many of the key global climate change-related risks (e.g. rising sea levels, increased extreme weather events), which are rapidly increasing. This is not limited to any one region: "risks, vulnerabilities, and impacts are increasing across the world in urban centres of all sizes, economic conditions and site characteristics" (IPCC 2013: II-5). On the other hand, urban areas are at the heart of successful adaptation to global climate change, which does not need to be limited to incremental development. In fact, the IPCC goes so far as to recommend "transformative development" to lead our societies towards resilience and sustainable development (Ibid.).

While we note above that urban centres in all regions of the world are facing climate risks, it is also important to understand that the sustainability, and indeed transition challenges, facing cities differ from region to region. For cities in so called developed countries, the challenge is essentially about decoupling from current resource-intensive urban forms and lifestyles towards more sustainable configurations of the city (Swilling and Annecke 2012). For cities in the global South on the other hand, the current challenge is to address existing underdevelopment i.e. rapid urbanisation, poverty and deficits in infrastructure and governance, in ways that avoid the resource-intensive paths taken by developed cities. As such, urban adaptation and sustainability trajectories will differ from region to region.

Nonetheless, the picture that emerges is one of cautious optimism as to the role that urban areas can play in resolving the persistent issues faced by today's and tomorrow's world – not least of which climate change and human poverty. For example, due to their high contribution to GHG emissions and their particular vulnerability to the localized effects of climate change, cities are ideally placed to be at the source of significant mitigation and adaptation activities. Moreover, local governments have a large degree of authority over many of the policy domains influencing these factors – e.g., transportation, waste management, land use (Kamal-Chaoui and Robert 2009; Bulkeley 2010; Dimitrov 2010; Bai 2007). Urban policies (e.g., London's congestion charge) can also be effective in providing local co-benefits such as public health improvements or energy security (Kamal-Chanoui and Robert 2009).

Additionally, cities offer the opportunity to directly engage a large variety of actors involved at multiple scales (local, regional, national, international), to mobilise their support and influence their behaviour (Loorbach 2009; Bai 2007; Roorda et al. 2011). Alternatives can emerge and be nurtured – they can be inspiring to others, or be translated for implementation at higher levels of governance. Actions at the urban level can thus have a global impact and cities are seen as critical arenas for addressing sustainability issues (Wittmayer et al. 2015; Wolfram and Frantzeskaki 2016).

Scopes of Urban Sustainability

There is an additional challenge in approaching cities in relation to sustainability and transitions towards it: what is the scope of a city? How do cities relate to other levels of governance? Indeed, many (or most) analyses of urban areas in relation to sustainability limit themselves to understanding the ways in which cities are (un-) sustainable. This has the unfortunate tendency of placing them in a bubble, divorcing them from other scale levels (e.g., national, international).

Bulkeley and Betsill (2005) emphasize that viewing urban areas as largely independent from other levels and therefore abstracting them from their links to them seriously limits our ability to understand urban sustainability and its challenges: "multilevel governance perspectives can start to open up these divisions, and provide insight into the opportunities and contradictions which emerge in the interpretation and implementation of urban sustainability across a range of scales and spheres of governance" (Burkeley and Betsill 2005: 43).

As Cash et al. (2006) also points "[...] in a world increasingly recognized as being multilevel, solutions must be as well. The opposite poles of top-down approaches, which are too blunt and insensitive to local constraints and opportunities, and bottom-up approaches, which are too insensitive to the contribution of local actions to larger problems and the resulting potential for tragedies of the commons, are clearly inadequate in providing both socially robust information and viable management solutions."

A further issue related to questions of scope is the fact that many of the persistent problems affecting urban areas are not specifically linked to any one scale. Climate change, for instance, is a global issue, albeit with local manifestations. This implies that there is an ever-present tension between scale levels in addressing it. Solutions targeting one level can never be fully independent of actions undertaken at others, nor can interventions targeting only one level be capable of comprehensively addressing a persistent problem.

Defining the city and its scope is, however, not just an administrative or academic exercise: where and how we draw the limits of a city or urban area can also have significant impacts on how we understand current persistent problems, forge solutions, and put them into practice (Box 3). For instance, cities are frequently held to produce 75–80% of global greenhouse gas emissions (Satterthwaite 2008). While this may be correct if we attribute to cities all emissions linked to their consumption (e.g., from agricultural or industrial production occurring elsewhere), this also blurs their limits and exaggerates the part they play in global unsustainability. Indeed, Satterthwaite (2008) finds that globally less than half of greenhouse gas (GHG) emissions come from within the boundaries of cities.

Similarly, the GHG Protocol – an international accounting tool and standard – provides three scopes for measuring the emissions of cities: (1) includes direct emissions originating from sources owned or controlled by the reporting entity; (2) includes indirect emissions from the activities of the reporting entity, but occur at sources owned or controlled by another entity; and (3) includes indirect emissions, such as the extraction and production of purchased materials and fuels, transport-related activities, and electricity-related activities not covered in (2) (GHG Protocol 2012). Thus, underlying any effort to understand urban sustainability and its governance, is the equally important task of defining or at least delimiting the scopes of what is being considered.

Box 3: Cities as Sites of Sustainability (Adapted from Khare et al. 2011)

Cities can be seen as sites for sustainability and the addressing of persistent problems for many reasons, including:

- The majority of the world's population lives in them;
- They consume large volumes of resources and produce equally large amounts of waste – they are therefore susceptible to efficiency policies;
- They play a key role in making sustainable development relevant on a local scale;
- They can be powerful actors in promoting sustainable development towards other scale levels – e.g. national governments – and frequently have the resources necessary to pilot small-scale initiatives.

They tend to have considerable experience managing the complex systems that drive unsustainability (e.g. energy, transport), and are therefore ideally placed to enact strategies to improve them.

With the planet becoming more and more urbanized, the ecological footprint of cities reaches far outside their geographic boundaries. Resources from other continents are sourced for large cities to operate. This makes even greater the challenge to create sustainable cities, since it will have a great impact on the way environmental resources are utilized and impacts are lessened or not.

> urban sustainability and land change studies cannot focus solely on a place of fixed geographical locations, but should examine the complex set of dynamic processes that link distant and sometimes multiple locations—longstanding themes in the urban literature. (…) Through the lens of urban land teleconnections, new and surprising diverse urban

> forms and processes, such as periurbanization, can be better understood and foreseen. The urban land teleconnections concept could also be useful to the wider research community to anticipate implications for global land resource use. (Seto et al. 2012)

Governing the Urban

Having established that there is something distinctly urban and that it is a key issue to explore, we can turn our attention to governance. How can such complexity be governed and how has it been governed? How can governance for urban areas effectively address persistent problems?

The traditional view is that urban governance refers to "the broad constellation of social, political and economic forces that mold the process of urban development within modern capitalism". Lefebvre, the famous French philosopher and sociologist, argued that urban governance unfolds across geographical scales, as urbanization processes include cities, regions, cross-border agglomerations, as well as supranational hierarchies (Lefebvre 2003 [1970]).

For most of the past century, the key task of urban governance was to find ways of effectively managing these complex and multi-level spaces. In the US, there was a particularly virulent debate regarding the appropriate size and division of urban functions at the urban or metropolitan level (Dowding et al. 1994). In this regard, two main schools of thought have emerged (Ostrom 1972). On the one hand, those advocating polycentric governance (Ostrom et al. 1988) claim that competition among multiple local jurisdictions can make local public services more efficient. On the other hand, 'consolidationists' argue that limiting the number of local governments is beneficial (Frug 1999). Regardless of such debates, the recognition that urban governance is of a fundamentally multi-level nature is well established.

Urban Governance for Sustainability

> While municipalities were once viewed simply as providers of services such as waste collection and utility provision, a shift has occurred in which the municipalities act as leaders on sustainability issues. (Nevens and Roorda 2013: 112)

Early attention to the links between the urban areas and sustainability can be traced back to the 1970s, when it was considered that urban quality of life had to be improved if societies were to flourish (Bulkeley and Betsill 2005). These early reflections tended to focus on the poverty–environmental quality–urbanization nexus, and more often than not ignored the wider impacts of urban areas.

The first explicit linking of urban areas and sustainability came with the Brundtland report (WCED 1987), which devoted an entire chapter to "The Urban Challenge". The report highlighted the central role of urban areas in the roll out of sustainable development. Subsequent initiatives such as the Local Agenda

21 (LA21) sought to cement the connection between urban areas and sustainable development, by encouraging local authorities to enter into partnerships with local stakeholders, and engage in localized implementations of sustainable development.

Spurred by such initiatives, cities, primarily in the Western world, have become increasingly engaged with sustainable development since the early 1990s (Bulkeley 2010; Pattberg and Stripple 2008; Kern and Alber 2008). Cities like Portland in the USA, Toronto in Canada, and Freiburg in Germany are frequently cited as frontrunners in the area of urban sustainability. These cities have developed ambitious plans, committed to stringent targets, and have generally institutionalized sustainability as a guiding principle for their governance (Aylett 2011; Roorda et al. 2011).

In 2014, for instance, Freiburg announced that it would aim to become climate neutral by 2030. Urban sustainability governance, however, goes beyond the boundaries of individual cities. Transnational city networks such as the International Council for Local Environmental Initiatives (ICLEI), the European Covenant of Mayors, and the Climate Alliance have been established to foster cooperation and knowledge sharing across local and national boundaries. City networks play an important role in mediating and diffusing knowledge across cities and are ever growing (Bulkeley et al. 2003; Bulkeley 2010). Moreover, at the international level, cities' attempts to tackle climate change are given ever-increasing attention, for instance during the recent UN World Summit in Johannesburg in 2012 or in the IPCC's Fifth Assessment Report (Bulkeley and Broto 2012; IPCC 2013).

At the same time, urban governance for sustainability is frequently limited to climate mitigation and adaptation activities (Bulkeley 2010; Anguelovski and Carmin 2011). Whereas mitigation aims to limit the risk of climate change, adaptation strives to reduce the vulnerability of human and natural systems (Corfee-Morlot et al. 2009). In spite of the pervasive policy rhetoric, which would suggest high political commitment to urban sustainability governance, empirical research reveals that local governments frequently only target 'low hanging fruits' and fail to reach their goals, especially when it comes to reducing GHG emissions (Bulkeley and Broto 2012). The reasons for this are not necessarily a lack of will to enact changes: sustainability problems are (for the most part) persistent problems, as noted above, which makes them particularly challenging to tackle.

Local governments' focus on short-term goals, the fragmentation of policy initiatives across departments, the discontinuous nature of commitments due to political cycles (e.g. elections), and more contribute to sap even the best-laid plans (Maas et al. 2012). There is frequently a fundamental contradiction in the temporal, spatial and institutional scales between urban decision-making and the global environmental problems affecting cities (Eames et al. 2013). There is thus a discrepancy between persistent problems and the governance approaches used to solve them.

This situation is nonetheless not without hope. Cities at the forefront of urban sustainability governance have been tackling these problems through participatory governance (Aylett 2011). This approach recognizes that public policy "is formulated and implemented in dynamic contexts where multiple actors interact at multiple levels" (Driessen et al. 2012, p.143). Thus, the traditional hierarchical model of urban policymaking is expanded to include a horizontal dimension. In doing so, distinctions between formal and informal, public and private become blurred, and it becomes possible to recognize and involve the myriad actors who together shape the city's ever-changing context.

Participatory governance is particularly relevant to urban sustainability governance (Bakker et al. 2012; Aylett 2011), as the underlying persistent problems cannot be addressed without the inclusion of those affecting and affected by them. Moreover, taking such a holistic approach can help to create consensus between (or at least a mutual recognition of) opposing views and interests, and through increased participation, generate a higher degree of legitimacy and ownership, thus improving compliance (Mees et al. 2012; Huxham et al. 2009).

> This [participation] goes beyond conventional discussions of behaviour change. Key interventions like the densification of existing neighbourhoods, radical increases in the energy efficiency of private buildings, the implementation of local renewable energy systems, or fundamental shifts in patterns of urban mobility are not under top-down municipal control. The design and implementation of these policies requires communities to be actively involved. Aylett (2011: 7)

In Dar es Salaam as in many African cities, there are well-established bottom-up efforts as households, communities, the market and NGO's make up for the governance and infrastructure deficits on a daily basis (Mguni et al. 2015). Dar es Salaam is predicted to be at risk of increased occurrence of flooding and droughts due to projected climate change impacts (START 2011). Any consideration of resilience to these impacts and overall sustainable development in Dar es Salaam has to consider this departure point: of the 3.3 million residents (with a projected annual population increase of 4.3%), 70–80% of the population live in unplanned settlements thus implying a growing infrastructure and adaptation deficit (Dodman et al. 2009). Such cities present a unique challenge for governing sustainability; as post-colonial top-down governance approaches based on the rationalist approach have failed to create liveable and equitable urban environments. Performative participatory approaches such as transition management may be key to moving beyond top-down decision-making that frequently disqualifies existing informal urbanism by providing platforms for linking bottom-up efforts with local government management policies to ensure urban development that is context-specific and that avoids the resource-intensive urban form found in many cities in the Western world (Fig. 2).

However, the most feasible and effective configuration for such participation remains unclear – How much time is needed? Who should participate? In which capacity? Reconfiguring the roles of different actors therefore also requires new skills and the willingness of those involved to learn and experiment (Roorda et al. 2011). As such, the underlying problem is how to devise participatory arrangements

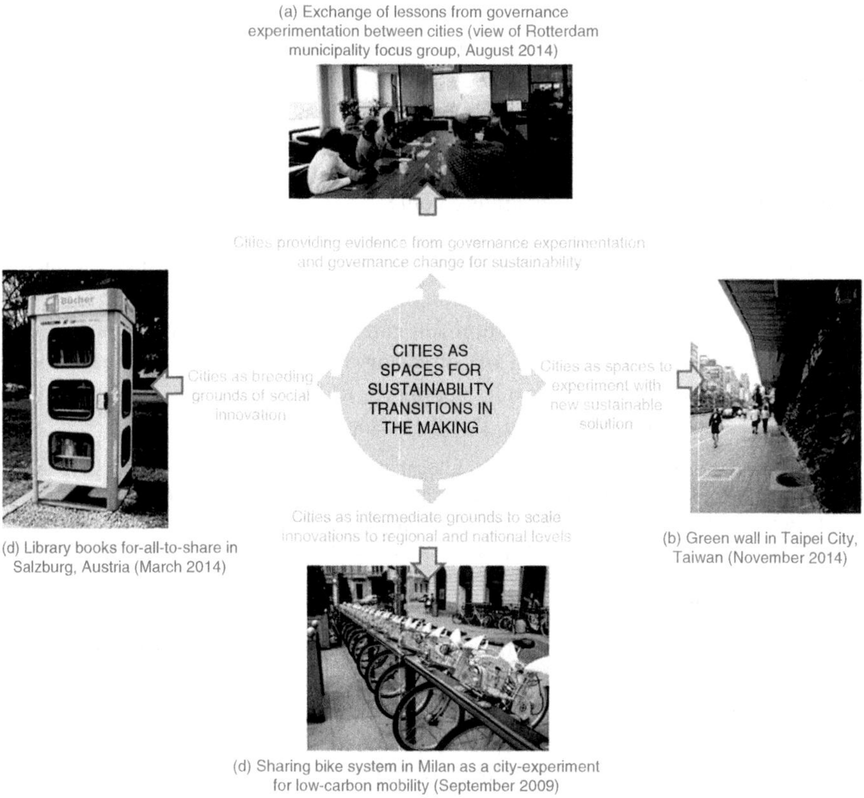

Fig. 2 A schematic of the basic reasons for exploring sustainability transitions in cities (Photo credits: Niki Frantzeskaki)

for specific purposes. However, before even determining the various parameters for participation in urban governance for sustainability, it becomes necessary to determine who the different actors are or could be. Khare et al. (2011) distinguish between three main groups.

First, local governments; they inevitably take decisions concerning economic efficiency, environmental quality, and social cohesion (Box 4). Second, business – companies and industries; they play a strong role through their high levels of resource consumption and waste production, as well as through their economic power. Third, citizens – “citizens desire environmentally healthy surroundings in which to live, work, play and bring up their families” (Khare et al. 2011: 229). Nevens and Roorda (2013) in their application of transition management at the urban level further disaggregate urban stakeholders. These include frontrunners (i.e., local leaders already involved in sustainability initiatives – see Wittmayer et al. (chapter “Transition Management: Guiding Principles and Applications”, this volume) for more information), city administrators, researchers, businesses, civil society organizations, and citizens.

Box 4: Cities as Sites of Innovation and Experimentation

Rotmans et al. (2000) elaborate an integrated city planning tool for sustainability, and argue that cities are motors for sustainable development.

For Ernstson et al. (2010), innovation is a key driver behind urban growth, and can be seen as a double-edged sword: innovation can both spur environmental degradation and offer solutions to improving the socio-ecological systems of a city.

Bulkeley et al. (2011) see cities as arenas for addressing sustainability issues.

Bulkeley and Broto (2012) argue that urban sustainability governance takes place through experimentation. They see this as giving rise to new forms of political space that blur the public-private divide.

Nevens and Roorda (2013) apply the transition management framework to the urban context through what they call "Urban Transition Labs" – "a hybrid, flexible and transdisciplinary platform that provides space and time for learning, reflection and development of alternative solutions" (115).

References

Amin A (2004) Regions unbound: towards a new politics of place. Geogr Ann: Ser B Hum Geogr 86(1):33–44

Anguelovski I, Carmin J (2011) Something borrowed, everything new: innovation and institutionalization in urban climate governance. Curr Opin Environ Sustain 3(3):169–175

Aylett ACE (2011) The urban governance of climate change: a comparative socio-institutional analysis of transformative urban responses to climate change in Durban (South Africa) and Portland (OR, USA), Ph.D. thesis, University of British Columbia

Bai X (2007) Integrating environmental concerns into urban management. The scale and readiness argument. J Ind Ecol 11:15–29

Bakker J, Denters B, Oude Vrielink M, Klok P-J (2012) Citizens' initiatives: how local governments fill their facilitative role. Local Gov Stud 38(4):395–414

Bulkeley H (2010) Cities and the governing of climate change. Ann Rev Environ Res 12:141–159

Bulkeley H, Betsill M (2005) Rethinking sustainable cities: multilevel governance and the 'urban' politics of climate change. Environ Polit 14(1):42–63

Bulkeley H, Broto VC (2012) Government by experiment? Global cities and the governing of climate change. Trans Inst Br Geogr 1–15

Bulkeley H, Davies A, Evans B, Gibbs D, Kern K, Theobald K (2003) Environmental governance and transnational municipal networks in Europe. J Environ Policy Plan 5(3):235–254

Bulkeley H, Castán Broto V, Maassen A (2011) Governing low carbon transitions. In: Bulkeley et al (eds) Cities and low carbon transitions. Routledge Taylor and Francis Group, London/New York, pp 29–41

Cash D, Adger W, Berkes F, Garden P, Lebel L, Olsson P, Pritchard L, Young O (2006) Scale and cross-scale dynamics: governance and information in a multilevel world. Ecol Soc 11(2):8–19

Collier MJ, Nedovic-Budic Z, Aerts J, Connop S, Foley F, Foley K, Newport D, McQuaid S, Slaev A, Verburg P (2013) Transitioning to resilience and sustainability in urban communities. Cities 32:521–528

Corfee-Morlot J, Kamal-Chaoui L, Donovan MG, Cochran I, Robert A, Teasdale P-J (2009) Cities, climate change and multilevel governance, OECD Environmental Working Papers No 14. OECD publishing, Paris
Dimitrov RS (2010) Inside Copenhagen: the state of climate governance. Glob Environ Polit 10 (2):18–24
Dodman D, Kibona E, Kiluma L (2009) Tomorrow is too late: responding to social and climate vulnerability in Dar es Salaam, Tanzania. Case study prepared for the Global Report on Human Settlements 2011 (UN Habitat, 2011)
Dowding K, John P, Biggs S (1994) Tiebout: a survey of the empirical literature. Urban Stud 31:767–797
Driessen PPJ, Dieperink C, van Laerhoven F, Runhaar HAC, Vermeulen WJV (2012) Towards a conceptual framework for the study of shifts in modes of environmental governance. Experiences from the Netherlands. Environ Policy Gov 22(3):143–160
Eames M, Dixon T, May T, Hunt M (2013) City futures: exploring urban retrofit and sustainable transitions. Build Res Inf 41(5):504–516
Ernstson H, van der Leeuw S, Redman C, Meffert D, Davis G, Alfsen C, Elmqvist T (2010) Urban transitions: on urban resilience and human-dominated ecosystems. Ambio 39:531–545
Frug G (1999) City making: building communities without building walls. Princeton University Press, Princeton
GHG Protocol (2012) Greenhouse gas protocol. http://www.ghgprotocol.org/. Used: March 13th 2013
Holland J (1995) Hidden order: how adaptation builds complexity. Addison-Wesley, Reading
Huxham C, Vangen S, Huxham C, Eden C (2009) The challenge of collaborative governance. Public Manag: Int J Res Theory 2(3):337–358
IPCC (2013) Climate change 2013: the physical science basis. Intergovernmental panel on climate change
Jhagroe S, Frantzeskaki N (2015) Politics of crisis: exceptional democracy in Dutch infrastructure governance, Critical Policy Studies, (https://doi.org/10.1080/19460171.2015.1066690)
Kamal-Chaoui L, Robert A (2009) Competitive cities and climate change, OECD Regional Development Working Papers No 2. OECD publishing, Paris
Kates RW, Parris TM (2003) Long-term trends and a sustainability transition. PNAS 100 (14):8026–8067
Kern K, Alber G (2008) Governing climate change in cities: modes of urban climate governance in multi-level systems. Prepared for OECD competitive cities and climate change, Milan, October 9th–10th, 2008
Khare A, Beckman T, Crouse N (2011) Cities addressing climate change: introducing a tripartite model for sustainable partnership. Sustain Cities Soc 1(4):227–235
Lefebvre H (2003) [1970] The urban revolution. U of Minnesota Press, USA
Loorbach D (2009) Urban transitions and urban transition management. The case of Rotterdam, Workshop on Urban Transitions, May 7–8, 2009, Manchester
Loorbach D (2010) Transition management for sustainable development: a prescriptive, complexity-based governance framework. Gov: Int J Policy Adm Inst 23(1):161–183
Maas S, Fortuin K, Frantzeskaki N, Roorda C (2012) A systems analysis methodology for exploring urban sustainability transitions. Exploring challenges and opportunities for urban sustainability transitions in Ghent and Aberdeen, conference paper presented at 3rd International Conference of Sustainability Transitions, Copenhagen, 26–28 August 2012
Markard J, Raven R, Truffer B (2012) Sustainability transitions: an emerging field of research and its prospects. Res Policy 41(6):955–967
Martens P, Rotmans J (2005) Transitions in a globalizing world. Futures 37:1133–1144
Mayntz R (2006) Governance Theory als fortentwickelte Steuerungstheorie? In: Schuppert GF (ed) Governance-Forschung: Vergewisserung u¨ ber Stand und Entwicklungslinien. Nomos Verlag, Baden-Baden, pp 11–20
Mees HLP, Driessen PPJ, Runhaar HAC, Stamatelos J (2012) Who governs climate change? Getting green roofs for stormwater retention off the ground. J Environ Plan Manag 1:1–24

Mguni P, Herslund L, Jensen MB (2015) Green infrastructure for flood-risk management in Dar es Salaam and Copenhagen: exploring the potential for transitions towards sustainable urban water management. Water Policy 17(1):126–142

Mulder K (ed) (2006) Sustainable development for engineers, A handbook and resource guide. Greenleaf Publishing, Sheffield

Nevens F, Roorda C (2013) A climate of change a transition approach for carbon neutrality in the city of Ghent (Belgium), Sustainable Cities and Society (2013), https://doi.org/10.1016/j.scs.2013.06.001

Ostrom E (1972) Metropolitan reform: propositions derived from two traditions. Soc Sci Q 53 (3):474–493

Ostrom V, Bish RL, Ostrom E (1988) Local government in the United States. ICS Press, San Francisco

Pan-African START Secretariat, International START Secretariat, Tanzania Meteorological Agency & Ardhi University (2011) Urban poverty and climate change in Dar es Salaam, Tanzania: a case study. Final Report

Parsons T (1977) Social systems and the evolution of action theory, vol 62. Free Press, New York

Pattberg P, Stripple J (2008) Beyond the public and private divide: remapping transnational climate governance in the 21st century. Int Environ Agreements 8:367–388

Roorda C, Buiter M, Rotmans J, Bentvelzen M, Keeton R (2011) Urban development: the state of the sustainable art. An international benchmark of sustainable urban development. Dutch Research Institute for Transitions (DRIFT), Rotterdam

Rotmans J (2005) Societal Innovation: between dream and reality lies complexity. Erasmus university of Rotterdam, inaugural address. Research in management series, 2005, reference number EIA-2005-026-ORG

Rotmans J, van Asselt M, Vellinga P (2000) An integrated planning tool for sustainable cities. Environ Impact Assess Rev 20:265–276

Rittel H, Webber M (1973) Dilemmas in a general theory of planning, Policy Sciences, vol 4. Elsevier Scientific Publishing Company, Inc., Amsterdam, pp 155–169

Satterthwaite D (2008) Cities' contribution to global warming: notes on the allocation of greenhouse gas emissions. Environ Urban 20(2):539–549

Seto KC, Reenberg A, Boone CG, Fragkias M, Haase D, Langanke T, Marcotullio P, Munroe D, Olah B, Simon D (2012) Urban land teleconnections and sustainability, PNAS, www.pnas.org/cgi/doi/10.1073/pnas.1117622109

Swilling M, Annecke E (2012) Just transitions: exploring sustainability in an unfair world. UCT Press/United Nations University Press, Cape Town/Tokyo

Tidball KG, Frantzeskaki N, Elmqvist T (2016) Traps! An introduction to expanding thinking on persistent maladaptive states in pursuit of resilience. Sustain Sci 11:861

UN (1997) Global change and sustainable development: critical trends. United Nations, Department for Policy Coordination and Sustainable Development, New York

UN (2005) The millennium development goals report. United Nations, New York

UNFPA (2007) State of world population 2007: unleashing the potential of urban growth

UN-Habitat (2016) Urbanization and development. Emerging futures, World Cities Report 2016. UN-Habitat, Nairobi

Voß JP, Kemp R (2006) Sustainability and reflexive governance: introduction. Reflexive Gov Sustain Dev:3–28

WCED, World Commission on Environment and Development (1987) Our common future. Oxford University Press, Oxford

Wittmayer JM, van Steenbergen F, Rok A, Roorda C (2015) Governing sustainability: a dialogue between Local Agenda 21 and transition management. Int J Justice Sustain 21(8):939–955

Wolfram M, Frantzeskaki N (2016a) Cities and systemic change for sustainability: prevailing epistemologies and an emerging research agenda. Sustainability 8:144. https://doi.org/10.3390/su8020144

Introducing Sustainability Transitions' Thinking in Urban Contexts

Niki Frantzeskaki, Matthew Bach, Katharina Hölscher, and Flor Avelino

Abstract This chapter describes the basics of transitions thinking, its rationale and relation to sustainability transitions. We outline the fundamental conceptual models and frameworks (multi-phase model, multi-level perspective, multi-path, power and agency) and make a quick introduction to the relation of the sustainability transitions field to environmental governance.

Introducing Transitions to Sustainability

As we have seen in the previous chapter, today's cities and urban areas face a host of challenges from poverty to climate change. These issues often find their roots in complex problems, which cannot be resolved through quick fixes. In fact, traditional modes of governance have failed to address them and at times even aggravated them. This has opened the way for experimentation with new approaches, for instance, focusing on participation and a greater delegation of authority (Frantzeskaki et al. chapter "Understanding the Urban Context and Its Challenges", this volume). At the same time, the widespread adoption of sustainable development as a frame of reference has helped to rebalance the different societal domains (environment, economy, society) and spread a recognition that change needs to be considered in and for the long term. However, simply adopting sustainability, altering who sits at the table, and giving them greater responsibilities will almost inevitably fall short, as by their very nature persistent problems require a deeper change: a transition. In this chapter, we introduce the central concepts, heuristics and frameworks underlying sustainability transitions research. This provides the conceptual basis for understanding and governing urban sustainability transitions.

N. Frantzeskaki (✉) · M. Bach · K. Hölscher · F. Avelino
Dutch Research Institute For Transitions (DRIFT), Faculty of Social Sciences, Erasmus University Rotterdam, Rotterdam, The Netherlands
e-mail: n.frantzeskaki@drift.eur.nl; bach@drift.eur.nl; holscher@drift.eur.nl; avelino@drift.eur.nl

N. Frantzeskaki et al. (eds.), *Co-creating Sustainable Urban Futures*, Future City 11, https://doi.org/10.1007/978-3-319-69273-9_3

What Are Transitions?

Asking "what is a transition?" yields no single answer, as the concept has been construed differently across several disciplines.

- Sociologists root it in biology and population dynamics (Davis 1945), using it to describe demographic transitions, which are the shift from high to low birth and death rates (UN 1997).
- Economists (see, for example, Rostow 1960 or Boulding 1970) view it relation to economic system changes, for instance, from a state-controlled economy to a market one.
- Those studying socio-technical systems (see Table 1) consider transitions as the outcome of developments unfolding at multiple levels (Rip and Kemp 1998; Schot 1998; Geels 2002): larger dynamics (e.g., macro-economic changes), the status quo (e.g., established firms, governments), and small-scale innovation (e.g., start-ups) interact with one another leading to structural changes.
- Finally, complex systems theorists think of transitions as system transformations from slow equilibrium dynamics to rapid development and instability, and reverting to relative stability (Rotmans 1994; Rotmans et al. 2000). The new equilibrium is a dynamic equilibrium, i.e. there is no status quo, because a great deal is constantly changing below the surface.

Regardless of one's theoretical inclinations, understanding structural change is a recurring theme across the social sciences, and one of their great challenges. A somewhat more unified field of study has emerged over the past decades to confront these questions head on – we will refer to these as transitions studies. They define transitions in different ways (see Box 1), though view them as nonlinear processes of social change, in which societal systems are structurally transformed. Going one

Table 1 Characteristics of transitions schools

Technical innovation systems	Approach focused on the emergence of new technologies and the related institutional and organizational changes
	Key references: Bergek et al. (2008), Hekkert et al. Hekkert et al. (2007), and Fuenfschilling and Truffer (2013)
Strategic niche management	Approach for deliberate niche creation to trigger transitions
	Key references: Hoogma et al. (2002) and Raven et al. (2012)
Transition management	Governance framework to influence transitions in a more sustainable direction
	Key references: Rotmans et al. (2001) and Loorbach (2007, 2010)
(Historical) transitions in socio-technical systems	Approach to explain (historical) technological transitions, which are seen as the result of dynamics at three levels (niche, regime, landscape)
	Key references: Geels (2002), Rip and Kemp (1998), and Geels and Schot (2007)

Adapted from Markard et al. (2012)

step further, they refer to a 'sustainability transition' as a "radical transformation towards a sustainable society as a response to a number of persistent problems confronting contemporary modern societies" (Grin et al. 2010:1). One of the central premises in transition studies is that persistent problems are symptoms of unsustainable societies, and that dealing with these persistent problems to enable more sustainable systems, requires transitions and system innovations (Box 1).

Box 1: Definitions of Sustainability Transitions

"A long-term process of change during which a society or a subsystem of society fundamentally changes" (Rotmans et al. 2000, 2001)

"A shift from an initial dynamic equilibrium to a new dynamic equilibrium" (Kemp and Rotmans 2001)

"A transition can be described as a set of connected changes which reinforce each other but take place in several different areas such as technology, the economy, institutions, behavior, culture, ecology and belief systems. A transition can be seen as a spiral that reinforces itself; there is multiple causality and co-evolution caused by independent developments" (Rotmans et al. 2001)

"A transition can be defined as a gradual, continuous process of societal change where the structural character of society (or a complex sub-system of society) transforms. [. . .] A transition can be described as a set of connected changes, which may reinforce each other but take place in several different areas, such as technology, the economy, institutions, behavior, culture, ecology and belief systems" (Martens and Rotmans 2005: 1136)

"A transition is a shift from one socio-technical system to another i.e. a system innovation." (Geels 2005a)

"A transition is a structural societal change that is the result of economic, cultural, technological, institutional as well as environmental developments, which both influence and strengthen each other" (Rotmans 2005)

"A transition denotes a long-term change in an encompassing system that serves a basic societal function" (Elzen and Wieczorek 2005: 651)

"A transition emerges out of co-evolutionary processes in which institutional, technological, behavorial, ecological, economic and other processes intertwine and reinforce each other" (Loorbach et al. 2009)

"Transitions are understood as processes of structural change in major societal subsystem. They involve a shift in the dominant 'rules of the game', a transformation of established technologies and societal practices, movement from one dynamic equilibrium to another – typically stretching over several generations (25–50 years)" (Meadowcroft 2009: 324)

Transitions Studies

Transition research has its intellectual roots in innovation studies as found in social studies of technology and evolutionary economics (Rip and Kemp 1998; Geels 2005). These areas of scholarship focused on technological transitions, including changes in user practices and institutional structures (e.g., regulatory, cultural), in addition to the technological dimension. The underlying notion is that socio-technical transitions do not just change the structures of existing systems (e.g., transportation); they also affect related societal domains (e.g., housing, planning, policymaking).

While originally the focus was on transitions in socio-technical systems (e.g., mobility, energy, agriculture), recent developments have broadened the focus towards societal systems more generally (e.g., regions, sectors), and to 'reflexive' governance for sustainable development (Voß et al. 2009). Either way, the primary object of transition studies remains societal (sub)systems, whether (sub)sectors or regions. This systemic perspective requires a holistic view that acknowledges the interaction between human and non-human aspects. The influence on societal systems is not only social, cultural, institutional and political, but also economical, ecological, and technological. Analytically, transitions studies draw on several core concepts, such as co-evolution – society and technology mutually shape one another. Then, they have developed several conceptual frameworks to better analyze transitions, the most widespread of which is the multi-level perspective (Table 1).

Core Principles

The core principles of transitions' thinking are about the conceptual foundations that inform the theory and tools (approaches to govern, to experiment) of sustainability transitions' studies. For getting a grasp of transitions' thinking and all associated conceptual models thereof, it is important to be introduced into these core principles as the foundations of the thinking behind these models.

Co-evolution

Although the term co-evolution has different meanings in different disciplines, in transitions studies it refers to the interactions between different systems, which influence the dynamics of the individual systems and thereby lead to irreversible patterns of change. Co-evolution has primarily been viewed in terms of science-technology interactions, though also those between culture and technology, and technology and society (Geels 2002).

Coordinated Emergence

A transition is conceived as a process of fundamental change that has both an emergent and coordinated aspect. The transitions approach builds on concepts from complex adaptive systems, such as coordination and emergence (see Box 2). The contradiction comes from the combination of two seemingly exclusive characteristics: coordination refers to the stimulus-response pattern of complex adaptive systems and emergence refers to the autonomous response that appears to emerge. Transitions are therefore societal processes of fundamental change including emergent and coordinated characteristics. Coordinated emergence describes the principle of radical change in incremental steps. In this way, a transition can be coordinated, while having an emergent character.

Evolutionary Revolution

> Evolution is change within the framework and the limits of the system. Revolution is a radical transformation of a system; the substitution of one principle or basis of organization for another. (White 1959: 281)

A transition is further conceptualized as a process of fundamental change that includes revolutions in different times and domains that jointly contribute to the evolution of the system in the long term. The tension inherent to the 'evolutionary revolution' concept comes from the juxtaposition of two mutually exclusive characteristics of change: evolution refers to continuous adaptation, mutation, and selection; revolution refers to drastic, immediate, and irreversible changes. The time horizons of the two change processes differ, as do the ways in which change is realized.

Revolutions are drastic changes that transform societal systems and are highly uncertain. Sanderson (2005: 1–2) perceives them as altering the basic values of a society, its social structure and institutions. Revolutions produce shifts in power, often through violent action, which makes the incumbent regime collapse. They occur as a conjunction of events that are "the necessary structural forces behind the revolutions" and that revolutions "are (. . .) the result of unintended consequences" (Sanderson 2005: 106).

The opposing element of evolution largely stems from the work of Darwin. In his view, evolution in nature is the outcome of a continuous process of adaptation, selection, and mutation (Mayr 2001). It is a random but guided process – a "directional change" (*Ibid*) – and while its realization is certain, its outcome is not. Evolution is therefore a complex process, in which the evolution of any one organism has an effect on that of others. This is referred to as co-evolution.

In the transitions approach, transitions are conceptualized as processes of change that are realized over the long term and are products of the combined effect of structural forces and developments in multiple sectors and levels (Rotmans et al. 2001). Loorbach et al. (2009) point out that a "transition emerges out of a co-evolutionary processes, in which institutional, behavioral, ecological, economic and other processes intertwine and reinforce each other." Evolutionary revolution thus refers to the process of fundamental change (revolution) as a sum of incremental changes (evolution) that constitutes a transition of the societal system (Frantzeskaki et al. 2012). Co-evolutionary processes occur in complex adaptive systems (Box 2) and societal systems are understood as complex adaptive ones.

> Radical structural change is needed to erode the existing deep structure (incumbent regime) of a system and ultimately dismantle it. (...) Incremental change allows the system to adjust to the new circumstances and to build up new structures that align to the new configuration. Radical change in incremental steps implies that the system heads in a new direction toward new attractors but in small steps. (Rotmans and Loorbach 2009: 189–190)

Box 2: Complex Adaptive Systems and Their Defining Characteristics
Complex adaptive systems are a special type of complex system that can adapt to change and learn from experience. They co-evolve, self-organize, and produce emergent patterns:

- Co-evolution refers to the process in which either (a) two systems or subsystems evolve over time while being interdependent resulting in mutual irreversible changes of both systems, or (b) the system and its environment mutually reinforce each others' change processes which results in an irreversible change of the system.
- Self-organization refers to the imminent capability of a complex adaptive system to organize itself without external involvement. Self-organization is inherent and distinctive to complex adaptive systems.
- Emergence describes the presence of processes at a high level of aggregation that are not related linearly to lower level decisions or structures.

Complex adaptive systems are also anticipatory, meaning that they have a memory that refers to past lessons and experiences (Holland 1995). Experience of the past is taken into account not due to their adaptive capacity, but due to their system memory that assigns them with anticipatory responses; meaning responses that are based on past experiences of the system as part of the system's memory.

Societal systems have similar characteristics and appear to behave as complex adaptive systems given that societal domains consist of numerous interlinked subsystems, there is a high degree of uncertainty about their interactions, they have an open and nested character in organization, and they tend to self-organize. Therefore, similar patterns can be observed, as for example emerging structures, co-evolving (policy) domains and self-organizing processes. Arguably, this complexity has increased over the last decades because of increased interaction, integration and interdependence.

Key Conceptual Frameworks

Transitions as highly complex societal phenomena, involving a multitude of actors, sectors, and governance levels, are particularly difficult to conceptualize and analyze. This has prompted transitions researchers to invest particular effort in developing appropriate conceptual and analytical frameworks to facilitate a better understanding of societal change and how it can be supported towards sustainability. We focus on the multi-level perspective, the multi-phase framework, the multi-path framework and reflections on agency and power.

Multi-level Perspective

The multi-level perspective (MLP) conceptualizes processes of change as taking place at (and in interaction with) multiple levels. Transitions can be viewed as outcomes of the continuous change of actor's practices and the interactions of practices and developments that take place at different levels. Changes at every level of interaction produce a convergent change can result in a system transition. The MLP is especially used in writings on socio-technological (Geels 2005a; Geels and Schot 2007). The three levels are defined as follows:

- The **micro level** where niches are located;
- The **meso level** where regimes are placed;
- The **macro level** that hosts trends and developments such as globalization.

Geels first put forward the MLP while researching socio-technological transitions and attempting to describe them (Fig. 1). According to him the levels "are not ontological descriptions of reality but analytical and heuristic concepts to understand the complex dynamics of socio-technical change" (Geels 2004: 33). These levels are neither hierarchically layered nor aggregated. This implies that a regime does not consist of niches and a landscape does not consist of regimes. Taken together, the micro, meso and macro levels form a descriptive framework of the constellations that interact during a socio-technological transition.

The MLP is therefore a valuable conceptualization of the nature of transitions. Scholars have used it as a descriptive framework to explain socio-technological transitions and to aid both the analysis and the suggestion of alternatives for innovation support (e.g. Kivits et al. 2010; Markard and Truffer 2008). Geels notes that it is an analytical tool that "provides an overall view of the multi-dimensional complexity of changes in socio-technical systems" (Geels 2010: 495). An MLP analysis helps to understand drivers and barriers to change and thus to open-up viable transition pathways (Frantzeskaki et al. 2010).

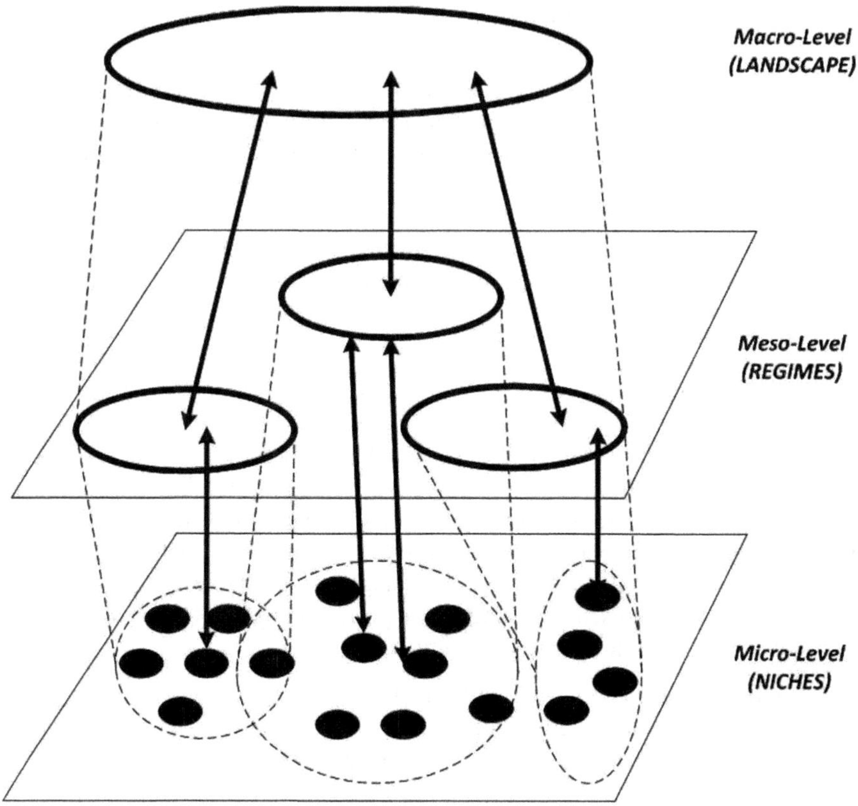

Fig. 1 The multi-level perspective of sustainability transitions (Geels and Schot 2007)

Macro Level

> The macro-level is formed by the socio-technical landscape, which refers to aspects of the wider exogenous environment, which affect socio-technical development (e.g. globalizations, environmental problems, cultural changes). The metaphor 'landscape' is used because of the literal connotation of relative hardness, and to include the material aspects of society, e.g. the material and spatial arrangements of cities, highways, and electricity infrastructures. Landscapes are beyond the direct influence of actors and cannot be changed at will. (Geels 2005a: 684).

Landscape factors are difficult to influence, and consequently difficult to change. This however does not imply that every factor that is difficult to influence is placed on the landscape. The multi-level perspective places trends and developments that are mainly external, to regime and to niche, at the landscape level. Landscape processes include "environmental and demographic change, new social movements, shifts in general political ideology, broad economic restructuring, emerging scientific paradigms, and cultural developments" (Smith et al. 2010: 441).

Meso Level

> The meso-level is formed by socio-technical regimes. (Geels 2005a: 683)

The notion of regime has a defining role in the transitions field. It is the system unit in which transitions need to take place. Though definitions have purposefully been left vague to accommodate a wide range of analytical scopes and policy issues, Rotmans and Loorbach (2010: 108) see them as dominant "conglomerate of structure (institutional physical setting), culture (prevailing perspective), and practices (rules, routines and habits)". In sum, the regime refers to the dominant status quo and business as usual, which is mostly viewed as (actively) resisting change, in seeking to preserve itself. Transitions are therefore ultimately about regime change.

Micro Level

> The micro-level is formed by technological niches, the locus for radical innovations ('variation'). (...) Niches are important because they provide locations for learning processes and space to build the social networks which support innovations. (Geels 2005a: 684)

> Niches are protected experimental settings where norms and practices are developed which depart from those of an incumbent technological regime. According to niche-based understandings, regime changes begin when practices and norms developed in the niche become adopted more widely. (Berkhout et al. 2003: 48)

At the micro level, niches act as innovation incubators. They are protected spaces within which innovation can be nurtured (Smith and Raven 2012). While in earlier transitions writings, niches were seen as being of a primarily technological nature (Geels 2005a; Hoogma et al. 2002), Transition Management has expanded them to include institutional, cognitive, and behavioral innovations (Loorbach 2010; van den Bosch 2010).

Inter-level Interaction

Change occurs within each level of the MLP and from the interaction across levels. Changes taking place within the regime are described as tensions, e.g. when the regime is required to optimize its functioning. Tensions can produce 'windows of opportunity' for (radical) niche innovations to break through (Geels 2004). The causes of these tensions are rooted in the internal inertia of the regime, inefficiencies of current technology, negative externalities, perceptions change (e.g. changes in user preferences), and market conditions (e.g. inter-firm competition). Some of these causes can refer to developments at the landscape or niche levels, for example when climate change puts pressure on existing regimes (landscape stress) or when there is a new demand for sustainable products (niche pressure).

Multi-phase Model

As already noted, a transition is a long-lasting process of change, which makes it particularly difficult to chart. Delineating the phases of its change is therefore essential when attempting to explain its development. A transition progresses in definable episodes, where significant system changes signal the passage from one to another.

Transition phases are blocks of time in which developments occur that result in irreversible system changes. Hence, the passage from one phase to another indicates different types of changes and processes. As the transition moves forward, existing structures of the societal system (values, institutions, regulations, markets, etc.) fade away, while new ones emerge (Geels 2004; Loorbach et al. 2008).

Societal systems go through long periods of relative stability and optimization that are followed by comparatively shorter periods of radical change. Specifically, the multi-phase framework of transitions is rooted in the hypothesis that "the dynamics of transitions in time can be described as altering phases of relatively fast and slow dynamics, which together form a strongly non-linear pattern where there is a shift from one dynamic state of equilibrium to the other" (Rotmans 2005: 23). Martens and Rotmans (2005) and van der Brugge and Rotmans (2007) have described a four-phase framework of transitions (Fig. 2).

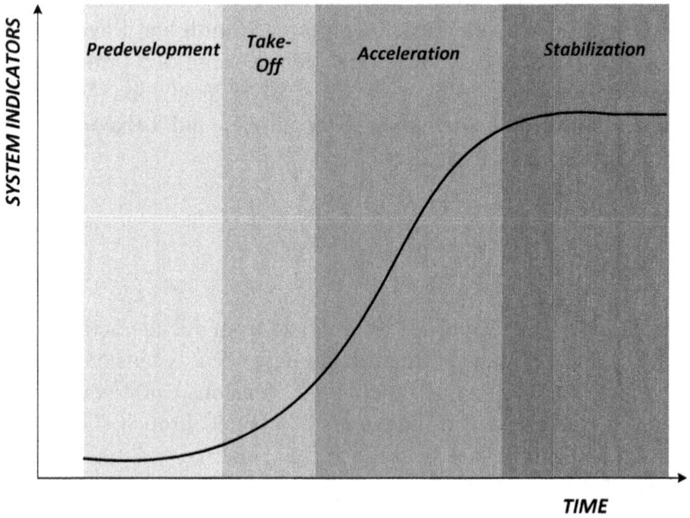

Fig. 2 The multi-phase perspective (Rotmans et al. 2001)

Predevelopment phase: "Co-evolutionary regime dynamics increase regime interdependencies"; "innovations are still isolated and fragmented, improperly embedded and insufficiently developed enough to compete with the existing regime" (van der Brugge and Rotmans 2007).

Take-off phase: "Triggering change and build up of a new regime", "innovations start acting as perturbation of the status quo", "regime dynamics collapse when the systems key functions fall out", "build-up of innovation networks based on alternative ideas, concepts, theories, and technology" (van der Brugge and Rotmans 2007).

Acceleration phase: "The 'old' regime transforms and consequently the 'selection environment' changes and is primarily based upon selection rules from the up-scaled innovation network" (van der Brugge and Rotmans 2007). In this phase, change accumulate through positive feedback loops and the speed of change increases (Rotmans et al. 2001).

Stabilization phase: "The new regime settles down. Transformation processes turn into optimization processes. Regime dynamics now are equilibrium dynamics in order to enhance efficiency. If this equilibrium is not reached, the new regime can still breakdown" (van der Brugge and Rotmans 2007).

Multi-path

Throughout history, there have been numerous transitions – in the economy, agriculture, mobility, and energy, but also in areas such as education, health care, and social structure (Geels 2004; Rotmans et al. 2001). In these domains, long temporal stretches of stability have alternated with shorter periods of rapid social change. Sigmoidal curves, or transition paths, can be used to classify and map the evolution of these transitions (Fig. 3).

Since every system is a unique entity, its transition path will be similarly unique. As stated by Rotmans et al. "it is possible to have different paths to the same equilibrium level as well as it is possible for the same transition pattern to be realized in different ways" (Rotmans et al. 2001: 18). Although these paths lead the systems to their new state, the characteristics of the system end state are used as the basis for their classification.

When the alternating phases of a transition are aggregated, a 'master' curve takes shape (Rotmans 2005). This curve is frequently represented as an S-curve, which can be seen as the ideal transition pattern: the system adjusts successfully to its changing internal and external circumstances. However, other forms can also be found, and non-ideal or even reverse transitions are possible. Thus, transitions do not always follow the S-curve. It should be noted that given the complexity and the uncertainty inherent to any societal transition, the transition paths do not represent all changes taking place over time. They represent 'ideal-types' of development trajectories and are to inspire action rather than represent real-type change trajectories.

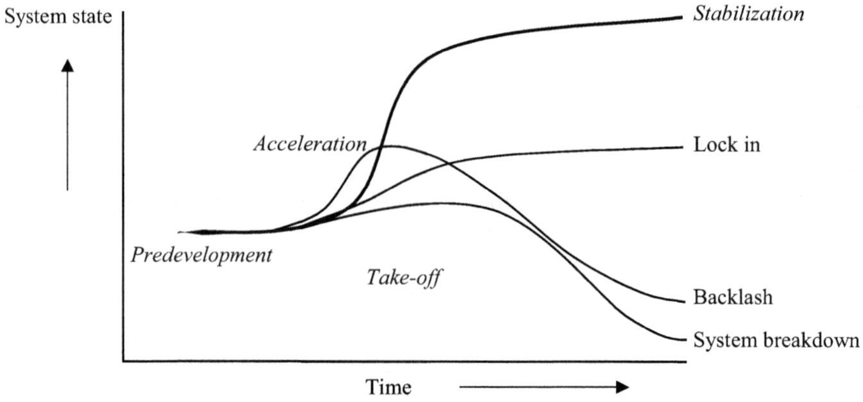

Fig. 3 The multi-paths conceptualization of transitions (van der Brugge and Rotmans 2007)

Power and Agency

The transitions approach in general, and transition management in particular, are not without their critics. The most prevalent point has been the lack of attention to agency dynamics, power and politics in transitions (Shove and Walker 2007; Hendriks 2007; Smith and Sterling 2008). Indeed, the transitions approach tends to overlook the political processes behind decision-making and societal goal setting, as well as issues such as values, identity, and conflict. However, as noted by Meadowcraft (2007), the governance of sustainability transitions impacts all spheres of life, making them impossible to ignore.

Overall, sustainability transitions approaches recognise agency and governance as central drivers for influencing system change processes into desirable directions. This builds on a recognition that societies are made up of networked connections between all kinds of individual actors that interact across scales, sectors and societal spheres to influence change (Grin 2010: 232). While recognising that sustainability transitions cannot be controlled, transitions scholars developed approaches like transition management to provide analytical and operational tools targeted at specific niches, actors or transformative innovations at a relatively small scale to understand and develop specific interventions to support sustainability transitions (Loorbach et al. 2015; Wittmayer et al. this volume, chapter "Transition Manage ment: Guiding Principles and Applications"). A particular focus has been on frontrunners who are aware of a complex persistent problem and take the lead in the change process by creating social and technological innovations (van Buuren and Loorbach 2009). While regime actors are often viewed as reinforcing dominant system configurations (Schuitmaker 2012), the role of change-inclined regime actors in connecting niche innovations with existing structures and processes is acknowledged as critical to gain support and legitimacy.

Still more attention needs to be paid to issues of contestation, conflict, power and vested interests. Gillard et al. (2016) criticises the lack of conceptualising the range of socio-political processes involved in transformative change, such as institutionalised ideas and values and intentions of actors that might resist social change. Rather than explicating the role of conflict, transition management proposes managerial strategies and purport the possibility of consensus across diverse self-organising systems. Another issue concerns the role of power and politics in transformative change and related questions about who is involved in decision-making processes (Avelino et al. 2016).

A specific issue refers to the ability of powerful actors from the incumbent regime to 'capture' a nascent transition, and thereby prevent radical change from taking place, or even steer a transition in an unsustainable direction (Shove and Walker 2007). The risk of capture is directly linked to what some critics see as the "managerial" or "technocratic" tendencies of transitions approach. This leads some to question its democratic legitimacy: "deliberations over structural transformation of socio-technical regimes affecting the lives of millions of people are led by a group of elite visionary forerunners" (Smith and Sterling 2008: 12).

Against the background of these and other criticisms a productive scientific dialogue has emerged addressing these challenges to the methodology (Jhagroe and Loorbach 2015; Avelino 2011; Jhagroe and Frantzeskaki 2012; Wittmayer and Schäpke 2014). The questions of power in urban sustainability transitions and transition management are taken up by Hölscher et al. (chapter "Context-Driven Transition Management as a Necessary Vehicle for Sustainable Urban Futures in Suriname", this volume) in relation to the extent actors become empowered to (re-) define and take up roles in contributing to a sustainability transition. Jhagroe (chapter "Urban Transition Management as a Democratic Practice? The Case of Rotterdam's Waterfront Regeneration", this volume) reflects on questions of democratic legitimacy in relation to transition management.

Approaches to Sustainability Transitions in the Urban Context

Geographical Approach (Cities as Transition Spaces)

The first writings of sustainability transitions from the geographical perspective noted that sustainability transitions need to consider the role of geography in the way patterns of change emerge and unfold (Truffer and Coenen 2012). Looking from the geographical approach, cities are places where transitions happen, cities are the places and spaces of sustainability transitions. So they offer empirical evidence to unpack, theoretically deepen and propose new conceptualizations of transition dynamics and measures, approaches for the governance of transitions through experimentation and not only.

Complexity Approach (Cities as 'Convergence Hubs')

Cities are the places that sectoral changes (e.g. energy, or mobility transitions) become irrelevant, since the interconnections between the different sectors are more evident and relevant. For example, when the mobility system a city changes towards a more ICT-advanced system, energy demands are changing. To accommodate these changing demands, the energy sector needs also to change. In this way, transformation in one sector, pushes for changes and transformations in other sectors. These interconnected changes build upon and reinforce urban complexity. These convergence points between different sectors are where synergies are to be searched for and exploited.

Governance Approach (Cities as Innovation Drivers)

Achieving urban sustainability and resilience are key governance challenges of our time. High pressures for action (esp. in response to climate change, resource depletion, loss of environmental quality, biodiversity, ecosystem services, and social segregation) require engaging with systemic change in urban areas upfront: Cities and their hinterlands are major concentrations of population that cumulate development drivers as well as negative impacts with heavy repercussions at all larger scales – while being genuine incubators of innovation and change.

The institutional proximity in cities makes it possible for innovations to cascade and for partnerships and synergies to happen. What cities offer is ground to experiment when applying innovative policies to tackle climate change, to reroute towards low-carbon development and as a result provide evidence for what works for national level policies to change. Cities are mediators for low-carbon sustainability transitions, transformations in policies, infrastructures, mindsets and skills.

Cities as Transitionscapes to Discover Processes and Actions of Agents of Change

Cities have social, ecological and technological assets that are essential for economic and social development locally and globally; there is thus an imperative for balancing their efforts for continuity and prosperity while ensuring a livable space for their citizens. Cities can be cradles for innovation and sustainability transitions. They are places where slow and fast transitions are initiated as well as test-beds and seedbeds for innovations.

Increasing policy attention is put on the processes of change that take place in cities and for cities driving the way to improved livability and resilience. This shows the diversity of opportunities and conditions in cities, where challenges such as urbanization and globalization contribute to the pressure for rerouting current development pathways towards sustainability and resilience.

With this volume, we move from understanding the complex dynamics of cities as transitionscapes to an inside view of strategic urban planning processes and the patterns of empowerment, mobilisation and activation they create. Specifically, we aim to shed light on strategic urban planning processes facilitated with transition management in order to understand how change agents can put their innovative potential for transformations in such processes.

Further Reading

Geels F, Schot J (2010) The dynamics of socio-technical transitions: a socio-technical perspective, part I. In: Grin J, Rotmans J, Schot J (eds) Transitions to sustainable development; new directions in the study of long term transformative change. Routledge, New York

Loorbach D, Rotmans J (2010) The practice of transition management: examples and lessons from four distinct cases. Futures 42:237–246

Wolfram M, Frantzeskaki N (2016) Cities and systemic change for sustainability: prevailing epistemologies and an emerging research agenda. Sustainability 8(2):144

References

Avelino F (2011) Power in transition. Empowering discourses on sustainability transitions, PhD thesis, Erasmus Universiteit Rotterdam

Avelino F, Grin J, Pel B, Jhagroe S (2016) The politics of sustainability transitions. J Environ Policy Plan 18(5):557–567

Bergek A, Jacobsson S, Carlsson B, Lindmark S, Rickne A (2008) Analyzing the functional dynamics of technological innovation systems: a scheme of analysis. Res Policy 37:407–429

Berkhout F, Smith A, Stirling A (2003) Socio-technological regimes and transition contexts, SPRU Electronic Working Paper, SPRU – Science & Technology Policy Research, University of Sussex, Brighton BN1 9RF, UK

Boulding KE (1970) A primer on social dynamics: history as dialectics and development. The Free Press, New York

Davis K (1945) The world demographic transition. Ann Am Acad Polit Soc Sci 237(4):1–11

Elzen B, Wieczorek A (2005) Transitions towards sustainability through system innovation. Technol Forecast Soc Chang 72:651–661

Frantzeskaki N, Slinger JH, Vreugdenhil H, van Daalen E (2010) Social ecological systems governance: from paradigm to management approach. Nat Cult 5(1):84–98

Frantzeskaki N, Loorbach D, Meadowcroft J (2012) Governing societal transitions to sustainability. Int J Sustain Dev 15(1/2):19–36

Fuenfschilling L, Truffer B (2013) The structuration of socio-technical regimes – conceptual foundations from institutional theory. Res Policy 43(4):772–291

Geels F (2002) Understanding the dynamics of technological transitions. A co-evolutionary and socio-technical analysis. PhD-thesis. Twente University Press, Enschede

Geels F (2004) From sectoral systems of innovation to socio-technical systems. Insights about dynamics and change from sociology and institutional theory. Res Policy 33:897–920

Geels F (2005) Technological transitions and system innovations, a co-evolutionary and socio-technical analysis. Edwar Elgar, Cheltenham
Geels FW (2005) Processes and patterns in transitions and system innovations: refining the co-evolutionary multi-level perspective. Technol Forecast Soc Chang 72:681–696
Geels F (2010) Ontologies, socio-technical transitions (to sustainability), and the multi-level perspective. Res Policy 39:495–510
Geels FW, Schot J (2007) Typology of sociotechnical transition pathways. Res Policy 36 (3):399–417
Gillard R, Gouldson A, Paavola J, van Alstine J (2016) Transformational responses to climate change: beyond a systems perspective of social change in mitigation and adaptation. WIREs Clim Chang 7(2):251–265
Grin J (2010) Understanding transitions from a governance perspective, part III. In: Grin J, Rotmans J, Schot J (eds) Transitions to sustainable development; new directions in the study of long term transformative change. Routledge, New York
Grin J, Rotmans J, Schot J (2010) Transitions to sustainable development, New directions in the study of long term transformative change. Routledge, New York
Hekkert MP, Suurs RAA, Negro SO, Kuhlmann S, Smits REHM (2007) Functions of innovation systems: a new approach for analyzing technological. Technol Forecast Socail Chang 74:413–432
Hendriks C (2007) "Making democratic sense of socio-technical transitions for sustainability", Paper presented at International Workshop: Politics and Governance in Sustainable Socio-Technical Transitions, 19–21 September 2007, Blankensee Berlin. Later published as:
Holland JH (1995) Hidden order: how adaptation builds complexity. Addison-Wesley, New York
Hoogma R, Kemp R, Schot J, Truffer B (2002) Experimenting for sustainable transport futures. The approach of strategic Niche Management. EF&N Spon, London
Jhagroe S, Frantzeskaki N (2012) A state of exceptional democracy: unravelling the politics of crisis in Dutch infrastructure governance, Paper presented at the 7th International Conference on Interpretive Policy Analysis Understanding the Drama of Democracy: Policy Work, Power and Transformation, 5–7 July 2012, Tilburg
Jhagroe S, Loorbach D (2015) See no evil, hear no evil: the democratic potential of transition management. Environ Innov Soc Trans 8. https://doi.org/10.1016/j.eist.2014.07.001
Kemp R, Rotmans J (2001) The management of the co-evolution of technological, environmental and social systems, Paper for the Conference 'Towards Environmental Innovation Systems', Eibsee, 27–29 Sept 2001
Kivits R, Charles MB, Ryan N (2010) A post-carbon aviation future: airports and the transition to a cleaner aviation sector. Futures 42:199–211
Loorbach D (2007) Transition management, New mode of governance for sustainable development, PHD Dissertation, Erasmus University Rotterdam, International Books
Loorbach D (2010) Transition management for sustainable development: a prescriptive, complexity-based governance framework. Governance 23(1):161–183
Loorbach D, van der Brugge R, Taanman M (2008) Governance in the energy transition: practice of transition management in the Netherlands. Int J Environ Technol Manag 9(2/3):294–315
Loorbach D, van Bakel JC, Whiteman G, Rotmans J (2009) Business strategies for transitions towards sustainable systems. Bus Strategy Environ., Published online in Wiley InterScience 10:1002
Loorbach D, Frantzeskaki N, Huffenreuter LR (2015) Transition management: taking stock from governance experimentation. J Corp Citiz 2015(58):48–66(19)
Markard J, Truffer B (2008) Technological innovation systems and the multi-level perspective: towards an integrated framework. Res Policy 37:596–615
Markard J, Raven R, Truffer B (2012) Sustainability transitions: an emerging field of research and its prospects. Res Policy 41(6):955–967
Martens P, Rotmans J (2005) Transitions in a globalizing world. Futures 37:1133–1144
Mayr E (2001) What evolution is, Orion Publishing Books, ISBN:13 978-0-7538-1368-3
Meadowcraft (2007) "Steering or muddling through? Transition management and the politics of socio-technical transformation", Paper presented at Internation Workshop: Politics and

Governance in Sustainabile Socio-Technical Transitions, 19–21 September 2007, Blankensee Berlin.
Meadowcroft M (2009) What about the politics? Sustainable development, transition management, and long term energy transitions. Policy Sci 42(4):323–340
Raven R, Schot J, Berkhout F (2012) Space and scale in socio-technical transitions. Environ Innov Soc Trans 4:63–78
Rip A, Kemp R (1998) Technological change. In: Rayner S, Malone EL (eds) Human choice and climate change, vol 2. Battelle Press, Columbus, pp 327–399
Rostow WW (1960) The stages of economic growth: a non-communist manifesto. Cambridge University Press, Cambridge
Rotmans J (1994) 'Transitions on the move', Global Dynamics and Sustainable Development-Working Paper, National Institute for Public Health and Environment (RIVM), Bilthoven
Rotmans J (2005) Societal innovation: between dream and reality lies complexity. Inaugural speech. Erasmus Research Institute of Management, Rotterdam
Rotmans J, Loorbach D (2009) Complexity and transition management. J Ind Ecol 13(2):184–196
Rotmans J, Loorbach D (2010) Towards a better understanding of transitions and their governance, a systemic and reflexive approach, as Part II. In: Grin J, Rotmans J, Schot J (eds) Transitions to sustainable development: new directions in the study of long term transformative change. Routledge, London, pp 105–120
Rotmans J, Kemp R, Van Asselt MBA, Geels FW, Verbong G, Molendijk K (2000) Transitions & transition management: the case of an emission-poor energy supply. ICIS (International Centre for Integrative Studies), Maastricht
Rotmans J, Kemp R, Asselt M (2001) More evolution than revolution: transition management in public policy. J Futur Stud Strateg Think Policy 3(1):15–32
Sanderson SK (2005) Revolutions, A worldwide introduction to political and social change. Paradigm Publishers, Boulder
Schot J (1998) The usefulness of evolutionary models for explaining innovation. The case of the Netherlands in the nineteenth century. Hist Technol 14:173–200
Schuitmaker TJ (2012) Identifying and unravelling persistent problems. Technol Forecast Soc Chang 79(6):1021–1031
Shove E, Walker G (2007) CAUTION! Transitions ahead: politics, practice, and sustainable transition management. Environ Plan A 39:763–770
Smith A, Raven R (2012) What is protective space? Reconsidering niches in transitions to sustainability. Res Policy 41(6):1025–1036
Smith A, Sterling A (2008) Social-ecological resilience and socio-technical transitions: critical issues for sustainability governance, STEPS Working Paper 8, Brighton: STEPS Centre
Smith A, Voss JP, Grin J (2010) Innovation studies and sustainability transitions: the allure of the multi-level perspective and its challenges. Res Policy 39:435–448
Truffer B, Coenen L (2012) Environmental innovation and sustainability transitions in regional studies. Reg Stud 46(1):1–21
UN (1997) Critical trends: global change and sustainable development. United Nations, New York
Van Buuren A, Loorbach D (2009) Policy innovation in isolation? Conditions for policy renewal by transition arenas and pilot projects. Public Manag Rev 11(3):375–392
van den Bosch S (2010) Transition experiments, exploring societal changes towards sustainability, PhD Thesis, Erasmus University Rotterdam, The Netherlands
van der Brugge R, Rotmans J (2007) Towards transition management of European water resources. Water Resour Manag 21:249–267
Voß J-P, Smith A, Grin J (2009) Designing long-term policy: rethinking transition management. Policy Sci 42(4):275–302
White LA (1959) The evolution of culture: the development of civilization to the fall of Rome. McGraw-Hill, New York
Wittmayer J, Schäpke N (2014) Action, research and participation: roles of researchers in sustainability transitions. Sustain Sci 9:483–496

Transition Management: Guiding Principles and Applications

Julia M. Wittmayer, Frank van Steenbergen, Niki Frantzeskaki, and Matthew Bach

Abstract This chapter describes a way of influencing sustainability transitions, namely transition management, and its strengths and limitations as a framework for strategic environmental planning in cities. This includes a description of the guiding principles of transition management, their translation into a governance framework and the operationalization thereof into process guidelines. For each phase of the process, rationale and methods as well as outcomes are outlined.

Influencing Transitions: The Transition Management Approach

Transition management, in a nutshell, is a new approach to deal with the persistent problems cities are facing that we have highlighted throughout the previous chapters. The underlying argument is that these problems are firmly rooted in the very structure of our societal systems, which implies that marginal changes (i.e., the optimization of existing measures or practices) cannot be effective and will lead to suboptimal outcomes. Transition management therefore advocates fundamental change, a transition, in order to address the root causes of persistent problems rather than their symptoms. Yet, unlike what its name might suggest, the approach is less about managing and more about influencing transitions through the creation of spaces for searching, learning, and experimenting. Through this process, transitions can be influenced, supported, and accelerated by playing into existing dynamics and embracing complexity and uncertainty as opportunities rather than as something to underrate or control.

Transition management is not about replacing other processes like policy formulating processes or policy implementation processes. It operates at strategic level

J.M. Wittmayer (✉) · F. van Steenbergen · N. Frantzeskaki · M. Bach
Dutch Research Institute For Transitions (DRIFT), Faculty of Social Sciences, Erasmus University Rotterdam, Rotterdam, The Netherlands
e-mail: wittmayer@drift.eur.nl; vansteenbergen@drift.eur.nl; frantzeskaki@drift.eur.nl; bach@drift.eur.nl

N. Frantzeskaki et al. (eds.), *Co-creating Sustainable Urban Futures*, Future City 11, https://doi.org/10.1007/978-3-319-69273-9_4

and is complementary to other processes, or as recently evinced, enacts a coordination function that brings inputs and streams of solution searching together. From our experience with transition management we can identify the similarities and differences with strategic urban planning. The similarities include the focus on the long-term horizon for identifying strategies and visions, the formulation of strategies and solutions that need to be operationalized further into a programme or action plan and the consideration of uncertainties and social complexities in the long-term orientation.

The differences between transition management and strategic urban planning include amongst others the following:

- Transition management focuses on transformative change, not on adaptive or corrective strategies to existing policy programs;
- Transition management facilitates co-creation processes to provide pathways to visionary futures. In their majority, strategic planning methods use trend-break or trend-mapping tools to project futures based on the current situation;
- Transition management links long-term visions to medium and short term actions positioning experimentation as the 'cornerstone' of open inclusive and reflexive planning, while maintaining an emphasis on the transformation of the current system (e.g., through system innovation).

From Understanding to Acting for Sustainability Transitions in Cities

Transitions as long-term processes of change involve a broad range of actors (Markard et al. 2012). Transition management offers a new governance approach capable of helping us understand and address the spiraling complexity of urban problems, as well as the entangled web of actors and processes that need to be reoriented towards a sustainable future.

> Transition management processes can be developed in any complex societal system in which individuals or organizations experience both persistent problems and the possibility of a transition, in other words, a context in which current common approaches and solutions do not suffice, future developments are highly uncertain, and major sustainability challenges are present. It therefore starts with people, individuals who seek more fundamental alternatives or are convinced these are necessary. (Loorbach et al. 2016)

Transitions thinking and transition management first appeared on the public policy scene in the Netherlands in 2001 through the fourth National Environmental Policy Plan (NEPP). The project put forward the concepts and principles that we will explore in this chapter as remedies to persistent problems. Amongst these are the combination of long-term thinking for short-term action and learning-by-doing or doing-by-learning. The project considered both transitions in general and in relation to the decarbonization of the energy system. Yet, why would the Dutch government opt for something so ambitious? There was a sense that something more radical was needed; the earlier incarnations of the NEPP have been largely ignored (Kemp and Rotmans 2009).

Not long after, a major research program on transitions was established (Grin et al. 2011). The KSI Knowledge Project, as it was named, had the mission to develop the knowledge and skills necessary to initiate and sustain system innovations and transitions towards a sustainable society (ibid). It focused on the interplay between fundamental knowledge (i.e., theory) and contextual knowledge, the latter of which was linked to specific sectors (e.g., agriculture, mobility, and health care). This is not to say that acceptance was easy to come by – some researchers believed that transitions and transition management were little more than hype (Shove and Walker 2007).

Since then, both transition management and the wider study of transitions have flourished with increasing applications in urban contexts. These have primarily been in the context of research and regional development projects funded by the European Union and other donors (see Box 1).

Box 1: Examples of Urban Applications of Transition Management

URBES (Urban Biodiversity and Ecosystem Services): Berlin/Germany, Rotterdam/The Netherlands, Salzburg/Austria, and Stockholm/Sweden (2011–2014).

MUSIC project (Mitigation in Urban areas: Solutions for Innovative Cities): Aberdeen/UK, Ghent/Belgium, Montreuil/France, Ludwigsburg/Germany and Rotterdam/The Netherlands (2010–2014) (see Roorda et al. 2014; Nevens et al. 2013; Nevens and Roorda 2014; Roorda and Wittmayer 2014).

InContext project (Individuals in Context): Carnisse/The Netherlands, Finkenstein/Austria, Wolfhagen/Germany (2010–2013) (see Wittmayer et al. 2011, 2014 and Nevens et al. 2013).

Melbourne, Australia Transition Management project(2012) (see Ferguson et al. 2013).

Port Vila, Vanuatu Transition Management project(2012) (see Poustie et al. 2016).

La Botija, Honduras Transition Management project(2014).

RESILIENT EUROPE project: Antwerp/Belgium, Bristol/United Kingdom, Burgas/Bulgaria, Glasgow/United Kingdom, Ioannina/Greece, Katowice/Poland, Malmo/Sweden, Thessaloniki/Greece, Potenza/Italy, Rotterdam/The Netherlands, and Vejle/Denmark (2015–2018) (see Frantzeskaki 2016).

Transition Management Basics

Having looked at transition management in general and its relation to the urban, we will now explore the approach in detail to understand what exactly it involves. To do so, we review its governance principles and the translation thereof in a governance framework outlining a number of related activities. We will then look at how transition management has been operationalized through specific phases organized in a cyclical format.

Governance Principles

The principles that have been developed for transition management form the basic elements for influencing societal transitions (Table 1). While not meant to be an exhaustive and immobile set of rules, they nevertheless offer some guidance for the governance of complex and adaptive societal systems (Loorbach 2010). The principles should be taken together, as they are intended to work as a whole, and complement one another. This also means that some can overlap.

Governance Framework

These governance principles have been translated into a governance framework, the transition management cycle, which distinguishes between four scopes of activities – strategic, tactical, operational, reflexive (Table 2). Transition management seeks to influence the interactions between these activities of different scopes through the creation of a facilitated and structuring process, which acts as a framework within which self-organization can take place (Loorbach 2010).

Strategic

Strategic level governance activities focus on the long term and the overall system. For example, in building a shared understanding of future energy systems to limit the harmful effects of climate change, we should consider all aspects of society, whether existing regulations, supply and demand projections, or emergent practices and technologies. Working at this level involves thinking together about ("structuring") problems, envisioning a desired future, and building bridges between the two. The focus on strategic level activities allows to counter-balance short-termism and reintegrate long-term governance into policymaking.

Table 1 Principles of transition management

Principle	Description	Example
System insight	It is essential to understand how a system works to effectively influence it in a sustainable direction. As such, managing change processes alone is insufficient, it must be linked to content	Deciding to implement a green house gas-reduction process cannot yield results unless it is connected to detailed knowledge about their sources, as well as the historical conditions which led to them in the first place
Long-term thinking	Radical societal change unfolds over a minimum of 25 years. Setting short-term policy therefore becomes a reflexive exercise, in which attention is continuously shifted to the future and backwards again	Developing a climate strategy for a city means thinking over the long-term and finding pathways to get there. Scenarios are a particularly useful tool in this context
Flexible objectives	Societal systems are complex, and rigid blueprints cannot offer adequate directions. As the system itself is constantly changing in response to interventions, objectives must also be adaptive	A long-term plan for expanding electric mobility in a city will need regular adjustment to account for developments: new technologies might appear, infrastructure might change, and unexpected behaviors from citizens might emerge
Timing is crucial	Crises offer possibilities for immediate and effective interventions, as they can help to overcome system inertia. At the same time, alternatives can be developed slowly while waiting for windows of opportunity	A food supply shortage can provide a unique opportunity to transform a city from a place of pure consumption to one of production, for example, through urban agriculture initiatives. These can also develop marginally until they spill over into the limelight
There is 'no outside the system'	It is not possible to influence a system from the outside: actors seeking to influence a system are an intrinsic part of it	A municipal government is as much part of the city (taken as the system) as any other actor
(Dis)equilibrium are both useful	Both equilibrium and disequilibrium offer opportunities for influencing a complex system in a desirable direction	Short-term disruptions, for example of fossil fuel supply, can make it possible to influence the system towards other, more sustainable energy sources. Stability can make it easier to create strategies that unfold over the longer term, for instance, the construction and connection of renewable energy capacity
Empowering niches	Agents need space to innovate, without the pressure of existing socio-economic structures	The development and spread of green buildings takes time, energy, and resources. They need to be shielded from the market and excessive regulatory requirements; the former would initially see them as uneconomical and the latter could cripple them. Support is therefore essential

(continued)

Table 1 (continued)

Principle	Description	Example
Focus on learning	Learning about the perspectives of different actors and the range of available options is essential for change: it cannot unfold without this	A city cannot reduce its environmental impacts by reproducing the policies that cause them: learning is the only means of discovering new possibilities
Participation	The involvement and interaction of stakeholders makes it possible to gain support for policies, reframe challenges, and derive solutions	Regenerating an urban area requires the support of many stakeholders, as well as the generation of new ways of seeing that area

Source: Adapted from Loorbach (2010)

Table 2 Overview of governance activities considered in transition management (Loorbach 2010)

Level	Key activities
Strategic	Problem structuring, vision development, strategic discussions, long-term goal formulation, collective goal and norm setting, long-term anticipation
Tactical	Achieving goals in a specific context, steering activities, negotiation, collaboration, agenda-setting, coalition forming etc.
Operational	Short-term, everyday decisions, specific projects and experiments
Reflexive	Learning, monitoring, assessment, evaluation

Tactical

Tactical level governance activities focus on the mid-term and on parts of the system. Actors designing actions with a tactical scope are less concerned with the overall direction of the system, than with specific goals unfolding over the mid-term (5–15 years), and tend to be in charge of program development, agenda-setting as well as regulation. For instance, a municipality might have a long-term vision to decrease its climate impacts; specific departments take up its various elements and prepare programs or regulations to reach the underlying strategic goal. Thus, in the pursuit of larger visions, actors at the tactical level negotiate, build partnerships and networks, and adapt wider goals to better fit reality. To prevent the fragmentation of visions for 'cross-cutting themes', such as climate change, it is important to find ways of integrating activities at the tactical level with the visions and problem framings arrived at in the strategic one.

Operational

Operational level governance activities focus on the short term and concrete activities – they are part of the constant flow of everyday decisions and actions. Projects, experiments, actions, and immediate decisions are all present here. This is therefore the level of practice from which new structures and routines can take shape that could eventually lead to system innovations. There is not only a flow from the strategic to the operational, but also the other way round. Nevertheless, action is guided by individual skills, ambitions and interests.

Reflexive

The reflexive level is inherently about learning about and consequently adapting activities at the different levels and between the different levels. It encompasses activities linked to the monitoring and evaluation of experiments, agenda, vision, policies and change processes. Such activities take place in a number of ways: through government-led evaluations, academic research, societal debates, the media, etc. This level should not, however, be separated from the other three, as it forms an integral part of governance processes and should continuously feed into activities taking place at the other three levels. It also helps actors to continuously readjust their goals and actions by learning about and taking into account changes occurring at all levels.

Putting Transition Management into Practice

The following operationalisation of the transition management principles and governance framework is targeted at urban policy makers. There are also other operationalisations which are more targeted at (action) researchers (Wittmayer et al. 2011; Ferguson et al. 2013). Policy actors use different types of interventions in their daily practice to influence the future of their city (Fig. 1). In the following, these interventions are described from a transition management perspective (rather than thinking about them from a strategic planning perspective). The categories are not fixed, necessarily sequential or tied to specific process phases; they go together, overlap at times, and diverge at others.

Orienting includes analyzing and positioning oneself vis-à-vis current societal developments and challenges – in the past, present and future – and other societal actors. Activities as part of this intervention type correspond to the strategic level activities described above.

Agenda-setting focuses on creating a shared sense of ownership and ambition for a sustainable future. Activities correspond to the tactical level governance activities, in which actors network and negotiate to give form to strategic insights.

Activating consists of experimenting towards a desirable future through projects and other (small-scale) activities. For example, by putting a tactical agenda into action, these activities (corresponding to the operational level activities) sow the seeds for transitions by multiplying pathways for fundamental change.

Reflecting is ongoing, as the primary means of learning from the different interventions and fostering a culture of reflexivity. Activities for this intervention type, such as monitoring, correspond to the reflexive level.

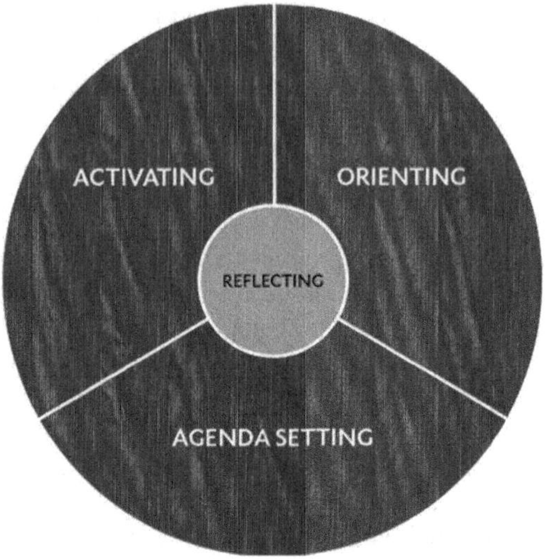

Fig. 1 Types of interventions in transition management process (Source: Roorda et al. 2014)

These types of interventions have been made part of a process through which transition management can be practiced in urban contexts (see Fig. 2). In this process consisting of seven phases, different activities and transition management instruments are deployed (see Table 3). For some of the phases there is a clear overlap in that that these can be assigned to different intervention types as is clearly shown in Fig. 2. The activities take place across different settings populated by different groups. These settings are also actors in the process: the Transition Team sets up the process, the Transition Arena carries out the process, and the Transition Experiments expand it outward and into broader networks.

Orienting

Before one can start this participatory searching and learning process that transition management is, a crucial process is taking place that we refer to as 'Setting the scene'. This is when the team is assembled which will arrange and facilitate the transition management process as well as the local situation explored and the overall focus negotiated, essentially this phase aggregates all work done ahead of the participatory process.

The **transition team** (not to be confused with the transition arena – see below) is the group that will facilitate the entire process, though its content role is strongest at the outset as it is tasked with carrying out the system and actor analyses, as well as identifying and selecting individuals for the transition arena. The team members are effectively the problem owners. To carry out its mission, the team needs both indirect and direct support: the former in terms of space for experimenting and

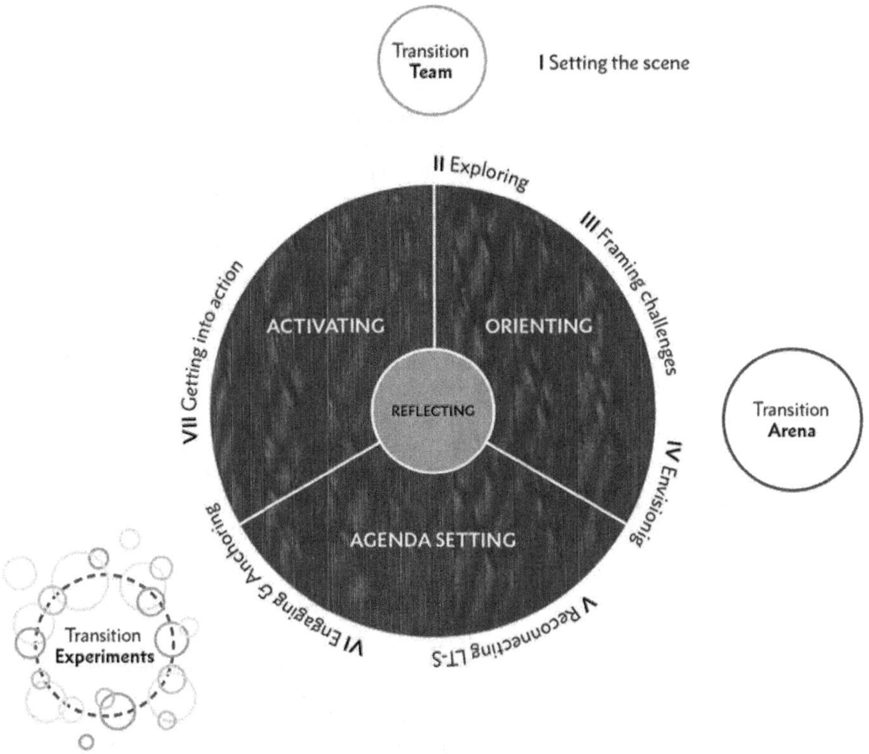

Fig. 2 Phases and settings of transition management (Source: Roorda et al. 2014)

learning to keep the process open-ended; the latter through resources and authority ('a mandate'). Moreover, they need to be highly motivated to withstand the inevitable barriers faced by any transition effort.

The composition of the group is a crucial point: diversity should be a defining feature to ensure an equitable transition management process and balance existing or emergent power dynamics. As the transition team selects the frontrunners (see below), it becomes a reluctant 'gatekeeper' either including or excluding actors from the process. This form of selective participation is important for an effective process, but also carries the risk of power asymmetries and biases coming to the fore.

Team members have tended to come from:

- The organization initiating the process (e.g. a municipality).
- The 'problem owner' (often a governmental organization in the urban context, though businesses and NGOs increasingly shoulder similar burdens).
- 'Experts' in the broadest sense of the word, capable of contributing to a specific transition management process.

Table 3 Overview of transition management process phases from MUSIC guidance manual 2014

Reflecting/ monitoring	I. Setting the scene for transition management	A **transition team** is formed to drive the process and embed it in the local context
	II. Exploring local dynamics	The transition team starts to explore the city's dynamics, performing interviews and desk research, and working towards a **system analysis** and **actor analysis**. Based on the actor analysis, a diverse group of change agents is invited to engage in a series of meetings as a **transition arena** group
	III. Framing the transition challenge	The change agents first explore what needs to change, the transition challenges and create a shared **problem framing**
	IV. Envisioning a sustainable city	Subsequently, they exchange and elaborate perspectives on a possible future, thereby creating **guiding principles** and **visionary images** for the future of the city
	V. Reconnecting long term and short term	As a final phase in the transition arena setting, the change agents elaborate **transition pathways**, indicating fundamental changes and corresponding actions needed to reach the envisioned future. The ideas brought forward by the transition arena group are summarized and published in a **transition agenda**
	VI. Engaging and anchoring	Actions are undertaken to make the transition agenda public and give others a chance to adopt and adapt it, and relate it to their own agenda and practices. This results in **transition networks**. A renewed transition team then takes the lead – these are change agents and others who assume ownership of the agenda's ideas
	VII. Getting into action	**Transition experiments**, radical short-term actions in line with the transition agenda, are initiated or adapted. Through these actions, more actors become engaged and learn by doing. Insights from these experiments can be taken to a more strategic level

Source: Roorda et al. (2014)

• Transition management experts (often scholars of sustainability transitions).

The **transition arena** on the one hand is a group of frontrunners who together develop a new vision for their city and on the other a setting in which a learning process amongst these frontrunners can take place.

– The transition arena as a group refers to the frontrunners who come together to delve into problems and generate visions of a sustainable future. In this sense, the transition arena is not an official 'body', but a network for societal innovation (van Buuren and Loorbach 2009).

– The transition arena as a space is more about the creation of a space for radical innovation processes to take hold. The spatial nature of the arena is, however, somewhat more complex: it is dependent on space at the outset – without it, learning, experimenting, and development would be jeopardized – while also creating space as an outcome of its activities.

Phase 1: Setting the Scene

The key activities in this phase include the formation of the transition team, the process design and leveraging support. The key outcomes of this phase include the transition team, the process plan and the support network.

Setting the scene is the first important phase of the transition management process: it is the work that precedes the participatory process that unfolds over the following phases. As in any project, this is a crucial phase, as it can largely determine the rest of the process. Depending on how the **transition team** is formed, the **process tailored** to a specific situation, and **support leveraged** from those assisting it (e.g., a city's administration), the transition management process can take root or remain marginal.

This risk is particularly salient for the approach, as unlike other projects, a transition management process is open-ended: it neither seeks to control the process, nor does it impose specific outcomes. Trust, readiness to learn and reflect, and a desire to go beyond business-as-usual are therefore important preconditions for a successful process. It is by adapting the process to reflect local priorities, opportunities, and dynamics that we can foster these.

In more practical terms, the elaboration of a process plan can help to better visualize the upcoming phases and, by making it more concrete, to maintain support for the process. Such plans can include elements regarding the local context (e.g., issues, relevance to pre-existing initiatives), organizational (e.g., resources, responsibilities), or procedural issues (e.g., rough schedule of phases, monitoring methodology). The process plan is, however, ever a work-in-progress – an evergreen document that grows with the process.

Phase 2: Exploring Local Dynamics

The key activities in this phase include the system analysis, the actor analysis and the identification and selection of frontrunners to participate in the process. The key outcomes of this phase include an insightful view on major issues and tensions to focus on, the identification and categorization of actors and the formation of the transition arena.

Once the transition team is formed and endowed with space and resources, it is ready to start exploring local dynamics. In urban areas, this can mean a city, a neighborhood, or any other spatial unit. There are three main aspects to this phase of the transition management process: the **system and actor analysis** (see Box 2 below) and the gathering of a group of active individuals ("**frontrunners**" or "change agents") in the **transition arena**.

Frontrunners or change agents are carriers of innovation, individuals who through their ideas, practices, and networks can play a leading role in influencing sustainability transitions. In the urban context, for instance, these could be community leaders, unorthodox thinkers, or social entrepreneurs. Regardless of their existing 'role', a frontrunner needs to be strongly connected to the issue at hand, while also being able to think beyond a specific worldview or set of interests. When invited into the transition management process, they are approached as individuals rather than representatives of whatever organization they might be members of. The methodological innovation, however, is that they are selected as a group and are meant to complement one another.

Box 2: System and Actor Analysis

The system analysis is a method used to acquire an integrated perspective and overview of the system under study (e.g., a city). Data can come from existing sources compiled through desk research, as well as from input provided by experts and individuals (e.g., interviews). Ensuring diversity in interviewees will ensure that the result truly has a bird's eye view of the system.

Loorbach (2010) summarizes the purposes of the systems analysis as follows:

- To gain an overview of the system in question: the main properties of the system, the persistent problems, dominant developments and innovative actors;
- To analyse the system from a historical perspective: describe and visualize changes in the system and changes in interactions of the system's properties over time;
- To provide the participants of the transition process with a common information base and concepts;
- To stimulate holistic thinking: enable participants to cross the limits of their work and/or personal expertise and understand the interconnectedness of the system;
- To serve as a provocative analysis to start the transition arena.

Through performing an ***actor analysis,*** the transition team gains an overview of relevant actors in the system, whether individuals, groups, or organizations. Broadly speaking, these actors correspond to the notion of stakeholder (their interests are affected by the issue, they have the potential to affect the issue, they have information or expertise related to the issue, they control elements linked to it), but additionally have new ideas, or, are part of initiatives regarding the issue.

(continued)

Box 2 (continued)

The actor analysis plays a recurring part throughout the transition management process. First, it will point to actors who can contribute to the system analysis. Second, it helps to identify the individuals who will be invited to the transition arena (i.e. frontrunners). Third, it will make visible potential partners for the diffusion of outputs from the arena (e.g. visions, narratives) at a later stage.

The following competencies can be expected of them (Loorbach 2010: 173–174):

- Ability to engage with complex and abstract problems
- Ability to go beyond the limits of their background or discipline
- Possess a degree of authority within one or more networks (i.e. be able to influence and mobilize others)
- Ability to share and make accessible sustainability visions within their networks
- Willingness to engage in collaborative thinking
- Openness to innovation (i.e. not pursuing a single option).

Based on these competencies, we can characterize urban frontrunners as:

- Actors that understand the complexity of current persistent problems;
- Actors who are innovators or active in (social, environmental, or economic) urban regeneration with an explicit or implicit interest in sustainability;
- Actors who are network brokers within their area of influence (e.g., a neighborhood, business network, or school);
- Actors with a genuine and passionate interest in society and a commitment to its sustainability.

Phase 3: Framing the Transition Challenge

The key activity of this phase is the participatory problem structuring. The key outcomes of this phase include the shared problem perceptions and the change topics.

Once the transition arena has been formed, the participatory transition management process can start. The first issue on the agenda is framing 'the problem' that the group and process will engage with. This search for a shared understanding of transition challenges usually starts from a discussion of the system analysis presented as an input to the process by the transition team and based on desk research and interviews (including those who are taking part in this process). In the process this analysis will be remolded until the group arrives at a shared problem perception. The value of this phase lies in the exchange of perspectives between frontrunners and the breaking down of institutional barriers. Another outcome from this phase is the formulation of core topics that the process can address.

Phase 4: Envisioning a Sustainable City

The key activity of this phase is the participatory vision building. The key outcome is a shared vision.

Departing from the platform of the shared problem perception, the arena group turns its attention to generating a sustainability vision – a contextualized understanding of sustainability, through which basic principles for long-term development are articulated. The vision is important in two ways. First, it orients the group towards a new 'attractor' – e.g., a car-free city, rather than a growing city – and, second, the intense discussions that lead to it allow the frontrunners to focus their efforts and move forward. Throughout the envisioning process, actors move from being passive problem watchers to realizing that their actions form a part of these phenomena. This is meant to provide an impetus for participants to reflect on their routines and, potentially, alter them.

Starting from exchanging future perspectives, the group agrees on a number of guiding sustainability principles – which can be based on the formulated transition challenges. These principles will be underlying the future vision that the group is working on. By focusing on a long-term vision made up of different (possibly competing/overlapping) visionary images at this point, the possibility for future debates is left open, while an overall future direction is set. Consensus is not necessary, most important is a collegial atmosphere so that participants can learn from one another and collaboratively imagine their common future. This is not to say that there is no place for conflict – it almost inevitably underlies transition visions, as these are frequently at odds with the vested interests of powerful actors who tend to oppose fundamental change and can challenge the rationale for a transition.

The elements that have been used to generate these visions include:

- Sharing perspectives on the future
- Drafting guiding principles for sustainability
- Creating visionary images
- Critically assessing visions.

Agenda Setting

After deploying 'orienting'-activities, the next two phases belong to the intervention type 'agenda setting'. The focus is here on reconnecting the long- and the short-term as well as anchoring the new visionary images in agenda's, program's and possibly regulation.

Phase 5: Reconnecting Long-Term and Short-Term

The key activities of this phase include the participatory backcasting, the definition of the transition paths and the formulation of the transition agenda and its specific actions. The key outcomes of this phase include the backcasting analysis and transition paths as well as the transition agenda and formation of possible experiments.

Now that the transition arena participants have envisioned a sustainable future, the challenge is to reconnect this long-term vision with the short term through more down-to-earth ideas about how to get there. In practice, this is done through **backcasting**, a method for working back from the future to the present (see Box 3 below). **Transition paths** are the first outcome of this phase – descriptions of possible routes to this future, or rather routes back from the future to the present. These are not blueprints or detailed scenarios, but narratives about how to get to the long-term through the mid and short term. The transition arena group tells and gathers these storylines, prioritizes them, and can split into smaller working groups to further flesh them out and brainstorm about ideas for short-term action.

The paths are then consolidated into a **transition agenda**, as the second outcome of this phase. It needs to be endorsed by all participants of the transition arena, as it is intended to present a common approach. Typically, a transition agenda would summarize transition paths, action ideas, topics for change, and visions, thereby elucidating the following points:

- "Why do we need a transition?" by mapping the flaws of the current system;
- "What needs to change?" by offering images of the desired future;
- "How is change possible?" by showing possible pathways to this future and immediate actions, projects or activities;
- "Who is involved?" by outlining key actors;
- "What is needed?" by determining necessary resources in their broadest sense.

The main purpose of this phase is to synthesize the work that has been done up to this point and to find ways of operationalizing it. In doing so, the transition arena participants can align their individual strategies and networks with the process, while also learning about possible opportunities (e.g., synergies with existing initiatives) and roadblocks for action (e.g., missing resources, conflicts).

Box 3: Backcasting in Transition Management Processes
Backcasting is a method to create pathways to an envisioned future, by starting from the future and working back in time. This reverse motion focuses on steps and milestones, highlighting how the short, mid, and long terms connect. This technique is especially helpful to formulate future-oriented strategies that reach outside the bounds of business-as-usual and look beyond the more immediate constraints of the short-term. A backcasting exercise will usually start with one of the visionary images (or a sub-topic within these) to draw up pathways as outlined in the previous phase. These can be defined in a "from – to" format, for instance, what fundamental changes are needed to go *from* persistent problems *to* a sustainable future? This type of reflection makes it possible to identify paths to the future, as well as potential barriers and opportunities.

Typical questions include: What changes would be needed to bring about the vision? What would be the three most important milestones on that path (e.g., in 2050, 2035, 2020)? What corresponding actions would be needed? What drivers and stepping-stones would support these changes? Which actors would play an important role in reaching these milestones?

Phase 6: Engaging and Anchoring

The key activities of this phase include the dissemination of the vision, pathways and transition agenda and the broadening of the network. The key outcomes of this phase include the broader public awareness and extended involvement to the transition agenda, the change agents' network and the spin-offs and new groups.

After having elaborated sustainability problems and challenges, future images and pathways in a small group of frontrunners from diverse backgrounds and worldviews – the resulting transition agenda should be challenging, motivating and desirable enough for the transition arena participants to get into action. This is the point in time, when the participants' leave the narrow bounds of the transition arena, which has sheltered the group from the pressures of the wider world, to influence and accelerate wider societal change. The underlying idea is that opening to ever more actors can generate support for a nascent sustainability transition.

The transition agenda of the arena participants helps them to share their vision with other individuals, organizations, or initiatives – these can then relate it to their own agendas and practices. In order to achieve a lasting impact, the group will need to reconfigure itself: the original frontrunners may disband, split into different groups, or take in new actors. One way of putting the transition agenda into practice is by creating "spin-offs" that emerge from the ideas or networks pioneered by the transition arena, but are only marginally related to the group itself. In any case, the number and nature of these activities will largely depend on the local context, its specificities and dynamics.

There are three primary types of activities that can take place at this stage:

- Organizing networking events;
- Getting exposure (e.g., in local media);
- Plugging the Agenda into other processes, such as existing sustainability initiatives, carbon reduction targets or networks such as 100 Resilient Cities, the Covenant of Mayors, or others.

Activating

The intervention type 'Activating' focuses on the actual practice, the short term and concrete activities. Rather than projects, transition management refers to **transition experiments**. Table 4 below details their characteristics and contrasts them with those of more typical innovation experiments.

> Transition management at this level focuses on creating a portfolio of related transition experiments that complement and strengthen each other, have a contribution to the sustainability objective, can be scaled up, and are significant and measurable (loorbach 2010: 176).

A transition experiment takes a "societal challenge as a starting point for learning aimed at contributing to a transition" (Van den Bosch 2010: 58) rather than a specific solution as is done for many innovation projects. As part of the transition agenda, transition experiments should be aligned with a specific transition path and a visionary image, as together they give an impulse to and influence the desired sustainability transition. That is not to say that all initiatives should be mutually reinforcing and coordinated: the ways in which they interact can follow

Table 4 Distinctive characteristics of transition experiments

	Classical innovation experiment	Transition experiment
Starting point	Possible solution	Societal challenge
Nature of problem	A priori defined and well-structured	Uncertain and complex
Objective	Identifying satisfactory solution (innovation)	Contributing to a transition (fundamental change in structure, culture, practices)
Perspective	Short- and medium-term	Medium and long-term
Method	Testing and demonstration	Exploring, searching and learning
Learning	1st order, single domain and individual	2nd order (reflexive), multiple domains (broad) and collective (societal learning)
Actors	Specialized staff (researchers, engineers, professionals, etc.)	Multi-actor alliance (across society)
Experiment context	(Partly) controlled context	Real-life societal context
Management context	Classical project management (focused on project goals)	Transition management (focused on societal transition goals)

Source: van den Bosch (2010), p. 63

any pattern – whether competition or cooperation. A diversity in transition experiments is not only desirable but necessary as it will increase the likelihood of successful action. Moreover, there is no need to reinvent the wheel; pre-existing innovative experiments that coincide with the above elements can be integrated into the process ('transitioned').

Transition experiments can contribute to sustainability transitions through three mechanisms: (1) deepening: learning as much as possible about an experiment in a specific context; (2) broadening: repeating the experiment in different contexts; and (3) scaling up– i.e. move from the micro (niche) level to the meso (regime) level (Van den Bosch 2010). This process is by no means straightforward or rapid. As experiments focus on exploring and learning, they tend to be costly in time and money, they need to be monitored closely for feasibility and their relation to physical, financial and institutional constraints of the system.

> Transition experiments are iconic projects with a high level of risk that can make a potentially large innovative contribution to a transition process (Loorbach 2010: 176).

Phase 7: Getting into Action

The key activity in this phase includes the implementation of transition experiments, policies and projects. The key outcomes of this phase include the learning and implementation of the transition agenda.

To put in practice the maxim of transition management to aim for long-term change through short-term actions, some of the transition experiments that have been devised during (or after) the arena should be further developed. Their implementation can bring together transition arena participants, as well as other actors. The transition agenda can also be linked to existing, new or planned activities.

Reflecting

Monitoring, as part of the reflecting activities, does not constitute a phase of the process per se as it is present throughout the transition management process and should therefore be integrated into all activities. This is particularly important as sustainability transitions are fundamentally searching and learning processes: there are no pre-established paths or blueprints. In fact, the system will adapt, as will institutions and the positions of actors, simply by orienting it toward a desired future. Thus, to influence the direction of such a sea of moving pieces, continuous monitoring and learning from it becomes essential.

What, then, are the objects for monitoring?

- The system itself – physical changes in the system, macro developments (slow), and niche changes (fast).
- The actors involved in the transition arena – their behavior and activities.

- The transition agenda – its goals and instruments.
- The transition experiments – for new knowledge and insights about what works, when and why (not).
- The transition process overall – its outputs, outcomes, and impacts; its rate of progress, barriers, etc.

As such, any effort to monitor a transition management process and its elements will have to pay attention to numerous system parts co-evolving at different paces. Doing so provides a richer understanding of the context in which transition management takes place, and makes it possible for learning to take place as a result of the interactions occurring between the different actors involved in both transition arena and transition experiments. On the basis of monitoring insights, adaptations can take place in either the desired direction of the system, in the different experiments or the transition agenda.

The results of the transition management process can be diffuse making them difficult to assess or quantify in a straightforward way. To provide insights into the results for those involved, the governance principles put forward earlier in this chapter can be used:

- System insight: has the process given actors insight into the system?
- Long-term thinking: are actors able to envision a sustainable future?
- Flexible objectives: can actors modify their objectives according to the situation at hand?
- Timing is crucial: do actors understand the barriers they face, as well as the opportunities afforded to them by crises?
- (Dis)equilibrium are both useful: are both disruption and stability seen as calling for different strategies for change?
- Creating niches: are actors able to create and maintain space for learning and reflection?
- There is no outside: do actors see themselves as intrinsic parts of the system they are trying to influence?
- Focus on learning: do actors understand the importance of learning for transitions?
- Participation: are actors willing to engage with participatory methods and carry on using them beyond the bounds of the Arena?

Transition Management in and for Cities

In practice, transition management comes down to a combination of developing around a common understanding of a transition challenge and a shared ambition to drive it toward sustainability. By using the transition perspective as a lens, transition teams develop an analysis of a particular persistent problem in a complex societal system. This could be a geographical area (from neighborhood to city, country, or region), sector, or societal issue. Based on the preliminary analysis, the transition team can identify regime and niche-actors that are potential contributors to a desired transition, who are brought together in so-called transition arenas or experiments. In and around these arenas and experiments, a shared discourse, ambition, and agenda is developed in such a way that it empowers participants and enables them to translate it to their own daily environments. Through such a social learning process, individuals start to relate to a broader common context and ambitions, creating the conditions for diffusion, self-organizational processes, and emergent innovation.

Transition management thus proposes to develop informal networks in which individuals and later, organizations, are provided the mental, social, and physical space to develop new ideas, common language, and ambitions, as well as new joint projects. In doing so over a longer period of time, participants will increasingly translate the transition perspective and ideas into their own operating context, while understanding the complexities, unpredictability and non-linear causalities involved. These transition management processes therewith indirectly influence and change regular policies in government, business, research, and civil society. In this chapter, the main ingredients of a transition management process are described, which in practice blend and are enacted often in parallel or at least in no specific sequential order.

References

Ferguson B, Frantzeskaki N, Brown R (2013) A strategic program for transitioning to a water sensitive city. Landsc Urban Plan 117:32–45

Frantzeskaki N (2016) Urban resilience: a concept for co-creating cities of the future. Baseline Study for URBACT project 'RESILIENT EUROPE', Licensed: Creative Commons, DRIFT: 2016

Frantzeskaki, N., J. Koppenjan, D. Loorbach, N. Ryan, M. Charles. 2012a. Theoretical and empirical contributions to an understanding of the governability of system transitions to sustainability—lessons and next-step challenges. Int J Sustain Dev 15 (½): 173–186

Frantzeskaki, N., D. Loorbach, J. Meadowcroft. 2012b. Governing transitions to sustainability: transition management as a governance approach towards pursuing sustainability. Int J Sustain Dev 15 (½): 19–36

Grin J, Rotmans J, Schot J (2011) On patterns and agency in transition dynamics: some key insights from the KSI programme. Environ Innov Soc Trans 1:76–81

Kemp R, Rotmans J (2009) Transitioning policy: co-production of a new strategic framework for energy innovation policy in the Netherlands. Policy Sci 42:303–322

Loorbach D (2010) Transition management for sustainable development: a prescriptive, complexity-based governance framework. Governance 23(1):161–183

Loorbach D, Wittmayer JM, Shiroyama H, Fujino J, Mizuguchi S (eds) (2016) Governance of urban sustainability transitions. European and Asian experiences. Springer, Tokyo

Markard J, Raven R, Truffer B (2012) Sustainability transitions: an emerging field of research and its prospects. Res Policy 41(6):955–967

Nevens F, Frantzeskaki N, Gorissen L, Loorbach D (2013) Urban transition labs: co-creating transformative action for sustainable cities. J Clean Prod 50:111–122. https://doi.org/10.1016/j.jclepro.2012.12.001

Nevens F, Roorda C (2014) A climate of change: a transition approach for climate neutrality in the city of Ghent (Belgium). Sustain Cities Soc 10:112–121

Poustie M, Frantzeskaki N, Brown R (2016) A transition scenario for leapfrogging to a sustainable urban water future in Port Vila, Vanuatu. Technol Forecast Soc Chang 105:129–139. https://doi.org/10.1016/j.techfore.2015.12.008

Roorda C, Wittmayer J (2014) Transition management in five European cities – an evaluation. DRIFT, Erasmus University Rotterdam, Rotterdam

Roorda C, Wittmayer J, Henneman P, van Steenbergen F, Frantzeskaki N, Loorbach D (2014) Transition management in the urban context. Guidance manual. Drift, Erasmus University Rotterdam, Rotterdam

Shove E, Walker G (2007) CAUTION! Transitions ahead: politics, practice, and sustainable transition management. Environ Plan 39:763–770

Van Buuren A, Loorbach D (2009) Policy innovation in isolation. Conditions for policy-renewal by transition arenas and pilot projects. Public Manag Rev 11:375–392

Van den Bosch S (2010) Transition experiments. Exploring societal changes towards sustainability. Ph.D. Thesis, Erasmus University, Rotterdam. http://repub.eur.nl/pub/20714

Wittmayer JM, van Steenbergen F, Quist J, Loorbach D, Hoogland C (2011) The community arena: a co-creation tool for sustainable behaviour by local communities. Methodological guidelines. Deliverable 4.1, InContext: EU ENV.2010.4.2.3–1 grant agreement n° 265191. Online available here: http://www.incontext-fp7.eu/sites/default/files/D4-1_Methodological%20guidelines_final.pdf

Wittmayer JM, Schäpke N, van Steenbergen F, Omann I (2014) Making sense of sustainability transitions locally. How action research contributes to addressing societal challenges. Crit Pol Stud 8(4):465–485

Urban Planning and Transition Management: Rationalities, Instruments and Dialectics

Marc Wolfram

Abstract This chapter juxtaposes the transition management approach with the rationalities and instruments of urban planning. Considering the historic evolution and current characteristics of urban planning in theory and practice, it identifies key challenges and opportunities for engaging with transition management. The discussion underlines the added value but also incompatibility of both approaches, and thus the need to create synergies and mutually compensate for critical deficits to effectively perform urban sustainability transformations. This implies to conceive of a dialectic relation between urban planning and transition management, deliberately striving for coordinated independence.

Urban Planning: A Brief Introduction

> *[. . .] conventional planning suffers from six key biases – all reflective of still being stuck in rationalism. Under these circumstances of a planning form limited by standardisation, disempowerment, economic rationalism, physical bias, exogeneity and formalism, transformation is unlikely.* (Karuri-Sebina 2014, p. 242)

Steering urban change into a desired direction is a core societal challenge that modern states have taken up and institutionalized rapidly since the beginning of the twentieth century. From the mid-nineteenth century, accelerated industrialization and new forms of urbanization were driven by multiple socio-technical innovations, fostering not only the accumulation of wealth and capital, but also of problems and conflicts in cities. Socio-economic segregation, along with deteriorating sanitary conditions and environmental degradation thus increasingly created pressures for responses and "sanitation" in the early industrial cities (Hays 2000). Such responses were devised mainly from two distinct angles:

M. Wolfram (✉)
Department of Architecture, Sungkyunkwan University, Suwon, South Korea

Dutch Research Institute For Transitions (DRIFT), Erasmus University, Rotterdam, The Netherlands
e-mail: marcwolfram@gmail.com

N. Frantzeskaki et al. (eds.), *Co-creating Sustainable Urban Futures*, Future City 11, https://doi.org/10.1007/978-3-319-69273-9_5

On the one hand, individual critics and professionals as different as Ebenezar Howard ('garden city' 1898), Patrick Geddes ('valley section' 1909), or, Le Corbusier ('contemporary city' 1922) developed radically new projects of the modern city, connecting designs for the physical and technical organization of urban space to social and economic reform, as well as to geographic and environmental conditions. These proposals for urban change were guided by a *vision* of a future society that radically differed from the conditions of the present, and that required to rethink the city. Depending on the values embraced, this led to plans for either disruptive or incremental change e.g. by rebuilding the city elsewhere from scratch (Howard), breaking down its fabric (Le Corbusier), or carefully transforming it through 'conservative surgery' (Geddes). Yet, such visions conceived first by individuals and increasingly also larger professional groups (e.g. Bauhaus) mostly fell short of also developing procedural orientation for *practical implementation*. Instead, their influence essentially relied on personal networks and the discursive reproduction of the concepts and imagery proposed – and thus the inherent alignment with existing ideologies, power positions and stakeholder interests (Hall 2002).

On the other hand, public authorities at local and increasingly also national levels, started to issue a steadily growing number of *regulations* meant to mitigate negative impacts and risks of urbanization on human health and living conditions. Rules and standards were defined to prescribe the position, dimension and design characteristics of buildings, the layout of infrastructures and provisioning networks, and the relative location of functions in the city (housing, industry, public facilities, green spaces, etc.). *The plan* became the key instrument for urban policy synthesis and communication: Spatial maps and schemes filled with coded information and accompanied by extensive textual specifications, data and analyses that define a desired future state of the city. In trying to cope with the increasing diversity of urban problems, plan preparation also involved a quickly growing number and range of *experts*: architects, engineers, economists, sociologists, ecologists, etc. In addition, procedures for *political decision making* were built into the plan-making process to provide plans with wider legitimacy and create liability for stakeholders. With this, urban planning quickly evolved from the military or civil engineering project it had been for centuries, to a highly institutionalized and interdisciplinary professional and political practice, closely intertwined with the articulation of the modern state and a wide range of sectoral policies concerning e.g. welfare, economic development, transport or environment (Albers 1997).

These different origins are constitutive of two implicit modes of urban planning that have maintained an inherently antagonistic relationship until today: Planning as a visionary exercise for radical societal change, and planning as a regulatory and technocratic practice (Davoudi 2001). While the former tends to conceive of integrated spatial images for urban futures, explicitly incorporating ideologies and norms, as well as independent and radical thought, the latter is more concerned with practical and specialized solutions for managing urban change through professionalization and institutionalization. With the growing complexity of sustainability problems linked to urban development, the relation between these rather different modes of urban planning has received increasing attention in urban planning theory since the 1990s. It is equally of importance here with a view to system innovations, since the regulatory mode of planning minds to *perform the stabilization* of socio-technical regimes embedded in cities, whereas the visionary mode aims to *inform their destabilization*.

Moving beyond these two general modes, attempts to theoretically define the characteristics of 'planning' coincide in identifying the following features that can serve here as a first baseline for a comparison with transition management. Urban planning is commonly understood to be (Faludi 1973; Forester 1989; Allmendinger 1992; Alexander 1992):

- *Future-oriented*: It deals with the prospective condition of a city, and with actions to be carried out in the future. It is therefore dependent on the use of some form of foresight.
- *Normative*: It sets priorities and defines goals related to a range of urban development characteristics, while also determining (the degree of) their liability for stakeholders.
- *Integrative*: It links diverse aspects of cities and decisions with the aim to formulate coherent and consistent overall concepts for urban development.
- *Preparing for action*: It provides guidance for decisions that will modify a city, and frameworks for corresponding actions implemented by diverse stakeholders. It does however not take such action itself.
- *Systematic*: It follows certain rules, procedures and methods – improvisation, trial-and-error, but also automatisms are opposites of planning.

Considering these five basic features, it appears that they are all clearly inscribed within the *regulatory* mode of planning. The visionary mode, however, does not have a strong affinity neither to a stringent 'systematic' approach, nor to 'preparation for action', while in turn putting emphasis on future-orientation, normativity and integration. As pointed out above, this trait has historically constituted a relative weakness of visionary urban planning in terms of implementation, thus forming an essential aspect to be considered in the (changing) relation between both modes.

Regarding transition management, these basic features are all shared in principal, but some important differences can also be discerned: Especially the central role of practical *experimentation* in transition management clearly exceeds and/or contradicts these orientations, while the related prominence of provisions for *social learning* has no adequate correspondence here. Moreover, transition management postulates particular requirements for *each* of the above features by drawing on systems thinking (e.g. 'arena', 'envisioning', 'pathways', 'agenda') that can only be traced in more recent urban planning debates – as will be discussed below.

Therefore, to compare between urban planning and transition management it is necessary to further unpack the interpretation of these basic features in theory, as well as their articulation in practice. To do so requires distinguishing between two perspectives on planning, respectively concerned with understanding (1) the very nature and *rationality* of planning as a form of state action and its role in societal development ('theory *of* planning'), and (2) the *instruments* and techniques used in planning practice ('theory *in* planning') (cf. Faludi 1973). Each perspective allows to further specify characteristics of urban planning that can then be juxtaposed with the particular rationality and instruments of transition management discussed in the previous chapters.

Evolving Rationalities of Urban Planning

Beyond the above basic features and co-evolving with wider societal changes and transformations of the modern state during the past century, urban planning has experienced continuous reinterpretations concerning the theoretical understanding of how its objectives should be defined, and what it should therefore practically do (Hall 2002; Fainstein and Campbell 2012). Although the scientific planning discourse may have been dominated by specific ideas for certain periods of time, this has in practice not necessarily resulted in a succession of radical 'paradigm shifts'. Rather, former interpretations have become superimposed, recombined and integrated with new ones, thus shaping complex patterns of urban planning rationalities adopted within national and local institutional contexts, and linked to different sets of (state) competencies, instruments and objectives. This implies that such patterns vary greatly in their composition and articulation between places and countries – and may thus serve to identify an *urban planning style* in a given city-region. It equally suggests that urban planning in practice pursues *different* rationalities simultaneously and independently, thereby inevitably contributing to development contradictions and conflicts (Sager 2001; Healey 2010). Without claiming comprehensiveness in terms of the historic evolution of urban planning rationalities, the following sections briefly outline key perspectives that continue to frame how urban planning is understood and approached today (Friedmann 1987; Hall 2002; Hillier and Healey 2010).

Rationalism in Urban Planning

Considering its historical origins, urban planning has unsurprisingly deep roots in rationalism. Rationalist thinking conceives of planning as a scientific method that helps to serve political or ideological goals. Based on knowledge acquisition, analysis and reason, rationalist planning thus claims to be able to achieve the 'rational mastery of the irrational' (Mannheim 1951), and create structure and order within urban realities perceived as messy and chaotic. Correspondingly, it is often also portrayed by attributes such as 'comprehensive' or 'synoptic'.

Rationalist planning strives to follow a linear process that leads from the specification of operational objectives and targets through information collection and analysis, towards option generation and assessment, plan implementation, and subsequent evaluation. The emphasis is on efficiency and on the optimization of goal achievement according to standardized assessment tools (e.g. cost/benefit). This implies a conception of 'the planner' as a central authority that provides the expertise(s) required and is capable of leading this process, while also making sense of the heterogeneous data gathered. Hierarchy and specialization within planning organizations are corollaries of this understanding. It is also essentially *the planner* who needs to learn from any failures occurring to improve future performance. Hence, from a rationalist point of view, urban change itself becomes framed as *the implementation of the plan.*

This conception of rationalist planning and its almost ubiquitous practical adoption have led to mounting criticisms since the 1950s, last not least in the light of the steadily accruing concentration of problems in cities that the approach proved incapable to solve (Jacobs 1961; Hayek 1967). Building on Simon's analysis of 'bounded rationality' (Simon 1947), first of all, the assumption of comprehensive knowledge and development control was questioned due to the limitations recognized in reality (data availability, knowledge and understanding, time for decision making, interest conflicts). Inflexible and time-consuming processes, the systematic exclusion of any alternatives that conflict with the optimization goal, and the lack of feedback mechanisms were identified as key obstacles for making adjustments to changing circumstances, or dealing with uncertainties. Moreover, also the inherently technocratic and authoritative character of rationalist planning was exposed since value orientations are merely 'adopted' from political decisions, thus removing all actors, their interest and strategies from the planning process. Apparently, these issues illustrate that rationalist thinking is also fundamentally at odds with key orientations of transition management.

Incrementalism in Urban Planning

Incrementalist ideas emerged in the 1950s from the recognition of the above limitations of rationalist planning, concluding that major radical change in society can and should not be planned for neither centrally nor comprehensively. Instead of clinging to the pursuit of an optimum future state, planning was advised to focus on 'second-best' and short-term actions for which the necessary consensus could be obtained. This understanding of planning as a 'science of muddling through' (Lindblom 1959) also implies to allow for decentralised expertise and steering, recognizing how stakeholders align their strategies through 'mutual partisan adjustment'. Planning should therefore remain open to continuous adaptations, using 'windows of opportunity' to purposefully connect means and ends (ibid.).

The method of 'disjointed incrementalism' suggested by Lindblom (1979) illustrates how such thinking translates into planning. It demands the following specific activities to obtain critical insights about change options and their feasibility (ibid. p. 517):

- Connect the analysis of goals/values with real problems observed;
- Focus more on problems to solve than on goals to reach;
- Analyse only a few familiar options;
- Explore only critical consequences of an option;
- Trial options and revise in case of failure;
- Let multiple stakeholders do this analysis.

This approach thus introduced new ideas into planning that are equally important for transition management. The aim to connect between normative goals and present day conditions, to engage in experimentation and learning-by-doing, as well as a concern for the plurality of knowledge and interests involved can be recognized. However, at the same time incrementalist thinking equally provides orientations that are highly problematic with a view to sustainability transitions. This concerns especially the intentional constraint put on the radicalness of options considered, the priority given to current problems over future goals, and the implicit reliance on existing institutions and actor positions. These aspects clearly contribute to reinforcing path dependencies while simultaneously diminishing the ability to pursue a long-term and integrated perspective. Therefore, while improving flexibility and adaptation, incrementalist planning practices have soon been criticized for their inherent opportunism and incapacity to move beyond short-term and piecemeal solutions (Rosenhead 1980; Weiss and Woodhouse 1992).

Participation and Advocacy in Planning

Since the 1970s, concerns referring to the legitimacy deficits and depoliticizing practices of both rationalist and incrementalist approaches became increasingly virulent. Planning was blamed for lacking the necessary *participation* of citizens in the decision-making process, and to consequently ignore their respective interests and concerns. In order to make planning more democratic and transparent, key demands were thus to properly inform and involve the general public, and to adequately incorporate the priorities expressed in this process into plans and programs (Arnstein 1969; Fagence 1977).

In addition, other critics recognized that even where planning practice included some form of public participation, certain stakeholders, their values and knowledge, remained underrepresented – typically the poor and ethnic minorities. Consequently, they suggested to deliberately create *advocacy* for such excluded communities through appointed professionals that could give them a voice in the planning process. By informing these communities in suitable formats and language, and by helping them to articulate their concerns and ideas, advocates should help to defend their values and interests vis-a-vis authorities and other powerful public or private actors (Davidoff 1965).

Over time, these criticisms and demands have informed the creation of a multitude of new approaches to participation and advocacy in urban planning, both formal and informal. From obligatory 'public consultation' procedures for official plans, or 'public hearings' and options for making petitions, to more interactive methods such as 'civic fora', including also foresight elements (e.g. visions, scenarios) – diverse efforts have since been undertaken to enhance both representation and democratic legitimacy in planning. While this is obviously achieved differently by the respective approaches and in different places, the general trend has been towards an increasing institutionalization of participatory and advocacy practices in urban planning, including the more recent proliferation of methods building on ICT, the diffusion of the internet and social media (e.g. 'participatory mapping', 'crowdsourcing') (Bryson et al. 2013).

Still, there is plenty of room for improvement here to consistently avoid distortions and manipulations in participatory practices. With a view to the broad deliberations and legitimacy required for sustainability transitions, however, the existing set of routines for involvement in policy-making and their institutional anchorage can be seen as an important asset to draw upon. In this respect, urban planning also clearly differs from transition management and its rather selective approach to participation and representation ('forerunners'), which has correspondingly been criticized for its neglect of the politics involved (Meadowcroft 2009).

Collaborative Planning

Beyond the concern for participation and equal representation, it is the specific qualities of communication processes and their implications that received particular attention in planning theory since the 1990s. Against the backdrop of an accelerating globalization and economic structural change, sustainability conflicts in cities started to aggravate, raising increasingly fundamental questions about how to reconcile economic growth with environmental health and social equity. The institutions of urban planning as such were thus called into question, pointing towards the need to base action on more diverse values, knowledge and practices. Inspired by Habermas' (1981) notion of 'communicative rationality', a *'communicative turn'* (Healey 1992) was postulated for urban planning that should imply especially to (Harris 2005):

- Recognize the social construction of knowledge both in science and practice;
- Acknowledge for different forms of developing and communicating knowledge within social contexts (e.g. including analysis as well as storytelling and artistic expression);
- Spread ownership and the range of knowledge and reasoning involved ('stakeholding');
- Shift from competitive interest bargaining to collaborative consensus building;
- Identify diverse interests concerned and their subordination through relations of power;
- Recognize planning activity as being embedded in everyday relations and place.

The intention was thus not only to democratize decision-making, but also the *knowledge production* involved in urban planning, thereby enabling processes of *social learning* and urban change that would affect values, practices and institutions (Innes and Booher 1999). Consequently, the role of the planner became reframed as that of a professional moderator, helping to ensure equity in the communication, develop critical reflexivity and translate ideas into actions. The tasks and procedures of planning were subject to revision and adaptation, leading to new practices of *'community planning'* and the creation of multiple *novel interaction formats* such as 'planning workshops', 'planning cells' or 'charrettes' in which diverse stakeholders could express themselves, develop shared problem frames and objectives, and the *social and political capital* required to move from reflection to practical implementation.

However, the normative aspiration of collaborative planning for guiding "[...] how political communities may organize to improve the quality of their places" (Healey 1997, p. xi) still turned out to be difficult to achieve in practice. Despite high efforts put into open communication processes, more substantial change remained constrained by existing institutions and dominant interests. Thus, the hopeful claim of collaborative planning to enable 'ideal speech situations' became widely criticized, underlining that the intended separation of discourse from power ultimately forms an illusion (Flyvbjerg 1996). Tactics, hidden agendas and power acts of stakeholders can still distort the desired equity in communications at any time, as well as the take up of results into formal plans and policies.

Nevertheless, with its focus on linking collective knowledge production, open processes, joint discourse development and social learning, collaborative planning has effectively contributed to establish key orientations equally highlighted by transition management. A key distinction, however, forms the strong emphasis collaborative planning puts on *place* as a central category for identifying problems, solutions and stakeholders to be involved (i.e. based on socio-spatial relations) – as opposed to transition management with its orientation at socio-technical system relations.

Strategic Planning

Complementary to the concerns of the collaborative approach, the uncertainties inherent to complex sustainability problems and possible planning responses to them received increasing attention since the 2000s. Drawing on earlier works in management-, policy- and organizational studies dealing with the role of *strategies* in purposive societal change (Etzioni 1967; Friend and Jessop 1969; Mintzberg 1987), some scholars suggested to reconceptualise urban planning as a form of strategy making in order to cope with the fundamental uncertainties involved in defining urban problems, solutions and priorities (Healey et al. 1997; Salet and Faludi 2000). New planning principles were thus identified: Explicitly accounting for *external conditions* and their influence on local developments (e.g. macro-trends, larger scales, higher authority tiers) should help urban actors recognize wider dynamics of change and their own room for manoeuvre. Distinguishing between *strategic* and *tactical* decisions (long-term value choices and directions of action as a prerequisite for short-/mid-term choices of the type and range of actions) should enable both flexibility to handle changing circumstances and yet provide orientation for diverse developments on the long run. Complementing formal planning methods through *cognitive and motivational tools* (e.g. mind maps, visions) should allow to elicit stakeholder values and steer through their self-commitment and in decentralized decision making contexts (public, private, civil society), rather than through command and control.

In practical terms, emphasis was thus put on the following set of strategy making activities (cf. Albrechts 2004; Cerreta et al. 2010; Hillier 2011):

- Develop a collective *long-term vision* as a normative framework for orientation, representing widely shared social values, priorities and meanings, and thereby increasing the transparency over actor interests and positions;
- Derive and implement *short-term projects and trials*, shaping coalitions of the main actors concerned, and focusing the available resources on key goals and places of the vision;
- Recognize *emerging patterns*, exploit windows of opportunity, and organise *learning processes for a continuous adaptation* of the strategy through monitoring and feedback. In this, strategic agility should remain more important than the strategy itself;
- Involve the *general public and key stakeholders* continuously to ensure inclusiveness and accountability, perform empowerment through *sharing knowledge resources* (deviating here from collaborative planning and its accent on *collective knowledge construction*), while developing actor networks, social and political capital, and capacity to act.

Apparently, all the above principles and activities are equally central to transition management. However, the strategic planning perspective conceives of them as an *integral part* of the public authorities' planning approach – not as necessarily *separated* from them in newly created formats ('transition arena'). The aspiration is to establish a 'permanent process' (Albrechts 2004, p. 753) that gradually incorporates any new formats into mainstream planning. Correspondingly, strategic planning embraces both wide involvement *and* coalition forming with selective key actors, thus oscillating between using *representation* and *interests at stake* as main criteria for participant selection – yet without articulated concerns for knowledge diversity and innovation potential as in transition management. Despite this difference, the delicate balance between participatory efforts and new forms of selectivity has therefore also attracted similar criticisms regarding the politics underplayed (Newman and Thornley 2011). Finally, with its explicit concern for scalar relations and focus on place-making, strategic planning necessarily incorporates the *spatial dimension* not reflected in transition management, but in turn lacks a comparable account for systemic configurations.

Multi-level- and Metropolitan Governance

Parallel to the rise of strategic planning, a major contribution to the evolving rationalities of urban planning resulted from the reception of relational geography (e.g. Murdoch 2006) and governance studies (e.g. Kooiman 2003) – apparently driven by the need to better understand the 'external conditions' of local development. Two aspects of a changing appreciation of stakeholder interactions can be distinguished here: On the one hand, notions of *multi-level governance* accounting for the shifting allocation of competencies and resources between levels of authority (Hooghe and Marks 2001) have brought the role of (supra-) national and regional governments in urban development to the fore (Healey 2007; Davoudi and Strange 2009). New types of state strategies have been identified, aiming to mobilize major urban territories as a key resource of national competitiveness e.g. through state-driven large-scale urban projects (Brenner 2004). On the other hand, recognition of the role of local private sector actors, as well as neighbouring authorities, suggested that new forms of *metropolitan governance* were emerging in response to both, pressures for urban competitiveness and/or for resolving diverse urban sustainability problems that escaped the city scale at large – regarding e.g. land use and transport patterns (Savitch and Kantor 2002; Kreukels et al. 2003).

In practice, a wide range of new approaches have thus been created for governance and planning, especially in metropolitan areas. These approaches usually target multiple *institutional voids* for coordinating action at a metropolitan scale: Between the city and its surrounding municipalities, between the public and private sector, and between various levels of government. To a much lesser extent this sometimes includes attempts for widening the participation and representation of civil society. This positioning of metropolitan governance as a new domain of collective action implies a relative weakness compared to the highly institutionalized planning practices for municipal, regional or national territories: All resources used and actions pursued in this context are necessarily subject to negotiation between the parties concerned, and thus largely dependent on identifying shared interests and creating new legitimacy. It is due to this constrained power position of metropolitan governance that it has turned into a major field of experimentation with new approaches guided by principles of collaborative and strategic planning, yet here relying essentially on informality, personal networks and leadership (Jouve and Lefèvre 2003).

Despite the broad variety of underlying motives and participating actors in practice, individual cases of metropolitan governance and planning may thus also reflect remarkable similarities with transition management. While still relying on territorial and administrative boundaries, they often involve a shift from a spatial towards a *systemic perspective*, considering e.g. mobility patterns, resource flows or industrial clusters. New interaction formats are created to elicit shared problem views and visions together with the stakeholders concerned in order to define an agenda for long-term change, as well as strategic projects for short-term action. However, key differences still reside especially in the extent to which such formats are deliberately opened-up for new knowledge and innovators, the reflexivity built into formats and projects (monitoring, social learning), and the overall ambition to achieve radical transformation rather than relative improvement.

Instruments of Urban Planning: Plans and Techniques

Despite of their successive emergence and partly also antagonistic conceptual orientations, in practice none of the urban planning rationalities sketched above has ever been fully 'replaced' by its successors. This includes even rationalism, although it is actually rejected by all planning theory developed since the 1960s. As emphasized in the introduction, the evolution outlined needs to be understood as a process of selective accumulation and superposition of planning rationalities in particular places. While the respective reorientations derived from a critical engagement with urban planning practice and interdisciplinary knowledge have always generated *new* approaches and instruments, many *existing* ones have remained unaffected or experienced only minor modifications. This patchwork has been favoured by the continuous differentiation and specialization of public administrations, which has multiplied the sites and tasks of urban planning. It has equally been shaped by the diverse cultural and socio-political context conditions of urban planning in different places, supporting the adoption of certain rationalities (and only for some tasks), while discarding others.

The resulting pattern of path dependencies as well as innovation opportunities relevant for sustainability transitions becomes manifest when regarding the *instruments* most commonly applied in urban planning today (cf. Newman and Thornley 1996; European Commission 1997; Ryser and Franchini 2015). In the following, this perspective will thus serve to further explore the institutionalization of different rationalities through particular types of plans and techniques, and the complementarities, synergies or frictions this implies regarding transition management. It equally allows to trace the role of the basic 'regulatory' and 'visionary' modes of planning identified above. Notwithstanding the multiple differences between countries and jurisdictions, four widely used types of urban planning instruments will be discussed here in the chronological order of their emergence: (1) Spatial plans, (2) Sectoral plans, (3) Environmental assessments and (4) Strategic plans.

Spatial Plans

Spatial plans are the basic instruments of urban planning. They encompass a wide range of urban development aspects and have the main purpose to provide a formal framework for changes in the urban structure and fabric, undertaken by public, private and/or civil society actors. Thus, they rely on regulations adopted at (sub-) national levels that specify the provisions these plans may contain. Spatial plans can focus on different scales (e.g. whole municipal territory or certain subareas) and time horizons (5–30 years), thus forming a *hierarchy* of urban development regulations. Although they are assumed to *derive from* a desired future urban development state, the process of formulating such a vision and translating it into spatial plans is usually not defined – or at best loosely described.

The most common type of spatial plans forms the *land use plan*, covering the whole municipal area and laying down directions for long-term development (20–30 years). It stipulates the permitted density, type and location of land uses through zoning codes (detailed for housing, industry, infrastructures, green space, facilities, etc.). Its elaboration is usually led by a specialized land-use planning unit that invokes multiple types of internal and external expertise (ecological, socio-economic, technological, etc.). Preparing a land use plan takes 5–10 years on average, including extensive quantitative analyses and modelling, as well as diverse informal stakeholder consultations and an obligatory public consultation on the final draft. Regardless these participatory add-ons, land use plans thus provide a striking example for the longevity of rationalism in urban planning. Whether and how these plans relate to a collective vision or transdisciplinary deliberation of alternative urban development pathways (including *strategic plans* – see below) remains subject to the willingness and initiative of the local government.

This rationalist bias is also inherent to spatial plans at smaller scales, but appears to be more frequently combined with certain collaborative efforts here. For instance, *area development plans* or *urban regeneration plans* can make manifold further detailed provisions regarding urban functions, form and design for a particular district or subarea (e.g. public space usage, building typology, street sections, construction features, etc.). In this, their more limited scope in terms of area covered, stakeholders affected, elaboration period (3–5 years) and time horizon (10–15 years) has also favoured the adoption of collaborative procedures that exceed the requirements of compulsory public consultation (such as planning workshops). However, at the same time this limited spatial and temporal scope also constrains the range of the (systemic) changes to be deliberated.

Spatial plans at the site scale reflect a broader diversity of rationalities. Standard *site plans* mostly follow a strict rationalist approach with technocratic procedures to ensure regulatory compliance and issue building permits. For larger sites with a strategic location (e.g. railway stations), however, the strong public and private (financial) interests involved have led to a proliferation of 'partnership' approaches. Often further enabled through amendments in national regulations, *public-private partnerships* (PPP) have thus been designed to accelerate particular types of urban change (prioritized by these interests) through a close coordination of investments and plan provisions. Since PPPs notoriously fail to ensure adequate participation and a balanced consideration of (long-term) impacts, more progressive cities have used formats such as charrettes for deliberating key site plans. In turn, *neighbourhood plans* display a very different kind of collaborative dynamic, rather driven by place-based issues of liveability, inclusion and cohesion. Correspondingly, methods for intensive local stakeholder involvement (e.g. citizen fora) and longer process timeframes are common, typically also linked to community development and empowerment policies. Therefore, neighbourhood plans provide for informality and considerable flexibility in terms of the measures incorporated, and have also given rise to *practical experimentation* as part of the planning process (e.g. new types of community centres).

Sectoral and Thematic Plans

To the degree that modernization has driven administrative specialization and policy fragmentation, urban planning has continuously developed new organizational sub-units and instruments with a *sectoral* or *thematic* focus. Examples range from transport, housing and cultural heritage to landscapes, retail and lighting. On the one hand, this trend has helped the perseverance of rationalist approaches in urban planning. The new forms of expertise developed and their institutional separation have favoured a focus on optimizing goal-subsets and legitimized reduced policy coordination, thus also fostering diverse sustainability problems – e.g. considering that highly interdependent components such as land use, transport and landscapes are usually dealt with in *different* plans. Moreover, in the absence of regulations specifying procedural requirements for sectoral and thematic plans, technocratic practices could further expand. For instance, even in urban transport planning, participation and collaborative approaches still form a recent innovation – as reflected in contemporary reform attempts (see e.g. the concept of Sustainable Urban Mobility Plans: http://www.eltis.org).

On the other hand, since the 1990s, several new thematic plans have been introduced in response to environmental problem pressures at global and local scales that have started to counter the rationalist undercurrent in sectoral urban planning. Formal plans have been driven through (supra-) national authorities and initiatives (e.g. EU directives and communications on air quality, noise, waste management, green infrastructures), and informal ones through local motives related to energy security, climate change adaptation and resilience. In both cases, such plans have implied an extension of *interdisciplinary coordination* among policies, but also *transdisciplinary deliberation* involving a wider range of stakeholders through *collaborative formats*. Moreover, especially in plans addressing climate change mitigation and adaptation, the long-term character, range and depth of urban transformations addressed have in addition multiplied instances of *strategic planning*, thus including the elaboration of *visions* and the initiation of *short-term projects and experiments* (Davoudi et al. 2009; Castán Broto and Bulkeley 2013).

Environmental Assessments

Environmental change has equally driven more institutionalized forms of policy coordination in urban planning through particular assessment procedures. While initially following a clear rationalist logic, these assessments have gradually incorporated elements of collaborative, strategic and governance rationality. Since the 1970s, nation states have adopted regulatory frameworks requiring (local) authorities to carry out *environmental impact assessments* (EIA) for larger projects, such as infrastructures, facilities or major urban developments. This approach obliges the promoting agency (public or private) to analyse all environmental impacts of a project through *systematic screening and scoping*, and to develop suitable mitigation measures. It also demands a consultation on the analysis results, involving key stakeholders and the general public, as well as a final public information report that is however not binding for the responsible planning authority (Glasson et al. 2011).

The effects of project EIAs are therefore rather ambiguous in terms of change for sustainability. On the one hand, EIA has certainly contributed to foster the involvement of environmental departments and of the general public in urban planning, helping to transfer ecological knowledge and to integrate ecological considerations in sectoral policies and private sector strategies. On the other hand, there are also critical deficits: EIAs are initiated on the basis of a given project proposal and focus on its narrow spatial and time scale, thus preventing consideration of alternative problem framings and solutions (e.g. "greening" the incineration plant instead of removing the very need for it). Implicitly they also format particular actor roles ("promotor" versus "assessor"), resulting in negotiations over minimum negative impact mitigation – rather than in a collective search for maximum positive impacts.

Such criticisms of project EIAs have motivated the creation of broader assessment approaches that seek to compensate for their shortcomings. Since the 1990s, procedures of *strategic environmental assessment* (SEA) have gradually been introduced through national legislation in many countries. Subject of SEAs are therefore *plans, programmes and policies* that significantly condition the environmental impacts of subsequent decisions and projects (e.g. land use plans or transport plans). The SEA does however only complement these existing planning instruments but its final report remains equally informational and non-binding upon actors as in an EIA (Sadler et al. 2011).

While the practice of SEA largely differs between countries in terms of formality, procedure and scope, it commonly tends to enhance the sustainability implications of urban planning in several important ways: First, it demands a broader public deliberation of *policy objectives* at an *early stage of plan-making*, involving a wide range of stakeholders and thus enabling a more transparent societal debate about alternative urban futures, often drawing on scenario techniques. Second, this implies consideration of *larger scales and longer timeframes*, thereby bringing regional urban metabolisms and global environmental change into local decision making, and enabling a more balanced account for (positive and negative) environmental, social and economic impacts.

Together, SEA and EIA thus constitute a repertoire of urban planning instruments of high relevance for TM. In conjunction with the spatial and sectoral plans concerned, they reflect the particular way in which current planning institutions frame and handle complex urban sustainability challenges and develop formally legitimized responses. Given the variation of practices between places, this may however range from strongly underpinning the status quo, to enabling the emergence of entirely new collective visions and strategies. Nevertheless, especially SEA provides critical regulatory anchorage for a stronger embedding of systems thinking and sustainability innovations in urban planning.

Strategic Plans

Strategic plans cover a broadening range of informal and voluntary arrangements created by local authorities in cooperation with diverse selected stakeholders to agree and practically pursue joint goals. While remaining coarse in terms of level of detail, their main purpose is to connect fragmented knowledge, approach different interests, set long-term objectives and ensure their uptake and consideration in diverse decision-making contexts. Already in the 1970s, *economic development plans* had emerged in response to growing inter-city competition pressures, indicating the location of key economic activities and the infrastructure developments deemed necessary for this. Such plans were elaborated by economic development agencies in narrow co-operation with local industry and chambers, thus privileging private sector interests in urban development. During the 1990s, these earlier routines have however conditioned and informed the wide-spread uptake of strategic plans to foster the global competitiveness of cities, now increasingly embedded in a wider metropolitan governance context. Despite limited participatory engagement and a clear prioritization of economic goals, the elaboration of these plans often included relevant methodological innovations. Especially scenario and creativity techniques, as well as qualitative analysis played an important role in aligning different stakeholder views over future development orientations (Healey et al. 1997; Salet and Faludi 2000).

More recently, a new generation of strategic plans has emerged in global cities that emphasizes "low-carbon", "resilient" and "smart" urban development as its principal long-term goal, thus ostensibly deviating from the previous focus on "competitiveness". Such shifts in terminology are also accompanied by efforts for achieving broader legitimacy through public information and international networking (e.g. http://www.c40.org). Yet, upon closer scrutiny these strategic responses to global environmental imperatives often have more in common with

Table 1 Urban planning rationalities and instruments compared to transition management

	Urban planning instruments		
Planning rationality	Regulatory mode	Visionary mode	Transition management instruments
Rationalism	E.g. land use plan, sectoral plan	E.g. 'garden city', 'contemporary city'	*None (rationale rejected)*
Incrementalism	E.g. site plan (PPP), EIA	*None (instrument rejected)*	(Partly) transition agenda, transition experiments
Participation and advocacy	E.g. public consultation, public hearing	E.g. future workshops	(Partly) engaging and anchoring
Collaborative planning	E.g. urban regeneration plan, community plan, planning workshop, planning cell, charrette		System analysis, transition arena, visionary images, pathways, agenda, experiments, engaging and anchoring
Strategic planning	E.g. SEA	E.g. "smart city" vision	
Multi-level and metropolitan governance	E.g. metropolitan body and plan	E.g. metropolitan vision	

their predecessors than the labelling suggests. Notably, their orientation at green growth, resource security and carbon control reflects preferences for urban development pathways that do not substantially question or alter the status quo, but strive for its socio-technical optimization. Moreover, these key ends are not necessarily subject to participatory deliberation or foresight, but to corporatist agreement among urban political-economic elites that set the stage for a more limited public debate of alternative means (Hodson and Marvin 2010, 2014).

Strategic plans thus represent a crucial instrument for urban transformations operating at the cognitive and normative levels of decision-making targeted by transition management (vision, pathways). However, their institutional legacy in terms of actor networks involved and routines developed appears to form a major burden for a stronger engagement with sustainability. Moreover, given their informal and strongly place-based character, huge differences exist in terms of the approach, scope, scale and liability of strategic plans; even though an increasing international peer-to-peer exchange and networking may play an important role in the future formation of standards (Table 1).

Dialectics of Urban Planning and Transition Management

The above analysis of rationalities and instruments in urban planning has allowed for a more nuanced account of similarities and differences regarding transition management. It has illustrated how urban planning evolved from a rationalist and authoritative approach relying on either regulation or vision development, towards a highly differentiated and heterogeneous set of practices that selectively incorporate a wide range of rationalities and instruments – including also features of transition management. As reflected in its evolution over the past decades, urban planning increasingly strives to (re-) shape the ability of stakeholders and institutions to enable, perform and/or accommodate innovations that *destabilize* and ultimately *transform* various systemic configurations. Yet, from a transition perspective urban planning may still be regarded as an articulation of the cultures, structures and practices that *stabilize* current regimes and resist radical change – and thus an essential part of the sustainability problems that transition management attempts to tackle (Malekpour et al. 2015). What does this mean for the future relationship between urban planning and transition management? Should their rationalities and instruments become confronted, connected, or possibly integrated in order to initiate and perform radical urban change towards sustainability?

Some fundamental differences between both approaches should be recalled in order to respond to this question. While urban planning is a deeply institutionalized and globally practiced form of state-led action for guiding urban development, transition management represents a conceptual framework for guiding system innovations that is currently shared by certain academia and policy networks, but has hardly been applied to cities in practice so far (see e.g. Nevens and Roorda 2014; Wittmayer et al. 2014). This provides for rather different starting points in terms of embeddedness, subject and ambition that will mostly require bridging huge gaps between their protagonists. Moreover, despite partial approximations and overlaps, significant divergences remain in terms of *governance modes* and regarding the *instruments and techniques* used. These distinctions will be briefly revisited here in order to clarify how urban planning may become "transitionized" while in turn "urbanizing" transition management (Table 2).

Planning approaches for steering urban development have undergone a shift from *government* to *governance*, gradually widening the range of stakeholders involved, and of interaction formats designed for this. While sustainability has hitherto remained a marginal guide for governance innovations in practice, the different planning rationalities discussed above have been decisive for this shift, resulting in particular types of process novelties: Earlier concerns for enhancing democratic legitimacy, representation and advocacy typically translated into *formal and informal* procedures of public consultation and citizen involvement, closely linked to *regulated territories* of intervention (e.g. municipality, district, site). However, later intentions to strengthen transdisciplinarity (collaborative planning) and/or the development of social and political capital (strategic planning, metropolitan governance) engendered only *informal* procedures that address especially *cross-boundary issues* between territories, sectors and/or levels, and tend to involve stakeholders more selectively. Governance modes in urban planning practice thus reflect an inherent divide between processes that provide spatially-bound interventions with strong legitimacy and power but weaken their plurality and coordinating role, whereas inversely cross-boundary processes develop strong effects on coordination, value and knowledge diversity, but lack democratic legitimacy and

Table 2 Comparison of urban planning and transition management – key features of governance, instruments and techniques

		Governance							Instruments										Techniques					
		government initiation and leadership	societal initiation and leadership	open public involvement	selected stakeholder involvement	multi-level coordination	multi-sectoral coordination	independent facilitation (science)	legally binding	voluntary commitment	regulation	economic cost/benefit	dialogue and negotiation	information and communicaton	spatial scope	systemic scope	elaboration period (years)	time horizon (years)	quantitative analysis	qualitative analysis	modeling and simulation	foresight techniques	creativity techniques	reflexive monitoring
Urban planning		●●		●●	●●	●	●		●●	●	●●	●	●●	●	●●	●	1-5	1-20	●●	●	●●	●	●	●
Transition Management		●	●●		●●	●●	●●	●●		●●			●●	●●		●●	1-2	1-50	●●	●●	●●	●●	●●	●●
Urban planning	Strategic plan	●●	●	●●	●●	●●	●●			●●		●●	●●	●●		●●	2	20	●●	●●	●	●	●●	●
	Land use plan	●●		●●	●●	●	●●		●●		●●	●●		●	●●		5	15	●●	●	●●			
	Area development plan	●●		●●	●●		●●		●●		●●	●●		●	●●		3	10	●●	●●				
	Site plans (PPP)	●●		●	●●	●●			●●	●	●●	●●	●	●●	●●		2	5	●●	●			●●	
	Building permits	●●		●	●●				●●		●●				●●		1	1	●●	●				
	Sectoral plans	●●		●	●●	●	●		●●		●	●	●	●	●	●●	3	15	●●	●	●●	●		●
	SEA	●●		●●	●●	●●	●●			●●			●●	●●		●●	1-5	10	●●	●	●	●		●
	EIA	●●		●●	●●		●●			●●			●●	●●		●●	1	5	●●	●				●
TM	Transition Vision	●	●●		●●	●●	●●	●●		●●		●●	●●	●●		●●	1	50	●	●●	●●	●●	●●	●●
	Transition Pathways	●	●●		●●	●●	●●	●●		●●		●●	●●	●●		●●	1	25	●●	●●	●●	●●	●●	●●
	Transition Agenda	●	●●		●●	●●	●●	●●		●●		●●	●●	●●		●●	1	10	●●	●			●●	●●
	Transition Experiments	●	●●	●	●●	●●	●●	●●		●●		●●	●●		●	●●	1-2	1-2	●●	●				●●
	Engaging and anchoring	●	●●	●●	●●	●●	●●	●●		●●		●●	●●			●●	n.a.	n.a.						●●
	Transition Monitoring	●	●●	●●	●●	●●	●●	●●		●●		●●	●●	●●		●●	n.a.	n.a.	●●	●●				●●

Pertinence of feature: ●● high; ● average; (blank) low

implementation capacity – a dilemma that requires continuous re-calibrations between these processes (cf. Allmendinger et al. 2015).

This is of equal importance for transition management since its approach essentially aims to add a layer of informal (meta-) governance guided by (cross-boundary) systems thinking, not by concerns for representation. It therefore confronts similar challenges concerning legitimacy and implementation that are further exacerbated by its lack of a *spatial* perspective. Moreover, transition management deviates from the existing mainstream of governance modes in four important ways: First, it places an explicit *normative and conceptual sustainability orientation* in the center, drawing on environmental ethics and scientific knowledge as a key source of legitimation for its governance approach. Second, it includes and encourages the option of *societal initiation and leadership*, thereby allowing for bottom-up processes, followed by authorities. Third, it aims to *selectively involve innovators* providing relevant future knowledge (forerunners), visibly privileging innovation capacity over democratic legitimacy. Fourth, it attributes a vital role to *(local) science and academia* as independent facilitators, contributing critical insights, skills and methods. Therefore, transition management cannot simply be performed by the existing planning institutions since it questions many of their constitutive assumptions. Forming itself a governance niche, it requires a "free space" for operation to effectively develop innovations for sustainability. But to do so for cities, transition management in turn needs to better resonate with the existing *spatial* patterns of urban governance and planning practices.

The sketched differences in terms of governance are mirrored by the instruments and techniques employed. Considering a basic spectrum of policy options, including regulatory, economic, dialogic and informational instruments (Lascoumes and Le Gales 2007), urban planning has accomplished a gradual shift from an exclusive reliance on binding regulations, taxation and incentives, all underpinned by quantitative analysis (rationalism, incrementalism), towards a growing emphasis on negotiation, communication and voluntary commitments, also attributing a more important role to qualitative reasoning (collaborative and strategic planning). Lengthy processes of "master plan" elaboration based on a spatial scope have become complemented by diverse shorter planning processes linking spatial and systemic scopes (e.g. sectoral plans, SEA, strategic plans), while extending the time horizon considered. This also implied an increasing use of modelling tools, creativity techniques, monitoring approaches and especially foresight. While this shift is supportive of central concerns of transition management, it has neither affected all planning instruments in the same way, nor does it fully embrace the more specific methodological requirements in transition management and their rationale (e.g. envisoning, backcasting, pathways).

This gap can only partly be explained by the high resource demands (human, time, skills, knowledge) of transition management processes and instruments, since planning authorities pressured by budgetary constraints and multiple legal obligations are less likely to deeply engage. More importantly, there is a strong emphasis in transition management on establishing *open processes* of collective searching, experimenting and learning, translated into particular instrumental choices (orienting, agenda setting, activating, reflecting), which makes it difficult for planning authorities and politicians alike to take responsibility when considering the risk of failure ("blame avoidance" strategies). Thus, from an instrumental point of view transition management is equally improbable to be eagerly embraced by existing institutions. However, the current spectrum of urban planning instruments does offer diverse vantage grounds that would allow to reinforce impacts on system innovations for sustainability – provided that these are actively identified, exploited and connected (e.g. strategic plans, strategic environmental assessment, neighborhood plans). This potential could be tapped by an urban transition management approach wary of embedding transformative instruments, while pursuing a complementary instrumental agenda: Reconnecting the regulatory and visionary modes of urban planning through a differentiated foresight approach, and especially adding novel techniques for reflexivity and social learning that have remained largely absent in urban planning.

These basic dissimilarities and interdependencies demand to conceive of a *dialectic relation* between urban planning and transition management in which both approaches remain *necessarily separate*, but engaged in a critical exchange and reflection. This would not only allow revitalizing democratic politics and challenging its legitimacy claims (Jhagroe and Loorbach 2015). A dialectics of urban planning and transition management suggests that their inherent tensions are not to be resolved, but rather *required* to continuously nurture sustainability innovations in urban governance and planning practices, while also enhancing wider capacities for transformative thinking and action among urban stakeholders. From the "outside", transition management can address critical gaps and frictions in planning institutions that will otherwise continue undisputed on their path. On the "inside", urban planning processes, instruments and techniques could become sharpened to foster and enforce radical systemic change, without demanding an institutional ground shift as a precondition. Therefore, establishing a dialectic relation of coordinated independence between transition management and urban planning may offer a promising future strategy for learning how to guide urban transformations towards sustainability.

References

Albers G (1997) Zur Entwicklung der Stadtplanung in Europa. Birkhäuser

Albrechts L (2004) Strategic (spatial) planning reexamined. Environ Plann B 31:743–758. https://doi.org/10.1068/b3065

Alexander ER (1992) Approaches to planning: introducing current planning theories, concepts, and issues, 2nd edn. Gordon and Breach Science Publishers, Philadelphia

Allmendinger P (1992) Planning theory, 2nd edn. Palgrave Macmillan, Houndmills

Allmendinger P, Haughton G, Shepherd E (2015) Where is planning to be found? Material practices and the multiple spaces of planning. Environ Plan C: Govern Policy. https://doi.org/10.1177/0263774X15614178

Arnstein S (1969) A ladder of citizen partizipation. JAIP 35:216–224

Brenner N (2004) New state spaces: urban governance and the rescaling of statehood. Oxford University Press, Oxford

Bryson JM, Quick KS, Slotterback CS, Crosby BC (2013) Designing public participation processes. Public Adm Rev 73:23–34. https://doi.org/10.1111/j.1540-6210.2012.02678.x

Castán Broto V, Bulkeley H (2013) A survey of urban climate change experiments in 100 cities. Glob Environ Chang 23:92–102. https://doi.org/10.1016/j.gloenvcha.2012.07.005

Cerreta M, Concilio G, Monno V (eds) (2010) Making strategies in spatial planning: knowledges and values, urban and landscape perspectives. Springer, Dordrecht/Heidelberg/London/New York

Davidoff P (1965) Advocacy and pluralism in planning. J Am Inst Plann 31:331–338. https://doi.org/10.1080/01944366508978187

Davoudi S (2001) Planning and the twin discourses of sustainability. In: Planning for a sustainable future. Spon Press, London, pp 81–94

Davoudi S, Strange I (eds) (2009) Conceptions of space and place in strategic spatial planning. Routledge, New York

Davoudi S, Crawford J, Mehmood A (eds) (2009) Planning for climate change: strategies for mitigation and adaptation for spatial planners. Earthscan, London/Sterling

Etzioni A (1967) Mixed scanning: A "third" approach to decision-making. Public Adm Rev 27:385–392

European Commission (1997) The EU compendium of spatial planning systems and policies, regional development studies. Office for Official Publications of the European Communities, Luxembourg

Fagence M (1977) Citizen participation in planning. Pergamon Press, Oxford

Fainstein SS, Campbell S (eds) (2012) Readings in planning theory, 3rd edn. Wiley-Blackwell, Malden

Faludi A (1973) Planning theory, Urban and regional planning series, 1st edn. Pergamon Press, Oxford

Flyvbjerg B (1996) The dark side of planning: rationality and "Realrationalität". In: Mandelbaum SJ, Mazza L, Burchell RW (eds) Explorations in planning theory. Center for Urban Policy Research, New Brunswick, pp 383–394

Forester J (1989) Planning in the face of power. University of California Press, Berkeley

Friedmann J (1987) Planning in the public domain: from knowledge to action. Princeton University Press, Princeton

Friend J, Jessop B (1969) Local government and strategic choice: an operational research approach to the processes of public planning. Tavistock Publications, London

Glasson J, Therivel R, Chadwick A (2011) Introduction to environmental impact assessment, The natural and built environment series, 4th edn. Routledge, New York

Habermas J (1981) Theorie des kommunikativen Handelns. Suhrkamp, Frankfurt/Main

Hall P (2002) Cities of tomorrow: an intellectual history of urban planning and design in the twentieth century, 3rd edn. Blackwell Publishers, Oxford/Malden

Harris N (2005) Collaborative planning: from theoretical foundations to practice forms. In: Allmendinger P, Tewdwr-Jones M (eds) Planning futures: new directions for planning theory. Routledge, London, pp 21–43

von Hayek FA (1967) Studies in philosophy, politics and economics. Simon and Schuster, New York

Hays SP (2000) A history of environmental politics since 1945. University of Pittsburgh Press, Pittsburgh

Healey P (1992) Planning through debate: the communicative turn in planning theory. Town Plan Rev 62:143–162

Healey P (1997) Collaborative planning: shaping places in fragmented societies, 2nd edn. Palgrave Macmillan, Basingstoke

Healey P (2007) Urban complexity and spatial strategies: towards a relational planning for our times. Routledge, London

Healey P (2010) Making better places: the planning project in the twenty-first century. Palgrave McMillan, New York
Healey P, Khakee A, Motte A, Needham B (eds) (1997) Making strategic spatial plans: innovation in Europe. UCL Press, London
Hillier J (2011) Strategic navigation across multiple planes: towards a Deleuzean-inspired methodology for strategic spatial planning. Town Plan Rev 82:503–527. https://doi.org/10.3828/tpr.2011.30
Hillier J, Healey P (eds) (2010) The Ashgate research companion to planning theory: conceptual challenges for spatial planning, Ashgate research companion. Ashgate Pub, Farnham, Surrey, England
Hodson M, Marvin S (2010) Can cities shape socio-technical transitions and how would we know if they were? Res Policy 39:477–485. https://doi.org/10.1016/j.respol.2010.01.020
Hodson M, Marvin S (2014) After sustainable cities? Taylor and Francis, Hoboken
Hooghe L, Marks G (2001) Multi-level governance and European integration, Governance in Europe. Rowman & Littlefield Publishers, Lanham
Innes JE, Booher DE (1999) Consensus building and complex adaptive systems: a framework for evaluating collaborative planning. J Am Plan Assoc 65:412–423. https://doi.org/10.1080/01944369908976071
Jacobs J (1961) The death and life of great American cities. Random House, New York
Jhagroe S, Loorbach D (2015) See no evil, hear no evil: The democratic potential of transition management. Environ Innov Soc Trans 15:65–83. https://doi.org/10.1016/j.eist.2014.07.001
Jouve B, Lefèvre C (2003) Local power, territory and institutions in European Metropolitan areas, The Cass Series in Regional and Federal Studies. Frank Cass Publishers, London
Karuri-Sebina JW (2014) Applying the systems of innovation approach to neighborhood planning: assessing local development analysis through an appreciative study of two South African townships (Ph.D. Dissertation). University of the Witwatersrand, Johannesburg
Kooiman J (2003) Governing as governance. Sage, London
Kreukels A, Salet W, Thornley A (2003) Metropolitan governance and spatial planning. Spon Press, London
Lascoumes P, Le Gales P (2007) Introduction: understanding public policy through its instruments? From the nature of instruments to the sociology of public policy instrumentation. Governance 20:1–21. https://doi.org/10.1111/j.1468-0491.2007.00342.x
Lindblom CE (1959) The science of "Muddling Through". Public Adm Rev 19:79. https://doi.org/10.2307/973677
Lindblom C (1979) Still muddling, not yet through. Public Adm Rev 39:517–526
Malekpour S, Brown RR, de Haan FJ (2015) Strategic planning of urban infrastructure for environmental sustainability: understanding the past to intervene for the future. Cities 46:67–75. https://doi.org/10.1016/j.cities.2015.05.003
Mannheim K (1951) Freedom, power and democratic planning. Routledge & Kegan Paul, London
Meadowcroft J (2009) What about the politics? Sustainable development, transition management, and long term energy transitions. Policy Sci 42:323–340. https://doi.org/10.1007/s11077-009-9097-z
Mintzberg H (1987) The strategy concept I: five Ps for strategy. Calif Manag Rev 30:11–24. https://doi.org/10.2307/41165263
Murdoch J (2006) Post-structuralist geography: a guide to relational space. SAGE, London
Nevens F, Roorda C (2014) A climate of change: a transition approach for climate neutrality in the city of Ghent (Belgium). Sustain Cities Soc 10:112–121. https://doi.org/10.1016/j.scs.2013.06.001
Newman P, Thornley A (1996) Urban planning in Europe: international competition, national systems, and planning projects. Routledge, London
Newman P, Thornley A (2011) Planning world cities: globalization and urban politics, Planning, environment, cities, 2nd edn. Palgrave Macmillan, Basingstoke

Rosenhead J (1980) Planning under uncertainty: 1. The inflexibility of methodologies. J Oper Res Soc 31:209. https://doi.org/10.2307/2581077
Ryser J, Franchini T (2015) International manual of planning practice: IMPP. Int Soc City Reg Planners (ISOCARP), The Hague
Sadler B, Aschemann R, Dusik J (eds) (2011) Handbook of strategic environmental assessment. Earthscan, London
Sager T (2001) Planning style and agency properties. Environ Plan A 33:509–532. https://doi.org/10.1068/a33170
Salet WGM, Faludi A (eds) (2000) The revival of strategic spatial planning. Royal Netherlands Academy of Arts and Sciences, Amsterdam
Savitch HV, Kantor P, 2002 Cities in the international marketplace: the political economy of urban development in North America and Western Europe. Princeton University Press, Princeton
Simon HA (1947) Administrative behavior: a study of decision-making processes in administrative organizations. Free Press, New York
Weiss A, Woodhouse E (1992) Reframing incrementalism: a constructive response to the critics. Policy Sci 25:255–273. https://doi.org/10.1007/BF00138785
Wittmayer J, Roorda C, Van Steenbergen F (2014) Governing urban sustainability transitions – inspiring examples. Dutch Research Institute for Transitions, Rotterdam

Part II
Applications of Transition Management in Cities

Chapter 6: Holscher et al. – Empowering actors in transition management in and for cities. The multi-actor nature of urban sustainability transitions challenges existing social fabrics and local governance settings, and raises questions about who takes decisions, with which agenda and to what end. A closer look is required for understanding who the actors involved in transition management processes are, what their roles and shifting roles are due to their engagement, what the responsibilities and relations of those actors and their affiliated networks entail. The chapter presents the notions of empowerment and disempowerment in view of transition arena processes, with the aim to support actor selection and analysis of their roles and the evolution of those thereof. The methodological propositions and reflections are further enriched with insights from empirical experiences with transition management in the cities of Ghent, in Belgium and Montreuil, in France.

Chapter 7: Maas et al. – Starting up transition management: A closer view on the system analysis and how it initiated transformative thinking in Ghent and Aberdeen cities. Maas et al. present methodological steps for conducting a systems analysis and reflect on how these steps were adapted during the transition management processes in Ghent, Belgium, and Aberdeen, UK. A systems analysis enriches the start of transition management by stimulating holistic and integrative thinking about a cities' problem domains and their interconnections, providing a common information base and pin-pointing areas and actors in the system both contributing to problems of unsustainability and solutions. In Ghent and Aberdeen, the systems analysis resulted in an overview of interconnected challenges and opportunities in both cities regarding their energy use, CO_2 emissions reductions and issues of sustainability and liveability. This revealed institutional barriers to sustainable development, highlighted opportunities for co-benefits and innovative actors and initiatives and broadened problem perceptions and integrated different viewpoints towards mutual understandings. While a systems analysis is thus critical for thinking about envisioned sustainability transitions in the later process phases, the authors emphasise that the system analysis methodology is not a cookbook and needs to be flexibly adapted according to the needs of a specific context and the involved actors.

Chapter 8: Wittmayer et al. – Transition management in urban neighbourhoods: The case of Carnisse, Rotterdam, The Netherlands. The case presented by Wittmayer et al. marks a socially disadvantaged community in the South of Rotterdam, the Netherlands: Carnisse. Transition management was applied in Carnisse in the context of decreasing public service provisions, closing of public facilities, high unemployment and crumbling social cohesion to facilitate the self-organisation of inhabitants to address persistent sustainability problems. Wittmayer et al. illustrate how the process implementation has been influenced by political, institutional and emotional factors related to the place-specific characteristics of urban neighbourhood transitions. While urban neighbourhoods mark opportunities for self-organisation – through for example emotional connectedness – the political playing field can provide support, or restrain it. The transition management application in Carnisse shows several advantages of applying transition management on the scale of an urban community. It provided a new perspective for the development of that community, increased social capital, supported reflexivity on roles and own behaviours and revealed opportunities for self-organisation.

Chapter 9: Hölscher and Wittmayer – A German experience: the challenges of mediating 'ideal-type' transition management in Ludwigsburg. Hölscher and Wittmayer review the transition management process phases and results in Ludwigsburg, Germany, from the perspective of how the 'ideal-type' process methodology gets mediated with respect to context-specific characteristics. Ludwigsburg represents a context that, before the transition management intervention, had already well-developed sustainability strategies and programmes and implemented a number of participatory processes to collect citizen input for strategic planning. The authors highlight several recurring challenges that occur from the need to differentiate transition management and position its added value to addressing the problems of a city. In Ludwigsburg, there was no perceived need for searching for alternative approaches to participation. In combination with a limited space to engage with the transitions thinking approach underpinning transition management this contributed to a framing of the process as resembling the other participatory efforts of the city. This led to a focus on the process on collecting new input to the city's strategies rather than (also) inspiring a co-creative process of jointly thinking and acting for a sustainable future of the city.

Chapter 10: Silvestri and Frantzeskaki - Transition Management for local sustainability: A case study from La Botija protected area, San Marcos de Colón, Honduras. This chapter presents two methods that we employed to facilitate the exploration for environmental stewardship in the La Botija area: Transition management was the transdisciplinary methodology that connected a co-created sustainability vision with the co-development and mobilization of local action. The ecosystem services framework was employed to systematize knowledge about resource use and ecosystem degradation and identify links between malpractices and ecosystems' deterioration. The transition management process activated participants to realise the co-created transition agenda by forging new partnerships, enabling self-organisation in community groups and empowering existing community groups. A highlighted learning outcome is the recognised need to develop own knowledge and community capacity to overcome local challenges deviating from the dependence on external aid.

Empowering Actors in Transition Management in and for Cities

Katharina Hölscher, Flor Avelino, and Julia M. Wittmayer

Abstract The multi-actor nature of urban sustainability transitions challenges existing social fabrics and local governance settings, and raises questions about who takes decisions, with which agenda and to what end. It demands close attention to who are actors involved in transition processes and what are the implications for (changing) roles, responsibilities and relations of actors and networks. This chapter provides an overview of different understandings of actors and roles in transition management in and for cities and how actors are empowered through transition management interventions to advance urban sustainability transitions. We first review different approaches and heuristics to understand actors and roles in transition management and in relation to the underlying goal to influence urban sustainability transitions. These enable to analyse actors and (changing) roles, support the actor selection and enhance reflexivity in transition management processes. We then discuss what the notion of 'empowerment' captures in terms of how actors (re-)define, develop and enact roles and relationships in the pursuit of urban sustainability transitions as a result of transition management interventions in cities. We exemplify our reflections with insights from our empirical experiences with transition management processes.

Introduction

The myriad of interlinked social, economic and environmental challenges and crises in cities prompted the recognition that local governments and their administrations are not able to address, let alone solve these on their own (Frantzeskaki et al., chapter "Introducing Sustainability Transitions' Thinking in Urban Contexts", this volume; Avelino and Wittmayer 2015; Loorbach 2014; Nevens et al. 2013). Sustainability transitions research understands transition processes to

K. Hölscher (✉) · F. Avelino · J.M. Wittmayer
Dutch Research Institute For Transitions (DRIFT), Faculty of Social Sciences, Erasmus University Rotterdam, Rotterdam, The Netherlands
e-mail: holscher@drift.eur.nl; avelino@drift.eur.nl; wittmayer@drift.eur.nl

N. Frantzeskaki et al. (eds.), *Co-creating Sustainable Urban Futures*, Future City 11, https://doi.org/10.1007/978-3-319-69273-9_6

be marked and carried out by a diverse set of actors and networks that formulate strategies, implement projects and mobilise resources (Farla et al. 2012; Grin 2010; Geels 2011). In cities, we witness numerous innovative projects and initiatives that are initiated by all kinds of public and private actors inside and outside of formalised policy channels (Bulkeley and Castán Broto 2013; Khan 2013; Nevens et al. 2013; Grimm et al. 2008; McCormick et al. 2013).

The multi-actor nature of transition processes can take the form of partnerships (Castán Broto and Bulkeley 2013; Frantzeskaki et al. 2014b; Wittmayer et al. 2015), but can also manifest in contestation, conflict and resistance to change (Hodson and Marvin 2010; Wittmayer et al. 2014a). The emergence of new actor configurations and partnerships challenges existing social fabrics and local governance setting (Wittmayer and Loorbach 2016). As such, accomplishing urban sustainability transitions "becomes a matter not only of policy, or of 'niche' experimentation, but of the reconfiguration of socio-technical networks – a process that is at once highly political and open to contestation and disruption" (Bulkeley et al. 2011: 30; cf. Avelino and Wittmayer 2015).

The multiplicity of actors and relationships in urban transitions demands close attention to who are the actors involved in transition processes, how do they interact and what are the implications for (changing) roles and responsibilities of actors and networks in the context of broader societal change (Wittmayer et al. 2017a). This ties in with questions about the kind of relations, power, politics and ethics involved in pursuing and governing sustainability transitions: Who is driving the initiatives, with which agenda, and to what end? How do the initiatives relate to incumbent and powerful actors? In this context, the re-organisation of what were considered governments' tasks vis-à-vis civil society and businesses through a privatisation paradigm raises questions about the (changing) roles of and the (possibly uneven) (re-)distribution of responsibilities between different actors that might have different decision-making authorities and resources bases (Swyngedouw 2005; Clarke 2005; Davies and White 2012). For example, governments increasingly call upon 'the community' to take over public services that they have become constrained or unable to provide due to financial pressure. Bottom-up initiatives that emerge to fill this gap often find themselves outside of standardised arrangements, i.e. outside of governance structures in terms of regulations, legislations and administrative organisations (Avelino and Wittmayer 2015).

In this context, transition management is a governance framework that seeks to "empower" individuals and communities "to shape sustainability in their own environments, and in doing so contribute to the desired transitions to sustainability" (Loorbach 2007: 284). In doing so, transition management aims "to create a societal movement through new coalitions, partnerships and networks [...] that allow for building up continuous pressure on the political and market arena to safeguard the long-term orientation and goals of the transition process" (Loorbach and Rotmans 2010: 239). The transition management process methodology offers operational guidance on how to set up "interactive and selective participatory stakeholder searching process aimed at learning and experimenting" (Grin et al. 2010: 140; Wittmayer et al., chapter "Transition Management: Guiding Principles and

Applications", this volume). Actors are selectively invited to engage in a co-creative process to address persistent unsustainability problems by defining desirable long-term visions and actively engaging in developing and nurturing alternatives to advance envisioned sustainability transitions (Loorbach 2010; Wittmayer et al. 2015; Kemp et al. 2007).

This contribution zooms in on the question of who are the actors in transition management processes in cities, what roles they (are expected to) take up in advancing urban sustainability transitions and how are they empowered to do so. Much of the past critique on transition management points to a neglect of conflict, politics, democratic legitimacy and power (Voß et al. 2009; Shove and Walker 2007; Smith and Kern 2009; Hendriks 2009; Meadowcroft 2009). Scholars identify questions for transition management in terms of who is initiating and who is invited to the process, whose framings count (about e.g. systems, problems, goals, sustainability) and what is the relationship with democratic institutions, incumbent regime actors and dominant discourses (cf. Wittmayer and Loorbach 2016).

For example, the focus on 'frontrunners' in transition management processes is criticised as possibly sustaining unequal power relations, supporting those already empowered and resulting in the *dis*empowerment of other actors (Smith and Stirling 2010; Wittmayer et al. 2015; Avelino 2009). Empirical examples of transition management show that different transition challenges and contexts require different approaches to actor selection (Kemp and Rotmans 2009; van den Bosch 2010; Avelino 2009). Meanwhile, the practice of transition management in cities and neighbourhoods has shifted towards the inclusion of the local community, at times to the detriment of other perspectives from government or business (Wittmayer et al. 2013; Nevens et al. 2013; Nevens and Roorda 2013). There is a risk to over- or underrate the abilities and power sources of different types of actors, particularly of so-called 'niche' actors, which might conceal the more deeper running changes needed regarding the roles and (power) relations between the diverse actors in order to achieve (urban) sustainability transitions (Avelino and Wittmayer 2015).

In response to these criticisms a productive scientific dialogue has emerged (Jhagroe and Loorbach 2015; Avelino 2011; van Steenbergen and Wittmayer 2012). A number of authors have explicitly theorised the role of power in transitions (Grin 2010; Avelino 2011; Hoffman 2013; Geels 2014), specified (roles of) actors in transition processes and transition management (Farla et al. 2012; Schuitmaker 2012; Bauknecht et al. 2015; Avelino and Wittmayer 2015; Wittmayer and Schäpke 2014; Wittmayer et al. 2014b, 2017a; Wittmayer 2016) and conceptualised empowerment outcomes targeted by transition management interventions (Avelino 2009, 2011; Hölscher et al. 2016, 2017). We draw on these insights to provide a critical perspective on who are the actors in transition management processes in cities and how are they empowered to take up what roles in advancing urban sustainability transitions.

In the following, we first review different approaches and heuristics to understand actors and roles in transition management and in relation to the underlying goal to influence urban sustainability transitions (section "Understanding actors and roles in transition management in and for cities"). We then discuss what the notion of 'empowerment' captures in terms of how actors (re-)define, develop and enact roles and relationships in the pursuit of urban sustainability transitions as a result of transition management interventions (section "(Dis)empowerment in transition management in cities"). We exemplify these sections with reflections from empirical experiences with transition management processes in cities – particularly drawing on the transition management interventions in five cities in the context of the MUSIC-project: Aberdeen (UK), Ghent (Belgium), Ludwigsburg (Germany), Montreuil (France) and Rotterdam (The Netherlands) (Nevens et al. 2013; Nevens and Roorda 2013; Roorda et al. 2014; Roorda and Wittmayer 2014; Frantzeskaki and Tefrati 2016; Krauz 2016; Hölscher et al. 2016; Hölscher and Wittmayer, chapter "A German Experience: The Challenges of Mediating 'Ideal-Type' Transition Management in Ludwigsburg", this volume; Hölscher et al. 2016, 2017). In the Interreg funded MUSIC-project (Mitigation in Urban Areas: Solutions for Innovative Cities) transition management was for the first time applied to cities to develop strategies and solutions for low carbon cities and to engage stakeholders in the process. The final section summarises the implications of these understandings of actors, roles and empowerment in transition management, points to issues of concern for implementing transition management processes in cities and identifies issues for future research.

Understanding Actors and Roles in Transition Management in and for Cities

We conceive an 'actor' as any social entity that is able to act (Hermans 2010). This refers to both individual actors (e.g. policy-makers, citizens, entrepreneurs) and organisational actors (e.g. public agencies, NGOs, community groups) (Avelino and Wittmayer 2015). An actor can take up different roles in the pursuit of urban sustainability transitions. The concept of roles is situated 'in between' the individual and society; roles refer to shared conceptions within a particular community and a change in roles can indicate changing interactions and relations between actors in such a community (cf. Wittmayer et al. 2017a). While some roles might be exclusive to some actors, some roles can be taken up by a variety of actors (ibid.).

There are various approaches to categorise, frame and analyse actors that highlight different roles in transition management processes and urban sustainability transitions. Firstly, they include a focus on the specific practical roles in prescriptive transition management interventions. These relate to the different levels of governance embedded in the transition management cycle as the process moves from more strategic towards more operational governance activities (Wittmayer et al., chapter "Transition Management: Guiding Principles and Appli cations", this volume). Approaches to actors and roles also include different heuristics to select and analyse actors and roles, and a discussion of roles of types of societal actors in transition management and urban sustainability transitions. Each of these approaches provides complementary insights on actors and roles that facilitate a better understanding of actor dynamics in transition (management) processes.

Practical Roles in Transition Management in Cities

Prescriptive transition management interventions take place in three types of (overlapping) actor configurations: transition team, transition arena and transition networks (Roorda et al. 2014; Wittmayer et al., chapter "Transition Management: Guiding Principles and Applications", this volume). On the one hand, these configurations offer delineated spaces for actors to reflect and engage in the different activities of the process and in that sense to take up different practical roles. On the other hand, they also represent three distinct collective actors that take up different sets of roles in transition management interventions (Wittmayer 2016; Roorda et al. 2014).

The creation of the transition team setting marks the start of the transition management process: the transition team initiates, prepares and guides the later process steps. In the transition arena setting, arena participants engage in a series of meetings to jointly formulate the transition challenges, develop future visions and transition pathways. The roles taken up by the actors in the transition team and transition arena hence refer to the more strategic and tactical activities of the transition management cycle (Wittmayer et al., chapter "Transition Management: Guiding Principles and Applications", this volume). After the transition arena, transition networks are formed that disseminate the new vision and transition agenda, connect the process results to other on-going initiatives and processes and implement transition experiments.

Actors can be members of all three settings, and in general there is a large degree of interaction between actors from different settings (Wittmayer 2016; Hölscher et al. 2016). The roles taken up in the respective settings should at best become overlapping as the process moves along and is co-created by all actors involved (ibid.). However, there might also be tensions between the different settings and roles. For example, on the one hand, the transition team is expected to initiate and guide the process, while allowing all actors in the process to meet on equal footing and take part in decisions about the framing, configuration and follow-up of the process (Hölscher et al. 2017).

Transition Team

The transition team encompasses a collection of actors whose role is to initiate and guide a transition management process. The roles of the transition team actors are particularly critical during the first stages of transition management to create space and an open mind-set for the intervention and to conduct the system and actor analyses (Roorda et al. 2014; Loorbach 2007). They have been described as follows:

> the team not only prepares, documents, analyses, monitors, co-ordinates, manages, facilitates and evaluates the whole process, but also chooses the participants and feeds them with background information and detailed knowledge. It brings together the various parties, is responsible for the internal and external communication, acts as intermediary in discordant situations and has an overview of all the activities in and between arena meetings. (Wittmayer et al. 2011: 31)

The transition team can in principle consist of any (combination of) kinds of actors – the actual composition depends very much on the respective context. To date, most transition management interventions have been initiated, coached, supported or accompanied by academic actors (Wittmayer et al. 2015; Wittmayer and Schäpke 2014). However, also local governments, businesses, citizens or entrepreneurs can take up the initiative for and guidance of the process. In many transition management processes in cities – such as in the MUSIC project – policy officers from the local governments were the key actors in the transition team (Roorda et al. 2014; Roorda and Wittmayer 2014; Hölscher et al. 2017). This might also link to a different scope of the transition management intervention. For example, when the scope of the process is strategic planning, agenda setting and coordination across planning processes, the transition team is likely to consist of both academic actors and policy officers. As transition management interventions usually build on flexible objectives, the scope for the intervention can also shift during the process.

Because of the influence of the transition team on the initial framing, direction and outcome of the process of the transition management process, who these individual actors are and what organisational background they represent must be treated with caution (Wittmayer et al. 2012, 2015). For example, a major task is to process the input from the transition arena meetings. While it is essential that the input is not substantially changed and that the arena participants verify the processing, this processing is critical because otherwise the process would take too long and might not arrive at a joint direction for the sustainability transition.

In the transition management process in the MUSIC-city Ghent, for example, the high level of political interest led to a rather controlled process implementation that did not leave much room for co-creation with the arena participants. This influenced the formulation of vision elements (e.g., the statement ‘car-free city centre’ was changed to ‘less cars’) and the extent the participants developed ownership over the process and engaged in its follow-up (Hölscher et al. 2016). In contrast, the process in Montreuil stands out in the sense that the transition team included the arena participants extensively in decisions about the process steps and intended results, which led to a high degree of ownership (Krauz 2016).

Such experiences highlight the importance of skilful facilitation by the transition team to indeed create a setting in which the roles of the transition team and the transition arena participants converge in the pursuit of an envisioned sustainability transition. On the one hand, the transition team members should feature a combination of skills and knowledge, including the ability to facilitate the meetings, adapt to the needs of the process and reflect on their role (Wittmayer and Schäpke 2014; Loorbach 2007). Furthermore, they should be able to go along with the innovative thinking that is advocated throughout the transition management process, while at the same time also 'playing the regime game'. This regime game includes strategically liaising with other mainstream (regime) actors and identifying windows of opportunity such as political momentum and leveraging financial resources (Wittmayer 2016). On the other hand, the transition team itself provides a setting for learning and a space for the involved actors to familiarise themselves with and embrace the open and co-creative approach of the transition management methodology (Wittmayer 2016; Roorda and Wittmayer 2014). Even when the team consists of researchers experienced with the methodology these researchers need to be able to contextualise it to take up their role in the transition team (Wittmayer and Schäpke 2014).

Transition Arena

The transition arena is a temporary setting that provides an informal but well-structured space to a small, mixed group of (15–20) participants to enter a joint learning process and develop a new narrative and discourse to frame, guide and pursue a sustainability transition (Wittmayer et al. 2017b, chapter "Empowering Actors in Transition Management in and for Cities", this volume; van Buuren and Loorbach 2009; Loorbach 2010; Roorda et al. 2014). A substantive outcome of the transition arena is a transition agenda for a city, which consists of (a) a shared integral problem statement, (b) a novel future perspective, (c) transition images and pathways that indicate concrete initiatives and actions (Wittmayer and Loorbach 2016). Through their participation in the transition management process, the arena participants should ideally become motivated, or 'empowered' to take up the developed transition agenda and engage in transition networks in the implementation of transition experiments (Roorda et al. 2014).

The transition arena participants are invited by the transition team on the basis of an actor analysis (Wittmayer et al. 2012). Central for the selection and composition of the arena participants is the notion of 'frontrunner'. Frontrunners refer to individuals who understand aspects of urban wicked problems, hold innovative ideas or engage in activities in urban development, are committed to their city and sustainability and have the potential to influence change (ibid., Wittmayer et al., chapter "Transition Management: Guiding Principles and Applications", this volume). Additionally, the group composition should be heterogeneous, bringing together actors from all kinds of societal spheres or sectors (i.e. government, business, civil society, intermediaries, science) and backgrounds (e.g. culture, mobility, social issues). This ensures diversity in perspectives and interests, which might enable richer debates on the problems, sustainability visions and pathways (Loorbach 2010; van Buuren and Loorbach 2009). The actor selection draws on the heuristics that are discussed in chapter "Transition Management: Guiding Principles and Applications", this volume.

Not all scholars agree with the choice of selective participation of frontrunners, conceiving it as 'elite group' (Smith and Stirling 2010), pointing to its legitimacy deficits (Hendriks 2009) and questioning the framing of frontrunners as an "enlightened" type of person (Jhagroe and van Steenbergen 2014: 2; cf. Wittmayer 2016; Avelino 2009). Researchers concerned with the democratic politics associated with transition management call for thinking beyond existing paradigms of democratic institutions and deliberative democracy (Hendriks and Grin 2007; Hendriks 2009; cf. Jhagroe, chapter "Urban Transition Management as Democratic Practice? The Case of Rotterdam's Waterfront Regeneration", this volume). Jhagroe (chapter "Urban transition management as democratic practice? The case of Rotterdam's waterfront regeneration", this volume) draws on post-foundational democracy to establish the democratic legitimacy of transition management in and for cities in that the process's goal is to challenge hegemonic discourses about cities and urbanisation and materialising alternative imaginaries in local and everyday settings.

Who is invited – and as such already empowered – and how the participation is framed (e.g. asking the participants to bring in their individual or organisational perspective) is important when applying transition management. In early years, transition management was often implemented in relation to sectors (e.g. energy, mobility), which led to the participation of mainly professionals in the transition arenas (Wittmayer 2016; Avelino 2009). The increasing focus on wider societal systems and questions and the application of transition management in cities coincided with increasing inclusion of social entrepreneurs, inhabitants and volunteers (Frantzeskaki et al. 2014a). In the MUSIC-project, the transition arenas were largely composed of such actors. While the heterogeneity of backgrounds and interests facilitated a broadening of perspectives and a move away from mainstream thinking, the perceived (financial, legislative) resource limitations also led to scepticisms by the participants themselves about their abilities to achieve sufficient leverage for the change they found necessary (Hölscher et al. 2017). In Ludwigsburg, this coincided with a framing of the process as a participation process to collect input from citizens rather than stimulating a reflection on roles of all actors in urban sustainability transitions (Hölscher and Wittmayer, chapter "A German Experience: The Challenges of Mediating 'Ideal-Type' Transition Management in Ludwigsburg", this volume). In contrast, in Rotterdam the transition arena mainly consisted of professionals. This did not contribute to a broadened view on different societal perspectives and to a momentum for co-creation of a sustainability transition (Wittmayer et al. 2012; Roorda and Wittmayer 2014).

Additionally, there is a focus away from (solely) frontrunners towards 'change-inclined' actors representing also mainstream or established institutions and interests (Loorbach 2014; Wittmayer 2016). This builds on the (empirical) recognition that the developed alternatives need to be connected to existing processes and structures to create greater leverage for embedding the ignited change (Brown et al. 2013).

Transition Network(s)

The transition network refers to the most loosely organised actor setting and it also is the least discussed in transition management literature - though there is an increasing discussion of the roles of networks in transitions (Frantzeskaki et al. 2014b; Hodson and Marvin 2010). The transition management process methodology foresees that after the formulation of the transition agenda the involved actors invite additional actors from their networks to join the pursuit of the transition agenda and "engage with the wider world" (Roorda et al. 2014: 35). This marks the start for nurturing 'networks of networks' that influence sustainability transitions by diffuse actions (Loorbach 2010, 2014) – it is also where the distinction between transition management interventions and the 'on-the-ground' and practical pursuit of urban sustainability transitions becomes blurred. These transition networks are not limited in time. They can refer to loosely organised actors that follow common ambitions towards sustainability and relate this to their own agendas and practices (Roorda et al. 2014). They can also represent more formalised partnerships that collaborate to implement concrete initiatives and projects or that share knowledge (Frantzeskaki et al. 2014b).

This approach of a transition network reflects the above stated recognition that urban sustainability transitions are influenced by a large variety of actors, who self-organise societal functions "in hybrid and decentralized, distributed forms" (Loorbach 2014: 38). In the transition arena it was necessary to have a limited number of participants to build trust, engage in in-depth exchange and think beyond business-as-usual. To keep the momentum of this temporary innovation impulse and maintain a lasting impact, the engagement of a growing number of actors and networks is needed to "increase the impact of the arena results and unleash the potential of others to contribute to the city's sustainable future" (Roorda et al. 2014: 35).

The self-organising transition networks consist of an increasingly diverse set of actors that operate in different, but often overlapping and nested network settings and engage in different types of activities (ibid.). They are aligned under the umbrella of the in the transition arena developed urban sustainability transition agenda. One cluster of transition network activities includes the dissemination of the transition agenda, actively engaging other actors, connecting with on-going initiatives and processes and starting groups to reach out to larger audiences. Secondly, transition experiments are implemented to put the transition agenda into practice. Moreover, the transition agenda is translated into changes of institutional structures and framework conditions – for example in new governance structures that facilitate collaboration and self-organisation, new relationships or new financing options. Finally, to align and support the diffuse nature of these self-organising activities coordination is suggested "as a driving force, [. . .] facilitating and expanding the network of engaged actors, and organizing space for reflection and inspiration" (ibid.: 35). Coordination can include identifying resource opportunities, providing space for collaboration, alignment, monitoring and connecting actors (ibid., Frantzeskaki et al. 2014b).

The move from the transition arena into transition networks has shown to be demanding in empirical transition management cases (Roorda et al. 2014; Roorda and Wittmayer 2014; Hölscher et al. 2016, 2017). For example, in the MUSIC-cities Ludwigsburg and Ghent transition networks were never really created or died out after a short time period. In Ghent, only few actors from the original transition team – i.e. involved policy officers – engaged in transition network activities. This is partially because the facilitation of the process generally ends after the transition arena – as does the project funding – and the transition management methodology has only recently proposed concrete steps of how to facilitate the follow-up of the transition arena (Roorda et al. 2014). There is still a need for a greater understanding of who takes up what roles in such networks, how they are enabled through transition management and how they can be coordinated in the long-term to sustain and facilitate the implementation of the transition agenda (Loorbach 2014). The experience of the transition management process in Montreuil offers a concrete example of how (a) transition network(s) can be institutionalised in the form of a recurring transition arena (Krauz 2016).

Heuristics to Understand Actors and Roles in Transition Management in Cities

Transition management draws on different heuristics to understand, analyse and frame actors and roles in transition management processes. This is for example necessary for a comprehensive system and actor analysis, to select arena participants and to support reflections on roles. The heuristics reflect different approaches of transition research to conceive of actors and to identify their roles in urban sustainability transitions.

One of these heuristics is based on the multi-level perspective (MLP) (Geels and Schot 2007) that broadly distinguishes between actors and corresponding roles according to niche and regime levels. Next to this, as part of the actor selection criteria, transition management literature categorises actors according to different backgrounds or societal spheres, to obtain a heterogeneous arena composition. Recently a Multi-Actor Perspective (MAP, Avelino and Wittmayer 2015) has been introduced to enhance the currently applied view on this, which has been based on the "societal pentagon" (Rotmans 2005: 46; Loorbach 2010: 174). Both approaches – categorising actors and roles according to the MLP or according to backgrounds – generate different types of understandings and framings.

Interestingly, in light of the current attention in particular to actor *networks*, it is striking that so far a conceptual network perspective is missing from the heuristics to understand actors.

The Multi-level Perspective

The multi-level perspective (MLP) is a central heuristic of transition studies to analyse how processes and dynamics at different levels co-evolve and what are the main factors that influence change (or resistance) (Frantzeskaki et al., chapter "Introducing Sustainability Transitions' Thinking in Urban Contexts", this volume; Rip and Kemp 1998; Geels and Schot 2007; Frantzeskaki and de Haan 2009; de Haan and Rotmans 2011). Though the MLP does not explicitly address actors and is criticised for paying too little attention to actors and roles (Smith et al. 2005; Genus and Coles 2008; cf. Geels 2011; Avelino 2011; Avelino and Wittmayer 2015), it is "shot through with agency, because the trajectories and multi-level alignments are always enacted by social groups" (Geels 2011: 29). It offers insights into the roles actors can play in facilitating, or, hindering sustainability transitions based on their positioning in the (structural configuration of the) system.

Following the MLP, actors can be broadly conceived of as pertaining to two archetypical, separated but interacting categories that have different options for agency and hence (can) take up different roles in urban sustainability transitions: While regimes generally have a very stable institutional structure that individual actors can influence only very limitedly and indirectly, the more loose institutional structure of niches offers niche actors more options to develop institutions and alternatives according to their own preferences (Raven et al. 2010; Grin 2010). 'Archetypical' regime actors take up dominant positions and roles in the incumbent regime. They are assumed to generally have an interest in and benefit from maintaining the status quo and to resist change (Raven et al. 2010; Avelino and Rotmans 2009). There can also be 'change-inclined' regime actors that can take a role in facilitating sustainability transitions by connecting niche innovations with existing structures and processes and to gain support and legitimacy (Bauknecht et al. 2015; Loorbach 2014; Apajalahti 2012). Niche actors refer to actors that develop and advance alternatives to existing structures, cultures and practices. The development of niches requires individual visionary actors, who take leadership, recognise persistent unsustainability problems, organise ideas and resources to grow the niche and identify opportunities for connecting niches with each other and to the regime (Bauknecht et al. 2015; Garud et al. 2010; Brown et al. 2013).

The identification of regime and niche actors is used to select arena participants, to map existing niche initiatives and reveal transition challenges. For example, in the MUSIC-cities the MLP was used in the system analysis to identify both change-resistant and change-inclined actors (Maas et al., chapter "Starting Up Transition Management: A Closer View on the Systems Analysis and How It Initiated Transformative Thinking in Ghent and Aberdeen Cities", this volume). These insights feed into the actor analysis and subsequent selection of arena participants that represent so-called frontrunners and niche actors. Furthermore, the identification of niche actors can also serve the broadening into transition networks, as it points to actors that might connect their activities to the transition agenda. For example, in Ghent the involved policy officers succeeded in linking up with other actors and their innovative initiatives (Hölscher et al. 2016).

The MLP is criticised for prompting somewhat ambiguous categorisations of who are regime and niche actors and creating black boxes of how regimes and niches work and interact. It is often taken to implicitly suggest that regimes have 'more power' than niches (Avelino 2011; cf. Avelino and Wittmayer 2015). For example, niche actors are often equated with citizens, social entrepreneurs and NGOs, while governmental and 'big business' actors are related to the regime (Avelino and Wittmayer 2015). However, one could also argue that trade unions and religious organisations might have both substantial power as well as an interest in maintaining the status quo. Such a generalisation causes bias towards bottom-up change and disguises the need for more top-down interventions (ibid.). What the MLP highlights, however, is that actors (can) take up different roles in influencing and facilitating urban sustainability transitions, and how roles shift, in accordance with their positioning in the system and interests (Bauknecht et al. 2015; Geels 2011).

Though transition management explicitly seeks to engage niche actors and frontrunners that are said to nourish innovative ideas, so-called 'regime' actors could nevertheless (unintentionally and intentionally) be part of the transition team or the transition arena. While there is a danger of 'regime capture' (Avelino 2009), insights from transition management applications in cities suggest that the processes benefit from actors who are able to connect alternatives to on-going dynamics, change institutional structures and identify opportunities in terms of leveraging financial resources or liaising with other actors (see chapter "Transition Management: Guiding Principles and Applications", this volume, Wittmayer 2016). Brown et al. (2013) conclude from their analysis of the historical transition to improved storm water quality treatment in Melbourne that the main focus of transition management to date, has been on the predevelopment phase of transitions, which accounts for the strong focus on empowering frontrunners and niches. More advanced phases in a transition, in which a strong niche, or niche-regime, is already developed, might require a different focus and a better understanding of and connection to the incumbent institutional and policy context (cf. Wittmayer and Loorbach 2016; Loorbach 2014; Loorbach et al. 2015).

A Multi-actor Perspective

Another important approach to understand and select actors in transition management is according to their background. A diverse arena group is expected to broaden problem perceptions and stimulate innovation and knowledge exchange (Loorbach 2010; Roorda et al. 2014). So far, transition management commonly applies the "societal pentagon" (Rotmans 2005: 46; Loorbach 2010: 174) to select "the most important societal perspectives relevant to the system under study" (Loorbach 2007: 140; cf. Avelino and Wittmayer 2015). This distinguishes between government, business, knowledge institutions, NGOs/societal organisations, and intermediaries.

While this distinction categorises actors according to sectors, transition management literature also claims that arena members are invited on a personal basis (Avelino and Wittmayer 2015). This suggests somewhat paradoxically that while being chosen for a certain background, the individual should not be seen as representing this perspective in the arena group (ibid.). A limited understanding of and attention to certain (groups of) actors can distort the arena group composition. For example, a number of Dutch national innovation programmes applying transition management ideas primarily used the 'triple helix' (i.e. government, business, science) to select actors (Avelino 2011). This resulted in a misrepresentation of the interests of business and government to the detriment of community or third sector (Avelino and Wittmayer 2015).

To address these concerns and to question the niche-regime bifurcation of the MLP, a Multi-actor Perspective (MAP) has been introduced (Avelino and Wittmayer 2015). The MAP distinguishes between individual and organisational actors that can take up different roles in relation to different sectors. The purpose of the MAP is not to categorise organisations or individuals per se, but to explore how they act within different sector logics, which sector logic tends to be 'dominant' in specific actions and discourses, and how the power relations between and within the sectors, organisational and individual roles shift over time (ibid.). The MAP emphasises that it is not about government vs. citizens, for example, but that one needs to be specific about what roles are important for influencing change from all sectors and about whom to empower.

The MAP draws on work by institutional scholarship, particularly the 'Welfare Mix' scheme by Evers and Laville (2004: 1740) and Pestoff (1992: 2537). This scheme distinguishes between four different sectors: state, market, community and third sector. The distinction of sectors is based on general characteristics and 'logics' of a sector (i.e. formal vs. informal, for-profit vs. non-profit, public vs. private). Notable is the category of 'third sector' as an intermediary sector between state, market and community. It includes the non-profit sector that is formalised and private, but also intermediary organisations that cross the boundaries between profit and non-profit, private and public, formal and informal (e.g. 'not-for-profit' social enterprises, universities, or cooperatives).

The consideration of the third sector enables to more sharply specify what is usually referred to as 'civil society' (Avelino and Wittmayer 2015). The transition management methodology needs to more clearly do so when it argues that the transition arena should involve 'civil society', 'societal' and 'non-governmental' organisations (Avelino 2011). Even if the arena includes actors from NGOs, citizens or grassroots initiatives can remain underrepresented (Avelino and Wittmayer 2015). Additionally, it is also important to recognise that the third sector can include rather powerful regime actors such as labour unions, large cooperatives and religious networks (ibid.).

The MAP takes the Welfare Mix scheme further and distinguishes between different individual and organisational roles in relation to which sector logic is

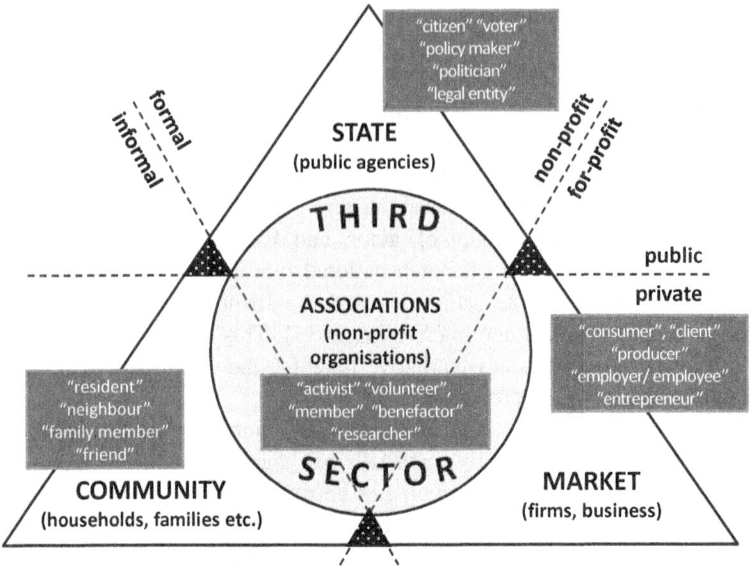

Fig. 1 MAP: level of individual roles per sector (Source: Avelino and Wittmayer 2015: 10)

applied (Avelino and Wittmayer 2015). As such, while sectors themselves can be viewed as actors, they can also be seen as institutional contexts, or, discursive fields, in which organisational or individual actors operate and with which they interact. Sectors can hence be viewed as sites of struggle and/or cooperation between different actors (e.g. the state as interaction between politicians, voters and different departments; the market as interaction between consumers and producers) (ibid.). Sectors can also determine how roles of actors tend to be constructed in a different manner following the specific sector logic.

Figure 1 illustrates how the MAP distinguishes individual roles according to sector logic, ranging from 'resident' or 'neighbour' (community) to 'citizen' or 'voter' (state) or 'consumer' (market). One single individual is constructed differently in different sector logics, e.g. a policy-maker is also a neighbour, consumer and possibly a volunteer in his/her free time. In this understanding, individual actors are performing social roles (Turner 1990; cf. Avelino and Wittmayer 2015).

The MAP helps to question underlying assumptions of the MLP about what type of sectors or actors are situated in niches or regimes and have more or less power (Avelino and Wittmayer 2015). On the one hand, within each sector broadly a regime and niches can be identified. For example, in the state sector, the regime refers to the dominant cultures, structures and practices that are utilised to fulfil the state function (e.g. centralisation, institutionalisation, short-term political cycles, formal democracy, bureaucracy). The state sector however also includes innovative practices and new pockets of culture and structure that deviate from this regime and reside in niches. The power relations between governments and citizens, politicians

Table 1 Use of MAP in transition management phases

Transition management phase	Use of MAP in transition management	
1. Setting the scene	Identifying and convening transition team actors;	*Analytical*
	Analysing and connecting to 'context' actors	*Prescriptive*
2. Exploring	System analysis: analysing actors and roles in existing governance processes;	*Analytical*
	Actor analysis: analysing and selecting actors for transition arena	*Prescriptive*
3. Framing challenges	Reflecting on and understanding actors and roles in existing governance processes	*Facilitative*
		Analytical
4. Envisioning	Identifying and framing actors and roles in future visions	*Facilitative*
		Analytical
5. Reconnecting	Assigning actors and roles in transition pathways	*Facilitative*
		Analytical
6. Engaging and anchoring	Connecting to actors for dissemination and broadening	*Prescriptive*
		Analytical
7. Getting into action	Engaging actors in the development of transition experiments	*Prescriptive*
		Analytical

and voters, national and local departments, are all part of the niche-regime dynamics within 'the state' sector.

On the other hand, the MAP enables to identify sectors that show most activity, dominant sector logics and reveal transition challenges in terms of where to search for alternatives. One could argue, for instance, that modern western societies have been dominated by a two-sector state-market logic during the past decades of welfare state development combined with privatisations and wide-spread 'neoliberal' discourses (ibid.).

The MAP can enrich transition management processes in various ways throughout the individual process phases (Table 1, Wittmayer et al., chapter "Transition Management: Guiding Principles and Applications", this volume). It can serve as an *analytical* tool to analyse and understand existing and envisioned governance processes. This could for example support the system and actor analysis by revealing the diversity of ascribed and constructed roles, facilitating reflexivity on who are 'dominant' actors and revealing current transition challenges and opportunities. Insights into the societal fabric visible both in the current situation as well as in possible future visions reveal which actors and roles are constructed as 'dominant' and how these might change as a result of envisioned role changes. The MAP can also facilitate (participatory) analyses of the arena composition and whether actors might be missing, what roles or organisational backgrounds individuals assume in the discussion, how actor involvement can be improved and to reflect on the vision and pathways.

In close relation to this, the MAP can also enrich transition management as a *prescriptive* tool that supports the implementation of a specific process phase. For example, insights gained from a MAP analysis at the outset of a transition management process can indicate which actors could be convened in a transition team. As stated earlier, who will participate in the transition team might depend on the scope of the transition management intervention (see section "Transition Team"). Similarly, the MAP analysis supports the selection of arena participants and of actors that are to be invited and engaged in broader transition networks. Again, the scope of the transition management intervention will determine which actors are invited or focused on, but the MAP helps to infuse reflexivity in this process, by enabling to be more explicit about which actor categories and roles the arena is supposed to represent in the particular context and to overcome a bias towards certain (groups of) actors and sector logics.

Finally, the MAP can be applied as a *facilitative* tool in transition management processes, in the transition arena sessions. For example, it serves to facilitate reflections of the involved actors on actor roles and (power) relations in relation to identified transition challenges and to discuss about what kind of actor relations are desirable in the future. Such reflexivity can also be critical for the development of transition pathways that are specific about what actors and roles are required to achieve sustainability vision. In setting up transition experiments, the MAP can be instructive in guiding whom to involve (and in which roles), in order to realise the sustainability vision.

Roles of Societal Actors in Transition Management in Cities

Actors can also be approached in relation to the specific (dominant) societal roles they take up in transition management processes in cities. This view connects the practical roles in prescriptive transition management processes and to pursue urban sustainability transitions (see section "Practical Roles in Transition Management in Cities") to actors' societal roles. This helps to understand what types of societal actors (are expected to) take up what specific roles in transition management processes in cities and their follow-up and to devise mechanisms to support actors in taking up their roles. We highlight few of those roles based on reflections of our experiences with transition management processes in cities. For future research, it might be fruitful to further investigate the roles of particularly market actors and to more critically differentiate between the roles of community and third sector actors in transition management.

Role(s) of Local Policy Officers and the Local Government

The role of local government and its administrative bodies is important, but also notoriously difficult to pinpoint, which is only partly due to the diversity of ways in which transition management is implemented in cities. In practice, representatives from local governments are involved as funders, initiators, facilitators initiators, participants, process leaders, or a combination of those roles (Wittmayer et al. 2015).

Especially when the local government is involved as initiator and facilitator of the transition management process the different logics and assumptions of transition management and the dominant daily practices and logics of local governments can clash. While the former is about empowerment, openness, accepting ambiguity and uncertainty, as well as encouraging reflection, the latter are inherently about providing stability, which often comes with an urge to plan and control (ibid.). This can lead business-as-usual and conventional governance practices to prevail the process implementation. For example, in the transition management process in Ghent the local government kept the process under tight control, which led to a limited ownership and uptake of the process by the arena participants (Hölscher et al. 2016). In contrast, in Montreuil the involved policy officers were free to engage in a flexible and co-creative process with the arena participants that resulted in open discussions about and inspired new roles and responsibilities (Krauz 2016).

In empirical experiences with transition management applied in cities, the local government has also been a central actor in the facilitation and support of the follow-up of the transition arena and the move into transition networks. It provides resources, connects actors to each other and mediates knowledge (Frantzeskaki et al. 2014b; Hölscher et al. 2016). This might partially relate to the challenges of stimulating ownership over the process and its results because of a too top-down process implementation, as in Ghent (Hölscher et al. 2016), or a framing of the process as 'conventional' stakeholder consultation (Hölscher and Wittmayer, chapter "A German Experience: The Challenges of Mediating 'Ideal-Type' Transition Management in Ludwigsburg", this volume). It also points to a new role that local governments might (need to) take up in the governance of urban sustainability transitions and what this implies on relationships towards other actors (Frantzeskaki et al. 2014b; Hölscher et al. 2017).

Importantly, while "local government" or "local administration" is used to refer to one seemingly coherent actor, it is more like a 'black box' in which different departments, individuals, and political interests need to cooperate, and ignore, compete, and struggle with each other. The policy officers who are involved in the transition team often act within a niche of the local administrative body. As such, transition management also asks for a new working attitude from local governments, that includes a willingness to reconsider one's role as an individual and as an organisation, and to create time and support for action and reflection (Nevens and Roorda 2013). To increase support for the transition management process and its results, including a new approach to governance, and to deal with tensions it is important that the policy officers involved connect with other actors from within the local government. As such, transition management requires "mediation between established (policy) institutions and policies and the developing transition projects, policies and networks" (Nevens et al. 2013: 8). In this sense, transition management also seeks to innovate and change the existing structures of how local governments work. This also involves new governance relationships between local government and citizens and the take up of a coordination role

(Wittmayer 2016). Empirical examples of transition management processes show that the experience of undergoing a transition management process led to the establishment of new ways of working within policy-making bodies (e.g. increased interdepartmental cooperation and an increased reflexivity regarding their own role) (Nevens and Roorda 2013; Roorda and Wittmayer 2014).

Role(s) of 'Civil Society'

The shift in the practice of transition management towards cities and neighbourhoods has led to particularly invite 'civil society' actors to participate in transition arenas, at times to the detriment of other perspectives from government or business (Nevens et al. 2013; Nevens and Roorda 2013; Wittmayer et al. 2013). This resonates with an increasing interest in the role of citizens, NGOs and social entrepreneurs in fostering sustainability transitions through the 'self-organisation' of services and products (Janssen and Hohman 2012; Seyfang and Smith 2007; Seyfang and Haxeltine 2012; Smith 2012).

Overall, civil society is a very heterogeneous actor category, containing for example NGOs, trade unions, political parties, environmental groups, interest groups and households (Fischer and Newig 2016). In transition management literature, it is so far unclear whether social entrepreneurs, that represent market sector logic, are considered as the same category. The term 'civil society' is here still applied rather vaguely and generalises the different types of roles individual actors take up in transition management and urban sustainability transitions. The previously outlined MAP (see section "A multi-actor perspective") helps to specify who those 'civil society' actors are and what roles they (are expected to) take up, differentiating for example between community and third sector actors and roles.

The role of civil society has so far not been adequately conceptualised or understood (Seyfang et al. 2010; cf. Fischer and Newig 2016). Several roles are ascribed to civil society actors (ibid.), including the diffusion of innovative niche ideas and practices, lobbying and protests to unsettle regimes, pushing and encouraging regime actors to seek new solutions from niches, and representing general landscape-level trends. Civil society actors are largely viewed as participating in transition arenas and the forming transition networks by taking up concrete roles in advancing urban sustainability transitions. While these roles usually remain rather open and up to the individual actors, how civil society actors are approached and how their roles in transition management processes are framed is critical. For the long-term pursuit of sustainability transitions transition management strongly emphasises critical reflection on and redefinition of roles and responsibilities. For example, the case of transition management in Ludwigsburg showed that a framing of the participation of civil society actors in the transition management process as 'collecting input from citizens' did neither stimulate such reflection nor empowerment (Hölscher and Wittmayer, chapter "A German Experience: The Challenges of Mediating 'Ideal-Type' Transition Management in Ludwigsburg", this volume). Similarly, sometimes framings and perceptions shift between participating as citizen or in relation to one's profession or organisational affiliation (Wittmayer 2016). This requires more attention in the practical implementation of transition management processes.

Role(s) of Researchers

Transition management engages researchers in various roles. It is often implemented in combination with action research that involves researchers in guiding and facilitating the process steps and/or in increasing the reflexivity of the involved actors on the process itself and their roles (Wittmayer 2016; Bartels and Wittmayer 2014). Additionally, the researcher is part of societal networks and is normative in his/her orientation towards sustainability (Loorbach 2014). This refers of a changing understanding of the role(s) of researchers in relation to society. Interdisciplinarity, transdisciplinarity and social relevance are key elements of a science that supports sustainability transitions (Wittmayer and Schäpke 2014; WBGU 2011). Researchers facilitate and participate in learning-by-doing and doing-by-learning in which fundamental research, theory development, participatory research and applied research are combined (Loorbach 2014).

The roles of researchers in transition management processes can be summarised as follows (cf. Wittmayer and Schäpke 2014). As *reflective scientists*, researchers perform activities closest to 'conventional' research, such as systematically collecting, analysing, interpreting and reporting from an observer point of view. Researchers can be involved as *process facilitator*, which refers to their involvement in initiating and facilitating the process implementation. They can also act as *knowledge broker* to mediate between different perspectives and providing space for critical reflection. In the role of *change agent* researchers explicitly participate in addressing real-world problems. Finally, as *self-reflexive scientists*, they reflect on their own positioning and motivations in the transition management process and change dynamics.

Action research in transition management contributed to transitions in local communities through small yet significant changes (Wittmayer et al. 2014b; Wittmayer and Schäpke 2014). Authors express concerns about the involvement of researchers in transition management when it moves beyond merely describing and analysing transition management interventions. For instance, taking up double roles in analysing and advocating transition management can obscure the analysis (Duineveld et al. 2007). Especially opening and maintaining an interactive space is hard work (Wittmayer et al. 2014b; Wittmayer and Schäpke 2014; Bartels and Wittmayer 2014). It requires facing the diverse, contested meanings of usable knowledge, dealing with existing power holders, mediate conflicts and induce reflexivity. Especially in the long-term there is limited knowledge about what can be expected from researchers in the facilitation of the follow-up process. Thus far, most research activity stops after the final transition arena meetings. This is often related to the end of project funding at this stage.

(Dis)empowerment in Transition Management in Cities

Transition management includes a portfolio of participatory tools directed towards enabling societal learning processes in which individuals become *empowered* to develop a direction for and pursue a sustainability transition (Loorbach 2010; Loorbach et al. 2015). Empowerment is a rather generic term and widely used to express "a family of somewhat related meanings" (Thomas and Velthouse 1990: 666; cf. Avelino 2011). It requires being specific about *who* is to be empowered, *how* and *for what* purpose. Ultimately, transition management in cities seeks to support urban sustainability transitions by empowering a diversity of actors to take up roles in influencing urban sustainability transitions. This is achieved through creating space for them to reflect on and (re-)define roles in relation to the envisioned transition, develop new social relations and become intrinsically motivated to work towards a common direction (Avelino 2009, 2011; Wittmayer and Loorbach 2016; Hölscher et al. 2016).

Who these 'empowered' actors are and what roles they (re-)define and take up strongly relates to the lens taken to study actors and roles (section "Understanding Actors and Roles in Transition Management in and for Cities"). For example, when looking at actors in their practical roles in transition management, empowerment includes considerations of how arena participants develop new ties towards other actors and how transition networks are formed and maintained. Drawing on the MLP, a focus of transition management has been on empowering so-called frontrunners and niches to stimulate innovative thinking and the development and connection of novelties (Loorbach 2007: 173). The MAP brings awareness to what kind of roles and logics are to be empowered – for instance, there are different underlying motives for empowering consumers, citizens or voters. Additionally, it reveals the development of new roles, relationships and power shifts between actors and roles, for example between citizens and policy officers.

It is critical to consider the potential *dis*empowerment resulting from transition management interventions, as they might result in an increased sense of powerlessness and/or reinforce existing power structures and relationships (Avelino 2009, 2011). Critical studies on empowerment that draw on postmodern and critical theory claim that processes geared at empowerment in fact re-enforce existing relations of power and worsen the patterns of domination and dependence, and as such are ultimately *dis*empowering (cf. Avelino 2011: 66). It has indeed been a constant criticism on transition management that it runs the risk of reinforcing incumbent power structures when being unaware of the politics underlying transition processes (Shove and Walker 2007; Hendriks 2009; Meadowcroft 2009).

In the following, we describe different kinds of empowerment targeted through transition management in cities and emphasise the risk of *dis*empowerment.

New Ties and Relationships

Transition management seeks to enable actors to question, develop, experiment with and nurture new and/or alternative social relations and to build collaborative transition networks (Wittmayer and Loorbach 2016; Wittmayer 2016). This can create a feeling of 'togetherness', facilitate cooperation and pooling of resources for implementing a project and stimulate sharing of knowledge (Loorbach 2010; Nevens and Roorda 2013; Roorda et al. 2014; Frantzeskaki et al. 2014b). These relationships are not necessarily 'new', but they can also change in quality, for example through developing a basis for cooperation and trust (Wittmayer 2016).

The new connections and relations between actors are often considered as one of the most valuable results of transition management processes in cities. These (new) relationships can take different forms – they can refer to new relations between arena participants, between transition team members, between arena participants and transition team members as well as towards actors that are engaged in transition networks (Wittmayer 2016; Hölscher et al. 2016).

Especially the new relationships between participants from the arena ('civil society' actors) and policy officers from the local governments were highly appreciated, e.g. for "knowing where to ask question". However, there is also a risk that the local government is reinforced as central actor and that other actors feel to have only limited leverage for change without its involvement (Hölscher and Wittmayer, chapter "A German Experience: The Challenges of Mediating 'Ideal-Type' Transi tion Management in Ludwigsburg", this volume). The (new) relations can differ in quality, ranging from more loose ties to deeper and newly institutionalised relationships. For example, in Montreuil the transition arena was established as a recurring process to institutionalise the communication between local governments and citizens (Krauz 2016; Roorda and Wittmayer 2014). In Rotterdam, formal and informal partnerships materialised following a transition management process to develop concrete transition experiments (Frantzeskaki et al. 2014b).

(Re-)defining Roles and Responsibilities

Roles are not static, but in constant development and different actors can assume different roles in different contexts (Wittmayer 2016; Wittmayer et al. 2017a). Transition management establishes a setting for actors to reflect on and (re-)define their roles and to help them to become aware of such on-going negotiations of roles and relations (ibid., Wittmayer et al. 2014b). The development of shared visions and transition pathways triggers the involved actors to locate themselves in the solution-seeking process and identify their (new) roles and responsibilities (Loorbach 2010; Frantzeskaki et al. 2012; Nevens and Roorda 2013). Such a questioning and (re-)definition of roles can correspond with a change in social relations (Wittmayer 2016).

For instance, a policy officer interacting with citizens and interested stakeholders in a new way might add this activity to his/her role understanding of what a policy officer should or should not be doing. Furthermore, such changes in a specific role understanding can have repercussions for the understanding of other roles (ibid.). As the understanding of what a policy officer is and does, related roles such as 'citizen' change with it.

The different process steps of the transition management framework have shown to facilitate a broadening of problem perspectives, identification and awareness of possible solutions and reflection on roles. For example, arena participants in Ghent identified how they could contribute to implementing the transition agenda by developing concrete initiatives (Hölscher et al. 2016). The experiences with transition management especially questioned the roles of policy officers and the city government, and the extent they engage in co-creation with and coordination of other actors and networks (Hölscher et al. 2017). It also triggered the 'creation' of new actors and roles, such as an institutionalised transition arena in which citizens and local policy officers take up new interactive governance roles (Krauz 2016), or the cross-departmental working group on energy in Ludwigsburg (Hölscher and Wittmayer, chapter "A German Experience: The Challenges of Mediating 'Ideal-Type' Transition Management in Ludwigsburg", this volume).

Overcoming engrained role patterns can pose major challenges (Roorda et al. 2014). In Ludwigsburg, the framing of the transition management process as resembling more 'conventional' stakeholder consultation hindered reflection on own roles, and resulted in mainly more specified roles for the local government (Hölscher and Wittmayer, chapter "A German Experience: The Challenges of Mediating 'Ideal-Type' Transition Management in Ludwigsburg", this volume).

Intrinsic Motivation

Actively engaging and taking up a role in urban sustainability transitions requires actors' willingness, or intrinsic motivation. Intrinsic motivation determines the commitment and willingness of actors to fulfil a task, create agendas and take action without being dependent on supervision, incentives or rewards (Avelino 2009, 2011). It depends on "interpretive processes through which individuals assess and evaluate their tasks, activities, and roles" (Avelino 2009: 377). There are four dimensions of positive task assessment: meaning (value attributed to task), competence (belief in capacities to fulfil task), impact (conviction to achieve intended effect of task), and choice (sense of autonomy and self-determination in initiating and completing task) (Thomas and Velthouse 1990; cf. Avelino 2011).

Through participating in transition management processes the involved actors attain a feeling of urgency for change, develop a direction for desirable change, reflect on their roles in this direction and experience a feeling of like-mindedness. Actors involved in the transition management processes in the MUSIC-project valued these experiences and insights and were inspired to take up a role in the urban sustainability transition (Hölscher et al. 2017). For instance, they engaged in working groups, dissemination activities and developed concrete initiatives.

Co-creation and the creation of ownership over the process and its results were identified as key factors contributing to intrinsic motivation (ibid.). For example, in Ludwigsburg the process stimulated only a limited reflection on and (re-)definition of roles because of the 'consultative' process framing (Hölscher and Wittmayer, chapter "A German Experience: The Challenges of Mediating 'Ideal-Type' Transi tion Management in Ludwigsburg", this volume). The experiences with Ghent show the challenges in sustaining longer-term commitment (Hölscher et al. 2016). The 'controlled' process implementation seems to have left insufficient room for developing ownership and hence also the follow-up was largely left in the hands of the city government. This also relates to the question of how the move into and coordination of transition networks can be facilitated (Loorbach 2014, section "Transition network(s)").

Conclusions

This chapter introduced perspectives and critical reflections on actors and roles in transition management in and for cities and on how transition management seeks to empower different types of actors to take up roles in advancing urban sustainability transitions. When implementing transition management processes in city contexts it is important not to be naïve about who is to be involved, in which role, what the relationships to 'outside' actors are, how roles and relations are reflected upon and what the implications are for changing roles, relations and governance structures in the city to advance an envisioned transition.

The heuristics and reflections introduced in this chapter help to obtain a better understanding of the actor dynamics in a particular urban setting – such as who are dominant actors and what are dominant role framings. In concrete transition management interventions this supports for example the system analysis, actor selection and guide reflections about actors and roles. It enables to be more explicit about who should be invited, what kind of (changes of) roles are needed (or desired/ expected) for achieving transitions and to develop facilitation tools for supporting particular roles in both transition management processes and the pursuit of urban sustainability transitions. The heuristics also support an understanding of actors and roles in sustainability transitions in general. Interestingly, in light of the current attention in particular to actor networks, it is striking that so far a conceptual network perspective is missing from the heuristics to understand actors. This also counts for transition management interventions, as there is limited attention in the operational process phases on what happens after moving into transition networks.

Attention to (dis-)empowerment in transition management processes illustrates how important actor and role framings are in relation to who should become empowered and how roles are changed and or taken up (for example as a result of transition management interventions) in the pursuit of the envisioned transition. The conceptualisation of (dis-)empowerment helps to reveal underlying power dynamics and interests in transition management processes as well as in the pursuit of sustainability transitions. Moreover, it supports an understanding of how actors can be motivated and connected to each other to pursue sustainability transitions. In this regard, the view on (dis)empowerment also provides insights into process techniques and setting designs that facilitate empowerment, such as giving more emphasis to a creation of space at the outset of the transition management process to familiarise the initiating actors with the process 'philosophy' (Hölscher et al. 2017; Hölscher and Wittmayer, chapter "A German Experience: The Challenges of Mediating 'Ideal-Type' Transition Management in Ludwigsburg", this volume).

In understanding (dis)empowerment as a result of transition management intervention, it is critical to remain attentive to the question of actors and roles. For example, it is important to understand the implications of central roles of policy officers from local governments in the transition team. It does matter who initiates and participates in processes of this kind – with regard to the actual meaning that sustainability acquires during the process, to the kind of change aimed for, and the reflexivity in the process/process design (Wittmayer et al. 2015).

These reflections on actors and roles and changes of roles (empowerment) in relation to empirical transition management applications in cities give way to some new directions for research and practice. There is a demand for being more specific on the types of actors and roles (needed, desired) in a particular transition management context – and by 'whom' they are needed or desired. Especially the move into transition networks and sustaining long-term commitment have been challenging. The heuristics in transition management literature do so far not pay attention to actor networks and the (process framework for) facilitation of transition management usually ends here – i.e. it is unclear how the transition arena connects to a broader transition network and how such a network should be facilitated (and by whom). This demands some further insights into how transition management can act as a framework for meta-governance of transition networks (Loorbach 2014), as well as what kind of coordinating roles are needed and who should take them up in navigating urban sustainability transitions.

References

Apajalahti E-L (2012) The role of large incumbent energy companies in energy system transition – The case of Finland's "coal coast". Paper präsentiert auf IST 2012 – International Conference on Sustainability Transitions. Lyngby, Denmark 29–31 August 2012

Avelino F (2009) Empowerment and the challenge of applying transition management to ongoing projects. Policy Sci 42:369–390

Avelino F (2011) Power in transition. Empowering discourses on sustainability transitions. PhD Thesis: Erasmus University of Rotterdam

Avelino F, Rotmans J (2009) Power in transition: an interdisciplinary framework to study power in relation to structural change. Eur J Soc Theory 12:543

Avelino F, Wittmayer JM (2015) Shifting power relations in sustainability transitions: a multi-actor perspective. J Environ Policy Plan. https://doi.org/10.1080/1523908X.2015.1112259

Bartels KPR, Wittmayer JM (2014) Symposium introduction: usable knowledge in practice. What action research has to offer to critical policy studies. Crit Policy Stud 8(4):397–406. https://doi.org/10.1080/19460171.2014.957335

Bauknecht D, Brohmann B, Grießhammer R (2015) Transformationsstrategien und Models of Change für nachhaltigen gesellschaftlichen Wandel: Gesellschaftlicher Wandel als Mehrebenenansatz. Bericht des AP2. Bundesumweltamt, Dessau-Roßlau. http://www.umweltbundesamt.de/publikationen/gesellschaftlicher-wandel-als-mehrebenenansatz. Letzter Zugriff: 01.03.2016

Brown RR, Farrelly MA, Loorbach D (2013) Actors working the institutions in sustainability transitions: the case of Melbourne's stormwater management. Glob Environ Chang 23 (4):701–718

Bulkeley H, Castán Broto V (2013) Government by experiment? Global cities and the governing of climate change. Trans Inst Br Geogr 38(3):361–375

Bulkeley H, Castán Broto V, Hodson M, Marvin S (eds) (2011) Cities and low carbon transitions. Routledge, London

Castán Broto V, Bulkeley H (2013) A survey of urban climate change experiments in 100 cities. Glob Environ Chang 23(2013):92–102

Clarke J (2005) New Labour's citizens: activated, empowered, responsibilized, abandoned? Crit Soc Policy 25(4):447–463

Davies AL, White RM (2012) Collaboration in natural resource governance: reconciling stakeholder expectations in deer management in Scotland. J Environ Manag 112:160–169

De Haan J, Rotmans J (2011) Patterns in transitions: understanding complex chains of change. Technol Forecast Soc Chang 78(1):90–102

Duineveld M, Beunen R, van Ark RGH, van Ssche KAM, During R (2007) The difference between knowing the path and walking the path: een essay over het terugkerend maakbaarheidsdenken in beleidsonderzoek. Wageningen University, Leerstoelgroep Sociaal-ruimtelijke Analyse, Wageningen

Evers A, Laville JL (2004) The third sector in Europe. Edward Elgar Publishing, Northhampton/Massachusets

Farla J, Markard J, Raven R, Coenen L (2012) Sustainability transitions in the making: a closer look at actors, strategies and resources. Technol Forecast Soc Chang 79(6):991–998. https://doi.org/10.1016/j.techfore.2012.02.001

Fischer L-B, Newig J (2016) Importance of actors and agency in sustainability transitions: a systematic exploration of the literature. Sustainability 8(5):476. https://doi.org/10.3390/su8050476

Frantzeskaki N, de Haan H (2009) Transitions: two steps from theory to policy. Futures 41 (9):593–606

Frantzeskaki N, Tefrati N (2016) A transformative vision unlocks the innovative potential of Aberdeen City, UK. In: Loorbach D, Wittmayer JM, Shiroyama H, Fujino J, Mizuguchi S (eds) Governance of urban sustainability transitions. European and Asian experiences. Springer, Tokyo/Heidelberg/New York/Dordrecht/London, pp 49–68

Frantzeskaki N, Loorbach D, Meadowcroft J (2012) Governing societal transitions to sustainability. Int J Sustain Dev 15(1):19–36

Frantzeskaki N, Bach M, Hölscher K, Wittmayer J (2014a) Applications of transition management: taking stock of 13 years of transition management and lessons for future research, 5th International Sustainability Transitions (IST) conference, 27–29 August 2014, Utrecht, The Netherlands

Frantzeskaki N, Wittmayer J, Loorbach D (2014b) The role of partnerships in 'realizing' urban sustainability in Rotterdam's City Ports Area, the Netherlands. J Clean Prod 65:406–417. https://doi.org/10.1016/j.jclepro.2013.09.023

Garud R, Kumaraswamy A, Karnøe P (2010) Path dependence or path creation? J Manag Stud 47:760–774

Geels FW (2011) The multi-level perspective on sustainability transitions: responses to seven criticisms. Environ Innov Soc Trans 1:24–40

Geels F (2014) Regime resistance against low-carbon transitions: introducing politics and power into the multi-level perspective. Theory Cult Soc 31(5):21–40
Geels FW, Schot J (2007) Typology of sociotechnical transition pathways. Res Policy 36:399–417
Genus A, Coles A-M (2008) Rethinking the multi-level perspective of technological transitions. Res Policy 37(9):1436–1445
Grimm N, Faeth S, Golubiewski N, Redman C, Wu J, Bai X, Briggs J (2008) Global change and the ecology of cities. Science 319:756–760
Grin J (2010) Understanding transitions from a governance perspective. In: Grin J, Rotmans J, Schot J (eds) Transitions to sustainable development. New directions in the study of long term transformative change. Routledge, New York
Grin J, Rotmans J, Schot J (2010) Transitions to sustainable development: new directions in the study of long term transformative change. Routledge, New York/Oxford
Hendriks C (2009) Policy design without democracy? Making democratic sense of transition management. Policy Sci 42(4):341–368
Hendriks CM, Grin J (2007) Contextualising reflexive governance: the politics of Dutch transitions to sustainability. J Environ Policy Plan 9(3/4):333–350
Hermans L (2010) Actor analysis. In: Enserink B, Hermans L, Kwakkel J, Thissen W, Koppenjan J, Bots P (eds) Policy analysis of multi-actor systems. Boom Lemma Uitgevers, The Hague, pp 79–108
Hodson M, Marvin S (2010) Can cities shape socio-technical transitions and how would we know if they were? Res Policy 39:477–485
Hoffman J (2013) Theorizing power in transition studies: the role of creativity and novel practices in structural change. Policy Sci: 46(3): 1–19
Hölscher K, Wittmayer J (2017) A German experience: the challenges of adapting 'ideal-type' Transition Management in Ludwigsburg. In: Frantzeskaki N, Hölscher K, Bach M, Avelino F (eds) Co-creating sustainable urban futures. A primer on applying transition management in cities. Springer, Tokyo
Hölscher K, Roorda C, Nevens F (2016) Ghent: fostering a climate for transition. In: Loorbach D, Wittmayer JM, Shiroyama H, Fujino J, Mizuguchi S (eds) Governance of urban sustainability transitions. European and Asian experiences. Springer, Tokyo/Heidelberg/New York/Dordrecht/London, pp 91–112
Hölscher K, Wittmayer J, Avelino F, Giezen M (2017) Opening up the transition arena: an analysis of (dis)empowerment of civil society actors in transition management in cities. Technological Forecasting and Social Change. Corrected Proof. https://doi.org/10.1016/j.techfore.2017.05.004
Janssen MA, Hohman N (2012) Addressing global sustainability challenges from the bottom up: the role of information feedback. CSID Working Paper Series, #CSID-2012-011
Jhagroe S (2017) Urban transition management as democratic practice? The case of Rotterdam's waterfront regeneration. In: Frantzeskaki N, Hölscher K, Bach M, Avelino F (eds) Co-creating sustainable urban futures. A primer on applying transition management in cities. Springer, Tokyo
Jhagroe S, Loorbach D (2015) See no evil, hear no evil: the democratic potential of transition management. Environ Innov Soc Trans 8. https://doi.org/10.1016/j.eist.2014.07.001
Jhagroe S, van Steenbergen F (2014) Van koploper naar koplopen: een pleidooi voor inclusief leiderschap. Drift essay 2014.04. Online at: http://www.drift.eur.nl/?p1/4262
Kemp R, Rotmans J (2009) Transitioning policy: co-production of a new strategic framework for energy innovation policy in the Netherlands. Policy Sci 42(4):303–322
Kemp R, Loorbach D, Rotmans J (2007) Transition management as a model for managing processes of co-evolution towards sustainable development. Int J Sust Dev World 14(1):78–91
Khan J (2013) What role for network governance in urban low carbon transitions? J Clean Prod 50:133–139
Krauz A (2016) Transition management in Montreuil: towards perspectives of hybridisation between 'top-down' and 'bottom-up' transitions. In: Loorbach D, Wittmayer J, Shiroyama H, Fujino J, Mizuguchi S (eds) Governance of urban sustainability transitions. European and Asian experiences. Springer, Tokyo/Heidelberg/New York/Dordrecht/London, pp 133–152

Loorbach D (2007) Transition management. New mode of governance for sustainable development. Erasmus University, Rotterdam

Loorbach D (2010) Transition management for sustainable development: a prescriptive, complexity-based governance framework. Governance 23(1):161–183

Loorbach D (2014) To transition! Governance panarchy in the new transformation, Inaugural lecture. Erasmus University of Rotterdam, Rotterdam

Loorbach D, Rotmans J (2010) The practice of transition management: examples and lessons from four distinct cases. Futures 42:237–246

Loorbach D, Frantzeskaki N, Huffenreuter L (2015) Transition management: taking stock from governance experimentation. J Corp Citizenship 58:48–66

McCormick K, Anderberg S, Coenen L, Neij L (2013) Advancing sustainable urban transformation. J Clean Prod 50:1–11

Meadowcroft J (2009) What about the politics? Sustainable development, transition management, and long term energy transitions. Policy Sci 42(4):323–340

Nevens F, Roorda C (2013) A climate of change: a transition approach for climate neutrality in the city of Ghent (Belgium). Sustain Cities Soc 10:112–121

Nevens F, Frantzeskaki N, Loorbach D, Gorissen L (2013) Urban transition labs: co-creating transformative action for sustainable cities. J Clean Prod 50:111–122

Pestoff V (1992) Third sector and co-operative services – an alternative to privatization. J Consum Policy 15:21–45

Raven R, van den Bosch S, Weterings R (2010) Transitions and strategic niche management: towards a competence kit for practitioners. Int J Technol Manag 51(1):57–74

Rip A, Kemp R (1998) Technological change. In: Rayner S, Malone L (eds) Human choice and climate change, vol 2 resources and technology. Batelle Press, Washington, DC, pp 327–399

Roorda C, Wittmayer J (2014) Transition management in five European cities – an evaluation. DRIFT, Erasmus University Rotterdam, Rotterdam

Roorda C, Wittmayer J, Henneman P, van Steenbergen F, Frantzeskaki N, Loorbach D (2014) Transition management in the urban context, Guidance manual. Drift, Erasmus University Rotterdam, Rotterdam

Rotmans J (2005) Maatschappelijke innovatie tussen droom en werkelijkheid staat complexiteit. DRIFT, Erasmus University Rotterdam, Rotterdam

Rotmans J, Loorbach D (2010) Towards a better understanding of transitions and their governance: a systemic and reflexive approach. In: Grin J, Rotmans J, Schot J (eds) Transitions to sustainable development. New directions in the study of long term transformative change. Routledge, New York/London, pp 105–220

Schuitmaker TJ (2012) Identifying and unravelling persistent problems. Technol Forecast Soc Chang 79:1021–1031

Seyfang G, Haxeltine A (2012) Growing grassroots innovations: exploring the role of community-based social movements in sustainable energy transitions. Environ Plann C 30(3):381–400

Seyfang G, Smith A (2007) Grassroots innovations for sustainable development: towards a new research and policy agenda. Environ Polit 16(4):584–603

Seyfang G, Haxeltine A, Hargreaves T, Longhurst N 2010 Energy and communities in transition – towards a new research agenda on agency and civil society in sustainability transitions. CSERGE Working Paper. https://www.econstor.eu/dspace/bitstream/10419/48803/1/636134792.pdf. Used: May 12th 2016

Shove E, Walker G (2007) CAUTION! Transitions ahead: politics, practice, and sustainable transition management. Environ Plan A 39(4):763–770

Smith A (2012) Civil society in sustainable energy transitions. In: Verbong G, Loorbach D (eds) Governing the energy transition: reality, illusion, or necessity. Routledge, NewYork

Smith A, Kern F (2009) The transitions storyline in Dutch environmental policy. Environ Polit 18 (1):78–98

Smith A, Stirling A (2010) The politics of social-ecological resilience and sustainable socio-technical transitions. Ecol Soc 15(1):11. DOI: http://www.ecologyandsociety.org/vol15/iss1/art11/

Smith A, Stirling A, Berkhout F (2005) The governance of sustainable socio-technical transitions. Res Policy 34:1491–1510

Swyngedouw E (2005) Governance innovation and the citizen: the janus face of governance-beyond-the-state. Urban Stud 42(11):1991–2006

Thomas KW, Velthouse BA (1990) Cognitive elements of empowerment: an 'Interpretative' model of intrinsic task motivation. Acad Manag Rev 15(4):666–681

Turner RH (1990) Role change. Annu Rev Sociol 16:87–110

Van Buuren A, Loorbach D (2009) Policy innovation in isolation? Conditions for policy renewal by transition arenas and pilot projects. Public Manag Rev 11(3):375–392

Van den Bosch SJM (2010) Transition experiments: exploring societal changes towards sustainability. Erasmus University, Rotterdam. http://hdl.handle.net/1765/20714

Van Steenbergen F, Wittmayer J (2012) Carnisse in transitie? Een verkenning van het verleden, het heden en de toekomst van een Rotterdamse wijk. DRIFT-uitgave, Rotterdam. E 2012.05

Voß J-P, Smith A, Grin J (2009) Designing long-term policy: rethinking transition management. Policy Sci 42(4):275–302

WBGU, German Advisory Council on Global Change (2011) World in transition – a social contract for sustainability, Berlin

Wittmayer JM (2016) Insights and lessons for the governance of urban sustainability transitions. In: Loorbach D, Wittmayer J, Shiroyama H, Fujino J, Mizuguchi S (eds) Governance of urban sustainability transitions. European and Asian experiences. Springer, Tokyo/Heidelberg/New York/Dordrecht/London, pp 157–173

Wittmayer JM, Loorbach D (2016) Governing transitions in cities: fostering alternative ideas, practices, and social relations through transition management. In: Loorbach D, Wittmayer J, Shiroyama H, Fujino J, Mizuguchi S (eds) Governance of urban sustainability transitions. European and Asian experiences. Springer, Tokyo/Heidelberg/New York/Dordrecht/London, pp 13–32

Wittmayer JM, Schäpke N (2014) Action, research and participation: roles of researchers in sustainability transitions. Sustain Sci 9(4):483–496

Wittmayer J, van Steenbergen F, Quist J, Loorbach D, Hoogland C (2011) In context. The community arena: a co-creation tool for sustainable behaviour by local communities. Methodological guidelines. http://incontext-fp7.eu/sites/default/files/Methodological%20guidelines_final.pdf. Used: December 17th 2012

Wittmayer J, Frantzeskaki N, van Steenbergen F, Roorda C, Henneman P (2012) Introducing a profile for urban change agents: experiences from five European cities, Drift Working Paper, Erasmus University Rotterdam, Rotterdam, Creative Commons

Wittmayer J, Mock M, van Steenbergen F, Baasch S, Omann I, Schäpke N (2013) Taking stock – three years of addressing societal challenges on community level through action research Pilot specific synthesis report. Deliverable 4.5. InContext: EU ENV.2010.4.2.3-1 grant agreement n° 265191. http://www.incontext-fp7.eu/sites/default/files/D4.5%20-%20Taking%20Stock-final_0.pdf. Used: 2 April 2015

Wittmayer J, Roorda C, van Steenbergen F (eds) (2014a) Governing urban sustainability transitions – inspiring examples. DRIFT, Rotterdam. 10.13140/RG.2.1.2367.7606

Wittmayer JM, Schäpke N, van Steenbergen F, Omann I (2014b) Making sense of sustainability transitions locally: how action research contributes to addressing societal challenges. Crit Policy Stud 8(4):465–485. https://doi.org/10.1080/19460171.2014.957336

Wittmayer JM, van Steenbergen F, Rok A, Roorda C (2015) Governing sustainability: a dialogue between Local Agenda 21 and transition management. Local environment: the international journal of justice and sustainability. https://doi.org/10.1080/13549839.2015.1050658

Wittmayer JM, Avelino F, van Steenbergen F, Loorbach D (2017a) Actor roles in transition: insights from sociological perspectives. Environ Innov Sustain Transitions 24:45–56. https://doi.org/10.1016/j.eist.2016.10.003

Wittmayer JM, van Steenbergen F, Frantzeskaki N, Bach M (2017b) Transition management in urban context. In: Frantzeskaki N, Hölscher K, Bach M, Avelino F (eds) Co-creating sustainable urban futures. A primer on applying transition management in cities. Springer, Tokyo

Starting Up Transition Management: A Closer View on the Systems Analysis and How It Initiated Transformative Thinking in Ghent and Aberdeen Cities

Suzanne Maas, Karen Fortuin, Niki Frantzeskaki, and Chris Roorda

Abstract In this chapter we present a systems analysis methodology for the first phase of Transition Management, the preparation and exploration phase, in an urban context. The methodology consists of the following four steps, which are followed in an iterative way: (i) system definition, (ii) system structuring, (iii) data gathering and (iv) data assessment and analysis. Characteristic of this methodology is the use of the SCENE sustainability model, in combination with a set of lenses that reflect a variety of ways of looking at the system from macro, meso, and micro level. Together, these lenses ensure an integrated perspective and thorough analysis of a city's government and governance. The systems analysis methodology was applied in the cities of Ghent and Aberdeen to identify the challenges and opportunities in their cities with respect to energy use and CO_2 reductions. The outcomes and reflections on their respective systems analyses show that creating an overview of the state of the system revealed interesting connections and that constructing a historical analysis of the city inspired and motivated participants. The systems analysis broadened the problem perception, revealed institutional barriers to sustainable development and served as a catalyst for mutual understanding and learning between the participants. Overall, the systems analysis methodology proved to be an effective method to create an overview of the city activities under study, and stimulated the creation of a common information base and holistic thinking amongst participants in the transition process.

S. Maas (✉) · K. Fortuin
Environmental Systems Analysis Group, Wageningen University and Research (WUR), Wageningen, The Netherlands
e-mail: suzanne.m.maas@gmail.com; karen.fortuin@wur.nl

N. Frantzeskaki · C. Roorda
Dutch Research Institute For Transitions (DRIFT), Faculty of Social Sciences, Erasmus University Rotterdam, Rotterdam, The Netherlands
e-mail: frantzeskaki@drift.eur.nl; roorda@drift.eur.nl

N. Frantzeskaki et al. (eds.), *Co-creating Sustainable Urban Futures*,
Future City 11, https://doi.org/10.1007/978-3-319-69273-9_7

Introduction

Many European cities have already committed themselves to reductions that go beyond EU *20-20-20 targets* that set the ambitions for 20% CO_2 emission reduction, 20% of energy consumption from renewable sources, and a 20% reduction in energy use by 2020 (European Commission 2010), for example through the Covenant of Mayors: a commitment between mayors of cities in EU countries to increase local sustainable energy production and consumption and reduce CO_2 emissions beyond the targets set by the EU (Covenant of Mayors 2011). Achieving these targets and realizing the necessary changes are difficult due to the complexity of these issues. A transition to a low carbon city requires a wide variety of actions (Swart and Raes 2007), because climate change has multiple causes and consequences and involves a wide range of societal domains, actors and scale levels (Loorbach 2009). Cities and their governments have to deal with these challenges, implement climate and energy policy and take leadership for their geographic area and hinterland (Newman and Jennings 2008).

Common barriers that prevent cities to deal with these challenges include: (i) policy fragmentation, referring to the phenomenon that local energy strategies are often developed in isolation from other policy domains such as spatial planning or mobility leading to inconsistent and incompliant strategies and policy plans; (ii) lacking, or, discontinuous political commitment due to political cycles and due to uncertain benefits from proposed actions; and related to this, (iii) a myopic focus on short-term goals and gains, instead of long-term sustainability goals; and (iv) the fact that the local government is just one of the actors in a city with (often) limited impact on the energy sector and market.

A local government cannot induce change alone; complex problems require mobilisation of a broad range of actors. An approach to bring together multiple actors to form a strategic agenda for sustainability transitions is Transition Management (Loorbach 2010). Transition management is characterized by long-term integrative thinking, involving multiple actors, and a focus on learning and system innovation next to system improvement (Frantzeskaki et al. 2012). An application of transition management in the urban context with a focus on the energy system is realised within the Interreg funded project MUSIC (Mitigation in Urban areas and Solutions for Innovative Cities). MUSIC was a transnational project aimed at reducing CO_2 emissions and energy use in five North-Western European cities. It applied transition management to formulate strategic climate mitigation action in urban areas in order to reach targets ranging from 50% CO_2 reduction by 2030 (MUSIC 2011) to climate neutrality in 2050 (Balthazar 2008). The project addresses an issue that local governments all over North-Western Europe struggle with: 'how to get from policy goals on CO_2 targets to reductions in practice?'

This chapter deals with a systems analysis methodology that was developed for the first phase of the transition management process in cities participating in the MUSIC project. It was expected that a systems analysis methodology could have a twofold contribution to the transition management process. Firstly, a systems analysis methodology can provide an overview of the system, in this case the city's status, characteristics and networks that are relevant from a sustainability perspective, integrating social, environmental and economic domains. This information is crucial for later phases of the transition management. Secondly, conducting a systems analysis can stimulate holistic thinking and provide a common ground among people participating in the analysis (Miser and Quade 1985; Rotmans and Loorbach 2009).

A systems analysis can create a shared language and common understanding of the problems and challenges in the city amongst project participants. As such, the systems analysis is expected to create a common information base to start the negotiation, collective learning and reflection by engaging people from different backgrounds, such as from the local government, business, science and civil society. The overall objectives of the system analysis methodology in this context are to enable the cities to identify key challenges and new opportunities with respect to energy and CO_2 reductions, and to initiate a dialogue between policy actors, scientists and community members.

Below we present the systems analysis methodology that was designed and applied in the MUSIC project as well as a reflection on the application of this methodology in two European cities: Ghent (Belgium) and Aberdeen (United Kingdom). More specifically, we address the following research questions: *What are the characteristics of the systems analysis methodology applied in Ghent and Aberdeen as part of a transition management process? What does the systems analysis methodology offer for unravelling sustainability challenges and for the identification of opportunities for change in European cities that face the challenge of reducing energy use and* CO_2 *emissions?*

We developed a systems analysis methodology fitting within the transition management process applied in the MUSIC project based on existing systems analysis methods and tools, which we identified through a literature review. As such the system analysis methodology that we developed has a modular configuration: it consists of adapted tools and techniques that in a step-wise application unravel challenges and opportunities for urban sustainability transitions.

In section "Systems Analysis Methodology" we elaborate on the objectives of performing a systems analysis and an elaboration of the systems analysis methodology. The third and last author have been involved as city coaches in Aberdeen and Ghent respectively, assisting and consulting the *transition team* – the team that organises and facilitates the transition process – on applying the systems analysis methodology and the wider transition process. The first author was involved in the development of the systems analysis methodology and its application in Ghent. The third author was involved in the application of the systems analysis in Aberdeen, United Kingdom. The context, outcomes and application of the systems analysis methodology in Ghent and Aberdeen are presented in section "Discussion". In order to gain information about the application and outcome of the systems analysis methodology the first, third and last author participated and helped with analysing the data gathered for the systems analysis. The results from reflection on the content and process goals in the two case studies by the different parties are presented in the discussion in section "Conclusions". The city coaches and the developer of the methodology reflected on the application and outcomes of the systems analysis methodology, and the first and last author interviewed three members of the transition team of Ghent to learn more about the experience of the transition team members. These transition team members and 12 transition arena participants – frontrunners that participated in a series of so-called arena workshops about the transition challenge in their respective city – in Ghent were asked to reflect upon the methodology by filling in a questionnaire which inquired about their experience. In section "References", we present our conclusions.

Systems Analysis Methodology

Systems analysis can be used to provide an overview of the system, to produce a baseline assessment of the system's state, to demarcate system boundaries, and to separate symptoms from problems (Findeisen and Quade 1985; Miser and Quade 1985; Rotmans and Loorbach 2009). An essential output of a systems analysis is a systematic and consistent analysis of the variables and elements that constitute its whole (Pahl-Wostl 2007; Saravanan 2008, p. 203). One of the important outcomes of the systems analysis methodology (that we present below) is that it should provide an integrated overview of the system with factual information as well as an understanding of the problems from different perspectives, from the viewpoint of different actors. This overview is needed to address the challenges related to energy and CO_2 emission reduction goals that cities face. It facilitates the identification of areas that require change as well as possible courses of action needed to meet these goals.

The systems analysis methodology should provide: (i) *an overview of the system and a detailed systematic analysis of its state and characteristics*. In this overview, the properties and elements of the system as well as the links and interdependencies between them should be identified (Saravanan 2008, p. 214); (ii) *an analysis of the history of the system including historical evolution pathways and events*. The timelines of elements and properties of the system can be used to provide a view of its historical development, explain historical path-dependencies, and create historical parallels (Rotmans and Loorbach 2009); (iii) *a problem identification and formulation from a system's perspective*. By analysing the system and its properties, problems can be revealed. The understanding of the interdependencies of system's elements provides an understanding of what the problem is, and gives insight into causes and symptoms (Rotmans and Loorbach 2009).

Next to providing an overview of the system under study, a systems analysis includes the perspectives and knowledge of those who are involved in the analysis. This is very important for the first phase of transition management, because it enables people from different (scientific) backgrounds to work together, and to combine objective data with subjective perspectives.

Therefore, the systems analysis methodology that we present below, should also meet the following objectives: (i) *stimulate holistic thinking among its participants*. Analysing the system in an integrative and holistic manner provides an information-rich and comprehensive base for negotiations and learning. Conducting a systems analysis in cooperation with experts and policy actors can stimulate holistic thinking about the system and the problem (Rotmans and Loorbach 2009); (ii) *provide a common information base and shared understanding*. Undertaking an integrated analysis of a system enables participants to create a common ground and a common perception of the system (Banathy 1988; Jansen 2009); (iii) *pinpoint areas and actors of the system contributing to problems of unsustainability and in turn to create a sense of urgency for action*. The presentation of persistent problems, showing the unsustainability of the current system, gives rise to a sense of urgency and justifies the need for change (Rotmans and Loorbach 2009; Saravanan 2008, p. 214).

The systems analysis methodology that we present below combines the provision of an integrated overview of the system with a process of inquiry that supports a holistic, participative approach to problem analysis, and forms the first step towards finding solutions.

Application of the Methodology

The basic structure of the systems analysis methodology we propose is built up in four steps: (1) system definition, (2) system structuring, (3) data gathering and (4) data assessment and analysis (Fig. 1).

Step 1: System Definition

The first step in the systems analysis is to determine the role of the systems analysis within the transition management process. Decisions need to be made on the working method, communication structure and project management. Thereafter, the system under study needs to be defined; the boundaries of the system have to be set in order to clarify the unit of analysis. In principle, there are no general inclusion or exclusion rules, so what is included depends on the problem the analyst(s) wants to address (van der Brugge 2009). The systems' boundaries can be defined in terms of space (the physical boundaries), time (how far to look back, how far ahead) and themes (the focus of the study and the understanding of relevant in- and outputs). The transition team decides on the system definition but, as the systems analysis is an iterative process, this can be challenged and further discussed in the subsequent steps of the systems analysis (see step 2, 3 and 4). Decisions on setting boundaries in space, time and themes require careful thought and discussion to grasp the implications of different choices. In the case of a local government with a policy plan, there are already some guidelines to steer the process. For example, the border of a city forms a natural boundary in terms of space, and the end term of a policy goal (e.g. climate neutral in 2050) forms a time boundary.

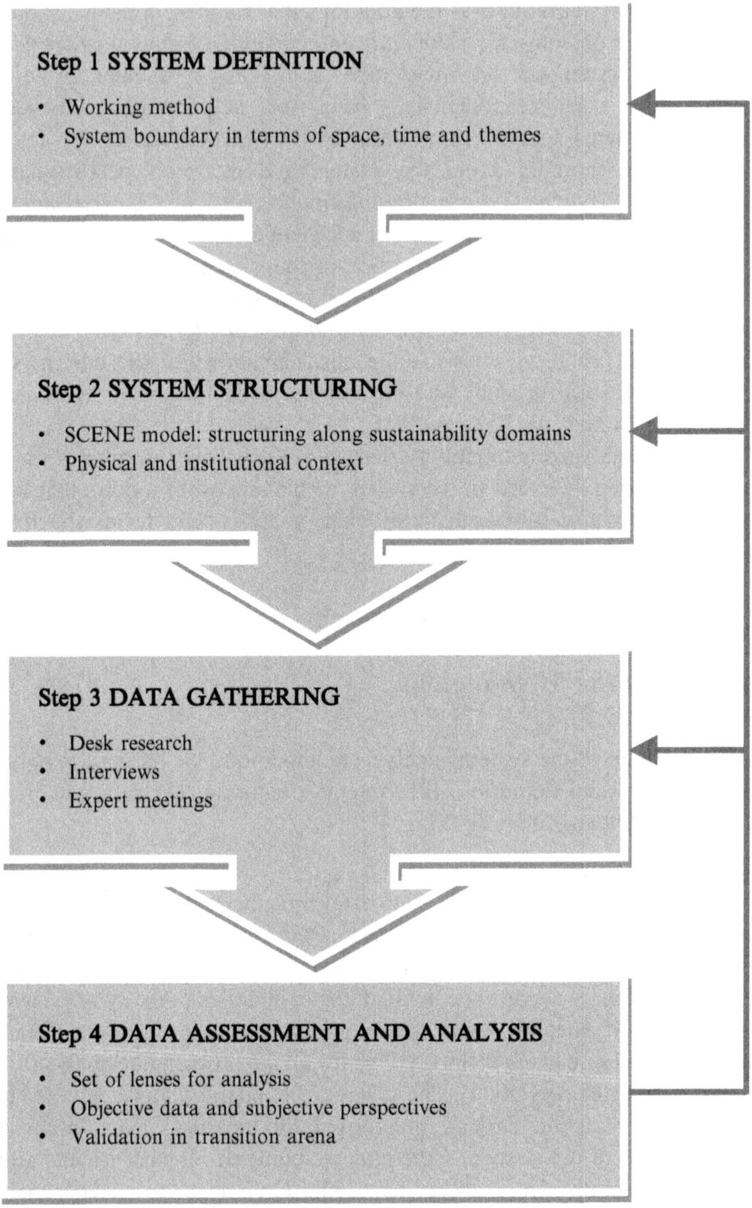

Fig. 1 Overview of the systems analysis methodology as applied in the transition management process in Ghent and Aberdeen cities as part of the MUSIC project

Step 2: System Structuring

The second step in the systems analysis is structuring the system, in order to be able to make an abstraction of the complex system. Such an abstraction is useful as a framework within which the system can be examined and discussed. In this systems analysis methodology, the SCENE model is used as a framework. SCENE is a model based on the sustainability triangle, consisting of three capital domains: SoCial, ENvironmental and Economic (Rotmans 1998). SCENE extends the traditional sustainability triangle, that solely portrays the interconnectedness between the three domains, with capital stocks that represent properties of the system and flows that define the relationships between them. Stocks can be quite generic terms, such as 'quality of life', 'environmental quality', or 'economic vitality' (Grosskurth 2008), but also more specific, such as 'education' or 'mobility'. Flows are aspects that change relatively quickly in the short-term and reveal the relationship between stocks (Rotmans et al. 2001).

Depending on the required level of detail the SCENE model with stocks and flows can be further specified to include characteristics and indicators. Because of the important role of the physical and institutional context in an urban environment, the stocks in the social, environmental and economic domains are placed within wider circles representing the physical and institutional contexts. Within the transition team, important stocks in the context of the specific city are defined, with the aim of coming to a good mix between the three sustainability domains. This can be done through discussions, but also more visually by means of placing sticky notes on the different sustainability domains. The flows between the stocks, as well as relevant physical and institutional factors, are also identified and described. The outcome of this step is an empty framework that will guide the transition team in the data-gathering step.

Step 3: Data Gathering

The third step in the systems analysis is gathering data. The systems analysis is conducted in an iterative, participative, trans-disciplinary way. Members of the transition team gather data for the systems analysis using a combination of desk research, interviews and expert meetings, to ensure input from different disciplines and perspectives. The systems analysis is done in an iterative manner: information from different sources is used to reconsider the findings from another source, so as to validate the results. Desk research is used to collect data and statistics of the city (e.g. demographic data, income distribution, historical data, spatial data and environmental data) and review and assess existing strategies, visions and white papers that the city has formulated on the themes of the systems analysis. Data and statistics collection and the reviewing of existing policy documents are important for the systems analysis since they bring to the foreground which information is available and how this information has been interpreted and used by e.g. policy makers. Interviews and expert meetings are used to complement the findings of the desk research with the perspectives of different actors and experts. In this way, the transition team gets a broader picture of what the properties of the systems are, how these are interconnected and which problems arise.

Step 4: Data Assessment and Analysis
The fourth step of the systems analysis entails data assessment and analysis. A set of lenses is presented that together reflect the variety of ways of looking at the system. The different lenses are structured according to the multi-level model (the landscape-regime-niche levels, or macro-meso-micro levels; Geels and Schot 2007) and together ensure an integrated perspective and analysis of the system (Table 1). The lenses are primarily used by the transition team, fed by information obtained from desk research, interviews and expert meetings. The systems analysis, including the presentation of the results of these different lenses, is presented in the first *transition arena*: a workshop involving frontrunners from different

Table 1 Systems analysis: lenses to examine contemporary dynamics at system's level

Levels	Lenses	Explanation
Macro Level	Landscape influences	A description of current landscape influences. From a system-environment view influences that occur outside, yet influence the system (such as global economic trends) are described
	Historical sketch	A description of historical change over a long time period (>100 years). Historical developments are described to unravel the characteristics of a city and draw historical parallels
Meso Level	State of the system	The SCENE model is completed, i.e. the stocks, flows and/or characteristics are identified and named, to get an overview of the properties of the system. Major influences from the physical and institutional context specific for the city are also described. An optional application is to assess the baseline of the system, to use for benchmarking
	Relations between stocks	An investigation and description of the relationships and interdependencies between stocks. This description can lead to the identification of important stocks and insight in feedback mechanisms
	Selection of main assets	Narrowing down of stocks to the most important assets for the city, to use for characterising the city, structuring the problem and further on in the envisioning process
	Trend analysis	A description of trends in stocks or main themes over the past 50 years, to identify capital losses or gains, turning points and possible path dependencies
	Dominant culture, structure and practices	Identification and description of the dominant structures, culture and practices in the city. Identification of these characteristics of the workings of the current regime is part of the problem structuring process
Micro level	Identification of niches	A description of niche developments within the main themes give an indication of what is happening outside of the regime and can be used as examples in the envisioning process

backgrounds. However, the systems analysis is not finished yet: the participants of the arena can include their input and perspectives by means of discussion and feedback, in order to adapt the systems analysis where necessary, create a 'common frame of reference', and ultimately validate the systems analysis.

Case Studies: Ghent and Aberdeen Cities

The systems analysis methodology was applied in the cities of Ghent and Aberdeen as the first step of the transition management process in the MUSIC project. The aim was to identify the challenges and opportunities in their cities with respect to energy use and CO_2 reductions as part of the first phase of transition management, the preparation and exploration phase. The cities adapted the analysis to their local circumstances. The outcomes of the systems analysis served as input for the second phase of the transition management strategy: the development of a joint vision and a transition agenda.

In both Ghent and Aberdeen, the initiative for taking action on the energy and climate goals of their local governments, and participating in the MUSIC project, came from within the local government. Therefore, the transition team was made up of policy officers, together with experts in participatory process management and transition management, as well as a city coach -the transition expert from DRIFT-who assisted the transition team. The transition team of Aberdeen consisted of a policy officer from the environment department, two policy officers with expertise on energy and community engagement from the environment department of Aberdeen's local government and the city coach – the third author of this chapter. The transition team of Ghent consisted of two policy officers from the environment department of the Ghent City, a transition expert and a process facilitator – responsible for project management – and the city coach – the fourth author of this chapter. The transition team was assisted by the developer of the methodology – the first author of this chapter (Fig. 2).

Figure 2 shows the actors in the transition management process in Ghent, who belong to three groups: the transition team, the transition arena and a group of experts. Black figures represent policy officers from within the local government. Grey figures represent individuals who work and/or live in Ghent and have been selected based on their innovative way of thinking: the frontrunners. The white-coloured participants represent the external professionals who assisted in the process by means of project management, consultation and coaching.

Figure 3 shows the actors in the transition management process in Aberdeen, who belong to two groups: the transition team and the transition arena. Black figures represent policy officers from the local government. Grey figures represent individuals who work and/or live in Aberdeen and have been selected based on their innovative way of thinking: the frontrunners. The white figure represents the external professional who assisted in the process by means of project management, consultation and coaching.

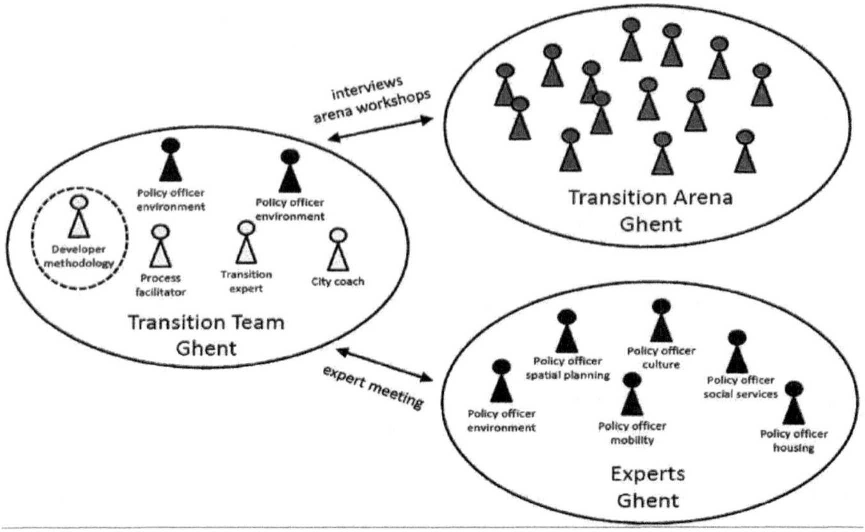

Fig. 2 Participants in the transition process in Ghent, Belgium in the MUSIC project

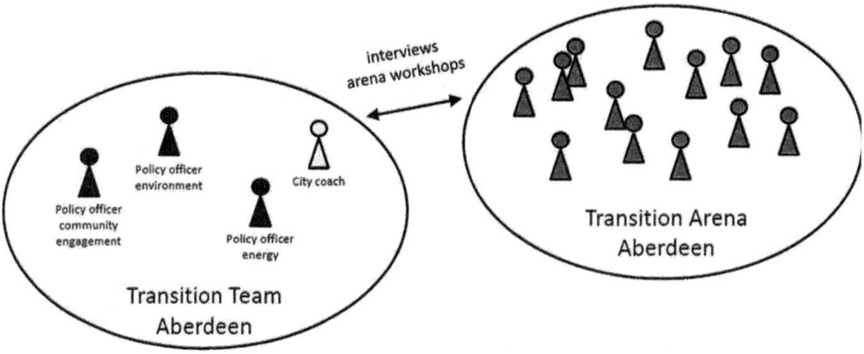

Fig. 3 Participants in the transition process in Aberdeen, United Kingdom in the MUSIC project

The systems analysis was performed in an iterative manner, together with experts and frontrunners (Figs. 2 and 3), by means of a combination of desk research, interviews, expert meetings and the transition arena workshops (Tables 1 and 2). In both cities, the system analysis methodology was meant to unravel the complexity of the city and to shift the local government's focus on initial solutions to systemic challenges and opportunities.

As Ghent, of all cities participating in the MUSIC project, was furthest ahead in the transition management process and had a tightly planned schedule for the different steps of the transition process, the systems analysis methodology was developed during the implementation of the systems analysis in Ghent, leading to an iterative cycle of development, implementation, results and evaluation to

Table 2 Application of the systems analysis steps in Ghent, Belgium in the MUSIC project

Step	Application in Ghent
1 System definition	January – February 2011: System definition based upon discussions in transition team meetings and first data gathering from policy documents
2 System structuring	January – February 2011: Analysis of existing policy documents of the municipality by the policy officers in the transition team
	March 2011: Discussion and sticky notes exercises to place stocks in domains by members of the transition team
	5 April 2011: Discussion and sticky notes exercises to place stocks in domains in expert meeting with policy officers from other departments within the municipality, in dialogue with the transition team
3 Data gathering	January – April 2011: Reviewing and analysing information from reports, statistics, newspaper clippings, websites, interviews and SWOT analyses by policy officers from the transition team
	March – April 2011: Interviews by policy officers from the transition team with 14 frontrunners, who were thereafter invited to participate in the transition arena sessions
4 Data assessment and analysis	March – May 2011: Discussion in transition team and working together to structure data according to the different lenses
	5 April 2011: Discussion about collected data for systems analysis between the transition team members and policy officers from other departments within the municipality in an expert meeting
	9 May 2011: First transition arena workshop, in which the (preliminary) systems analysis was presented to the participants of the transition arena, by the transition team. Enables an effective exchange of perspectives between the participants of the arena. Results from the first transition arena session are discussed within the transition team and are used to modify the systems analysis
	May 2011: Transition team processes input from the first arena session to improve systems analysis
	7 June 2011: Second transition arena workshop, with the same participants as the first transition arena session, together with the transition team. Results in in-depth expansion of systems analysis, priorisation of themes and a first draft of goals for the future
	June – October 2011: Transition arena workshops aimed at envisioning and agenda setting build upon the systems analysis

improve the methodology along the way. This means that the systems analysis was very much tailored to the specific needs in the local government of Ghent due to the fact that it was developed together with the people who would be using and applying the methodology.

Below, we use the results of the systems analysis in Ghent to illustrate possible outcomes of the systems analysis (based on Maas 2011; van Sande 2011). Section "Zooming In: Outcomes of the System Analysis in Ghent City" presents the outcomes according to the lenses, following the gliding scale from macro to micro level, as shown in Table 1. Table 3 highlights the application of the four steps of the systems analysis methodology in Aberdeen. Table 4 illustrates the state of the city system in Ghent and Aberdeen, from the perspective of their respective

Table 3 Application of the systems analysis in Aberdeen, United Kingdom in the MUSIC project

Step	Application in Aberdeen
1 System definition	May 2011 – System definition based upon discussions in transition team meetings, by the policy officers and city coach. Transition team analyses data from existing policy documents and plans, and conducts first interviews with potential participants of the arena
2 System structuring	May–June 2011 – Collected data are systematically analysed to inform and form a historical analysis and a SCENE model. The system structuring is realised by the transition team
3 Data gathering	April–June 2011 – Data collection from existing policy documents, plans and databases existing in the municipality. First data collection is realised by policy officers from the transition team. Data collection involves 5 in-person interviews with potential participants following semi-structured open questions about problems, challenges, strengths and weaknesses of the city and the energy sector
4 Data assessment and analysis	June 2011 – During the first transition arena workshop, the transition team presents the systems analysis to the participants. Discussions and inputs from the participants' results into adapting the system structuring
	September 2011- During the second transition workshop, new participants point at themes and issues revealed by the system analysis resulting in expansion of system definition and boundaries: inclusion of mobility into the broader system definition

transition teams. It provides an overview of the stocks and characteristics that are identified as most important in the social, environmental and economic domains, relevant to the energy and climate context in Ghent and Aberdeen. Information on these stocks and characteristics was gathered in a background document and served the further exploration and analysis of the system, in terms of relations, trends and main assets (section "Zooming In: Outcomes of the System Analysis in Ghent City"). To fuel the discussion on the relations in Ghent, a few stocks were highlighted in the synthesis by the transition team: *mobility and infrastructure*, *housing*, *economic vitality*, and *health*.

Furthermore, a system model was developed to map the relations between the stocks, and to investigate linkages between the different components of the system. This system model generated insight in (unintended) effects that can be expected when changing one part of the system and in linkages that provide opportunities for change. An example from Ghent was the relation between social housing and energy use. Ghent has a large social housing stock, with houses of relatively poor quality and low energy efficiency (e.g. lack of double glazing, poor insulation, and lack of central heating). The systems analysis revealed – that there might be a great potential for increasing energy efficiency and living quality by renovating and retrofitting these houses using low-carbon technologies. This improvement could lead to both increased energy efficiency and living quality, as well as to lowering energy bills, of major importance to the poorer segment of society that often live in social housing.

Table 4 Domains, stocks and characteristics of Ghent and Aberdeen as identified during the system analysis phase of the transition management process in the MUSIC project

Ghent		Aberdeen	
Stocks	Characteristics	Stocks	Characteristics
Water	Surface water	Environment	Drinking water
	Water quality		Water use
	Water use		
Green space	Surface forest and nature		Quality public space
	Public green space		Green infrastructure
	Green roofs		Urban parks and gardens
Energy	Energy prod./cons.		
	Renewable energy		Greenhouse gas emissions
	Energy infrastructure		
Waste, soil and nuisance	Municipal waste		Household waste
	Soil pollution		Industrial waste
	Nuisance		Air quality
Air pollution	Air quality		Flood prone areas
	Fine particulate matter		Ecological footprint
	CO_2 emissions		
Population	Population density	Demographics and settlements	Population density
	Demographics		Demographics
	Immigration/emigration		Suburbanisation
Housing	Ownership	Housing	Housing types
	Housing types		
	Energy performance		Land use
Social services	Education		
	Healthcare		
	Associations		
Liveability	Satisfaction of living	Liveability	Safety
	Social cohesion		Health
	Segregation		Social deprivation
			School attendance
			Cultural institutes
Leisure	Sports	Tourism and leisure	Sports
	Culture		Culture and art
	Lifestyle		Tourism

(continued)

Table 4 (continued)

Ghent		Aberdeen	
Stocks	Characteristics	Stocks	Characteristics
Participation and politics	Participation possibilities	Politics	Voter participation
	NGOs		Citizen participation
	Political attention climate		Social cohesion
Mobility and infrastructure	Car use	Mobility	Car ownership
	Public transportation		Public transportation
	Infrastructure		Mobility infrastructure
Economic vitality	Income	Economy	Income
	Unemployment		Unemployment
	Educational level		Educational level
Economic sectors	Harbour	Energy	Energy production
	Knowledge institutes		Renewable energy
	Services		Heating (district, central)
Knowledge and innovation	Knowledge economy	Knowledge economy	University-local government collaborations
	(Green) investments		Education level
		Low carbon economy	Governance responsibilities
			Green investments

The stocks were placed within the physical and institutional context, as they influence the developments in the city. Relevant physical factors in Ghent were for example the large number of protected monuments, limiting possibilities for adaptation and retrofitting of buildings, and flood vulnerability. A relevant institutional factor for Ghent was the challenging political reality in Belgium, which had led to a political deadlock in the past because of disagreement between the three political regions. This had resulted in a complex division of responsibilities between the federal government and the regional governments.

Zooming In: Outcomes of the System Analysis in Ghent City

Six main assets of Ghent were identified, on the one hand to emphasize the assets the city can build upon to reach a climate-neutral future, on the other hand to frame the challenges related to these assets. Finally, an inventory of existing innovative projects in the city was made, in order to investigate the possibilities and barriers for good practices, and look into ways to stimulate take off of these desired developments.

Macro Level

Landscape Influences

Many developments and trends at a higher level (e.g. national, European and global level) influence the institutional organisation and policy developments at city level. Worldwide, the attention for climate change has exploded in the past decades, both scientifically and politically. In order to combat climate change, CO_2 emissions from fossil energy sources have to be reduced. Furthermore, fossil-based energy supply is highly vulnerable to conflicts, scarcity of resources and rising prices. These factors combined give rise to the need to reduce energy use, reduce CO_2 emissions and to switch to renewable energy sources. Belgium is one of the most prosperous countries in the world. Subsequently, the ecological footprint of Belgian citizens is very large: they rank as the fourth in the world. If the entire worlds' population would live as the Belgians do, more than four times the surface of our planet would be needed: earth cannot sustain this level of consumption (van Sande 2011). The financial crisis and subsequent recession of the past years have revealed the vulnerability of the global economic system (Dahl 2009). This provides opportunities for rethinking the economic growth paradigm and deliberating alternatives more aligned with sustainability. The financial crisis and economic recession have shown that the 'green economy' is more resilient: green companies with green jobs have withstood the crisis well and represent a remarkably healthy economic sector.

Historical Sketch

Ghent is a historical city, and settlements on the site of the contemporary city reach back until Celtic times, around 500 B.C. Ghent was located in a marsh area, which made it unsuitable for agriculture, but where sheep breeding was possible. This led to a flourishing wool industry, and later on the production of other fabrics, which gave Ghent the name of "the Manchester of the European mainland". From around 1000 to 1500 A.D., Ghent was the biggest city of the low-lying countries and was bigger than Cologne and Paris. In the thirteenth century Ghent was home to around 60.000 inhabitants. Ghent has always been a rebellious and wayward city. Its inhabitants fought for hundreds of years against the repression of their rulers and strode for their freedom and privileges. Even now, rebellious people in Ghent are still referred to as *stroppendragers* ('knot wearers'), because the inhabitants of Ghent were punished by wearing a hangman's knot for revolting against Emperor Charles V.

From the sixteenth to eighteenth century, the city of Ghent suffered from economic decline. The population decreased from 60.000 to only half that size around 1630 and the city lost power over the access to the channel connecting to the North Sea. An economic upswing was brought about by the introduction of industrialized linen and cotton production. Ghent was the first city on the

European mainland that became industrialized during the Industrial Revolution at the end of the eighteenth century. Around that time, Ghent also became a true university city and opened its sea port, with a connection to sea through the newly dug Ghent-Terneuzen channel. In 1827 the first gas factory in Belgium was built in Ghent to provide the street lanterns with gas. Soon after, two others were opened.

From 1860 onwards, most of the industry settled outside of the city. Also, the last city gates were broken down, allowing for the construction of new neighbourhoods outside the historic city walls. Ghent was the first city to have labour unions, and was the birthplace of the socialist movement in Belgium. The Vooruit, the historic labour union stronghold and socialist party centre, is still a dominant feature in the centre of Ghent.

Because Ghent was not bombed during the two World Wars, many historical buildings have been preserved. From the 1950s and onward, Ghent was an industrious, but grey and polluted city. The streets and quays were clogged with cars, and the water in the canals was black and smelly. In the 1970s local politicians took some important decisions to improve the situation in the city. In general, there was a shift from a typical industry city towards a service- and knowledge-based economy. The local government demarcated part of the centre as a traffic-free zone, the most prominent quays in the centre became traffic-free and the water quality was greatly improved. The water became the lifeline of the city again and the quays, such as the *Graslei*, are a hotspot for inhabitants and tourists during summer to enjoy outside life alongside the canal. Furthermore, there were big investments in cultural activities, most famously the introduction of the *Gentse Feesten*: the Feasts of Ghents, a renowned festival.

Looking back, political decisions made 40 years ago (e.g. introduction of traffic-free zones, improvement of water and air quality, introduction of cultural activities) had a big impact, and are still visible when looking at the current situation in the city. They provide an inspiration for the political decisions that have to be made now to ensure Ghent will be a liveable city 40 years from now.

Meso Level

Main Assets

Knowledge city: strong presence of university and students, but link to practice is weak.

Prosperous city: strong high-tech sector, but also need for production of local goods and lower-skilled employment.

Water city: water is the lifeline for the harbour, tourism and citizens, but the transport potential is underdeveloped and climate adaptation can become a challenge.

Green city: Ghent has abundant green space, but many citizens do not have sufficient access to them. They live too far away.

Energy city: Some companies in Ghent are real frontrunners in renewable energy, but the presence of the fossil-energy sector remains strong.

Lively city: Ghent is a welcoming and buzzing city for citizens and tourists, but social cohesion is a weak spot in some areas and the main festival – De Gentse Feesten – has become so famous that people from Ghent do not feel it is their festival anymore.

Dominant Culture, Structure and Practices

Ghent's political structure is mostly aimed at securing short-term success, in order to get public support and get re-elected. This political short-termism makes it very difficult to introduce innovations and policies that take longer than a legislative period to prove their value.

Within the organization of the local government, the different departments do not always work together or differ greatly in their way of thinking on issues that involve multiple policy domains, leading to fragmented, inefficient and sometimes even contradictory policies.

Other characteristics of the city are: the shift from government to governance, the city's history of political consensus and dialogue, overregulation, stubbornness of the inhabitants, local pride regarding the city's heritage, and modesty about their frontrunner status.

Micro Level

Identification of Niches

Innovative niche developments are the Ghent bio-energy valley and the gasmetersite. *Ghent Bio-energy Valley is* a public-private partnership (e.g. Local Government and Province of Ghent, Port of Ghent, Ghent University and several companies) to stimulate the bio-based economy and the production of sustainable bio-energy. *Gasmetersite* is a new housing development on a former brownfield, with passive and low-energy houses and space for renewable energy generation. Other stimulating niche activities are the mayor coming to his office by bicycle, taking away prejudices about cycling to work being unprofessional or uncomfortable; and companies providing their employees with a foldable bike, instead of a (lease) car.

Systems Analysis in the Transition Team

In Ghent, the systems analysis methodology was developed in close cooperation with the transition team, since they were the first city within the MUSIC project starting on the systems analysis. Therefore, most dialogue was about *how* to implement the methodology and little effort was made to zoom out and discuss the *purpose* of the analysis, which led to some confusion from the side of the policy officers who were conducting the systems analysis. Working on an overview of relevant stocks and characteristics in the social, environmental and economic domains, led to a useful discussion on what is relevant to the energy and climate context. The overview thus constructed helped to select and structure relevant insights from the abundantly available data and reports, this was collected in a background document that served the further exploration and analysis of the system, in terms of relations, trends and main assets. Furthermore, a system model was developed to map the relations between the stocks, and to investigate linkages between the different components of the system. This system model was discussed in an expert meeting. It was perceived as too complex for the limited time devoted to it in the expert meeting, but appeared useful for the transition team to have an insight of the expected effects when changing one part of the system and in linkages that provide opportunities for change. An example was the relation between social housing and energy use.

For the members of the transition team, the methodology was useful to give meaning to the large amount of data and studies available. They also highly appreciated that the interviews gave them insights in other perspectives, illustrated by one of the members through the following quote: "I realized how little I knew of what is going on in my own hometown". The insights from the systems analysis were presented in the first arena session by a policy maker from the transition team. The presentation included a synthesis of the analysis and specific quotes. It was structured in three parts: the first part discussing the urgency of action towards a low-carbon city, the major section discussing strengths of the city and how they're under pressure and finally a discussion on the need for a "climate for change".

Systems Analysis in the Transition Arena

From the first two transition arena sessions, in which the systems analysis plays a central role, it appeared that the participants perceived Ghent as a city with many assets and attractions and a very pleasant place to live and work. However, the current situation was perceived as unsustainable, and the social, economic and environmental challenges that the city faced (e.g. in terms of climate change and rising energy prices) ask for a pro-active position. The local government has no single key to success: many different actors made up this current unsustainable situation and influenced the development. Therefore, the local government needs to work together with industry, organizations and inhabitants. The local government should set the framework and function as a lever, to enable public-private partnership to flourish and initiate the necessary changes. It is really a matter of will-power and showing leadership and commitment. One of the participants gave the following example: "Kennedy said: We'll be on the moon within 8 years. And they were".

The participants openly supported relating energy, climate and sustainability to the broader concept of liveability. When taking a holistic perspective and relating different domains or sectors, it became easier to perceive that a measure that will lead to costs in one domain can lead to benefits in other sectors. One of the participants said: "Taking transport as an example, what are the potentials for synergy between climate neutrality, air quality, noise pollution and safety? We often talk about the costs, but we also have to look at the benefits!" In the first meeting of the arena group (May 2011), a setting was created for an open exchange of perspectives: not only so that the participants could get to know each other, but also they could reflect on the draft of the system analysis. The most dynamic moments in the discussion, according to the transition team, were the discussion on the urgency, about the scope of climate neutrality (including consumption or not?) and the social aspects of a sustainable city (the difference between the ability to pay for sustainability measures and the role of social cohesion).

The work of the transition team – with a focus on a climate neutral city – already integrated many domains, the discussions in the transition arena seemed to go even broader: the link of the sustainability ambitions to the liveability of the city and the health of citizens was stressed. The participants experienced a lack of attention to green in the city, and to economic aspects. After the first arena meeting, the transition team used the insight from the discussion to create a new synthesis of the analysis, again in the form of a powerpoint presentation. This was recapitulated and once more sharpened in the first half of the second arena meeting (June 2011) by the arena participants. The arena group built upon the analysis in their meetings: it fuelled the discussion on what basic principles a city should comply with in order to be sustainable, and especially the discussion on the strengths of the city was reflected upon for the later drafting of visionary images. The systems analysis was not consolidated in a document: the transition team has not elaborated the synthesis further. Neither the transition team nor the transition arena sought to summarize the problem statement. Elements of the synthesis have though been included in the *transition magazine* that was created after six arena meetings.

Discussion

In this section, we reflect upon the application of the systems analysis methodology in Ghent and Aberdeen and draw lessons for future applications. In both Aberdeen and Ghent, the climate and energy goals of the local government were the reason to get involved with the MUSIC project. Therefore, the identification of stocks was strongly focused on this realm. However, the SCENE model greatly aided in forcing the transition team and arena to take a wider perspective of the system, and getting an overview of the whole.

In Ghent, the stock 'energy' was placed in the environmental domain; despite the importance it has for the economic domain, since Ghent historically has been a big steel- and electricity producer. This can be explained by the energy and climate focus that the local government took for this project: energy is mainly viewed in terms of emissions (harmful to the environment), or in terms of the switch to renewable energy (beneficial to the environment); it is mainly framed in terms of its environmental impact. At the start of the systems analysis the policy officers showed some surprise at seeing the proposed stocks suggested by the developers of the methodology mainly in the social domain, such as 'housing' and 'social services'. It was felt that these had no clear relationship to the focus of the project: climate and energy. However, later in the process, after data gathering, interviews and discussions, it became apparent that there are some very interesting links between housing, social services and energy, such as possibilities for sustainable social housing and a social benefit scheme to prevent energy poverty among poorer groups in society.

In Aberdeen, the application of the SCENE model for mapping stocks into three core sustainability domains revealed the policy logic behind understanding the operation of each stock for specific domains. Policy officers initially placed 'energy' as an economic asset of the city. The rationale behind this perception was that the energy sector is providing employment to the city and it is coupled to its economic vitality. During the discussions and reflections on this energy-employment coupling, the officers shifted their understanding in two ways. Firstly, the policy officers recognized that energy assets (oil, wind, etc.) are part of the environmental assets and provide economic services instead of only being perceived as part of the economic fabric of the city. Second, energy initially was the asset of the city that was connected directly or indirectly with all characteristics and services of the city. During discussing what constitutes the fabric of the city, policy officers addressed additional assets such as education, trade, agriculture and art as important characteristics of the city that are not directly relating to the energy sector per se.

The Analysis of the History Inspired and Motivated Participants

In both cities, the historical analysis of the city gave insight in the importance of historical events and developments, and created a feeling for the possibilities in the future. In Ghent, the historical sketch, depicted major events and developments in the past of the city provided some interesting perspectives for the future. The characterisation of the inhabitants of Ghent as being rebellious and taking up a role as frontrunners can be used to position and frame the ambitious policy choices made today. Radical policy choices made in the 1970s – including the creation of a traffic-free zone in the centre of the city and the improvement of air and water quality – provide an inspiration for the political decisions that should be made now to ensure Ghent will be a liveable city 40 years from now and show how much can change in this relatively short time span.

In Aberdeen, the presentation of the history of the city created a feeling for its future. During the transition workshop, the participants shared their (family) stories about the past, and in this way, they connected personal memories with historical events and path-dependencies. Stories about the cultural heritage and activities of the community in different seasons and the changes in culture that were brought by the newcomers that are employed in the energy sector have been shared and expressed. The participants shared a feeling for the need to regain 'ownership' of the future of their city and "to live to its true potential".

The Systems Analysis Broadened the Problem Perception, Which Was Considered Valuable by the Participants

In Ghent, dialogue within the transition team, discussions between the transition team, experts and participants of the transition arena, and the execution of the systems analysis enabled broadening the problem perception from the narrow focus on climate and energy towards a more holistic approach to the problem. This led to the broadening of the initial problem formulation from "How can Ghent become climate neutral in 2050?" to "How can Ghent become climate neutral in a context of becoming a social, green and liveable city in 2050?" Historical characteristics and current assets of the city of Ghent, such as its social and green character and strong cultural sector, were thus linked to the ambition of CO_2 emission reduction and a transition to renewable energy.

Widening the problem space, by adopting a holistic perspective on the system was deemed very valuable by the members of the transition team and the participants of the transition arena in Ghent. For example, the inclusion of social aspects while talking about climate and energy goals led to insights and opportunities that would otherwise have been missed. Having an integrated perspective shows that climate and energy issues do not exist in a vacuum, but need to be considered in relation to other policy domains in order to look for the deeper lying problems and opportunities for improvement. As a city officer of the transition arena expressed: "This is a method for participation where the level of knowledge is increased and complexity of the subject is acknowledged" (Roorda et al. 2012).

The Systems Analysis Revealed Institutional Barriers to Sustainable Development

Differences in institutional logic of every local government were revealed during the dialogues about the domains of change and persistent problems in the transition teams of both Aberdeen and Ghent. Aberdeen's governance is characterized as pragmatic, meaning that policy agenda's points have to relate one to one with the issue at hand, and only realistic top-down measures can be included in policy action plans. Any innovation that requires 'creative' arrangements between departments that do not pre-exist are screened out as infeasible. The pragmatic approach to governance yields fruits on the short term. However, long term changes that may be required to address an issue such as climate mitigation are confronted with a strong avert governance reality.

Ghent's political structure is mostly aimed at securing short-term success, in order to get public support and get re-elected. This political short-termism makes it very difficult to introduce innovations and policies that need a time span longer than a legislative period to prove their value. Besides a long-term perspective, sustainable development requires an integrated, holistic approach. The different departments do not always work together or differ greatly in their way of thinking about issues that involve multiple policy domains, leading to fragmented, inefficient and sometimes even contradictory policies.

In Ghent, the systems analysis has forced the policy officers involved, as well as the experts who participated in the expert session, to think more holistically about energy policy at local level and promote cooperation between different departments. Discussing the systems analysis in a trans-disciplinary group greatly aids in the process of stimulating holistic thinking. However, there were also difficulties with thinking holistically in the session with the transition team, the expert session and the transition arena sessions. There is a clear need for a transition from compartmentalized thinking towards holistic thinking. The fact that data gathered through desk research and was complemented with insights and perspectives of experts (in the expert session) and actors (through interviews), was considered very valuable by the members of the transition team of Ghent. It made them realize how little they know about the city they live and work in, and in how many possible ways one can look at an issue.

In Aberdeen, systems analysis discussions between policy officers and the city coach, as well as the transition workshops with a variety of frontrunners, stimulated holistic thinking about the problem and its root causes. More specifically, during the systems analysis discussions, a first step to decouple the city's growth from carbon growth emerged in the policy officers' perceptions. During the dialogue with the policy officers, the problem perceptions have been shifted from anchoring the oil-industry for city's prosperity towards seeking possibilities to maintain and expand the city's prosperity. The understanding that the city consists of more economic assets than the oil industry alone and that different assets also contribute to the city life was a first step in decoupling the city's growth from carbon growth. A second step to decouple the city's growth from carbon growth was proposed by workshop participants. Two participants started questioning the reality that "oil rules the city" and argued about the need to start a new social fabrication of the city's growth looking outside the energy-economic growth couple. They mentioned Tim Jackson's *Prosperity Without Growth* (2010) ideas and growth perspective as indicative for what can create a shift in thinking and understanding.

The Systems Analysis Served as a Catalyst for Mutual Understanding and Learning Between the Participants

The difference in backgrounds and perspectives of the people involved in the systems analysis could have posed problems in the discussions. However, in Ghent, the systems analysis helped to overcome this. The participants recognized their perspective was integrated in the systems analysis, so they recognized it as "theirs" and gained ownership of the results of the systems analysis. Moreover, discussing the systems analysis served as a catalyst for mutual understanding and learning between the participants. It made the discussions go beyond platitudes and symptoms. The level of knowledge on sustainability issues ranged widely, but still all participants say to have learned from other participants and most say to have gained new insights with respect to sustainability.

The transition arena did however inhibit a delicate balance: some participants wanted to have more in depth discussions and research, others asked for more speed in the process and were eager to get more concrete. In general, the participants of the transition arena sessions highly valued the diverse background of the arena participants and the variety of technical information and perspectives brought to the foreground by the systems analysis. Some illustrative quotes are: "I learned a lot due to the mixed background of the arena participants" and "I valued the mix of technical knowledge and different backgrounds" (Roorda 2012).

The Systems Analysis Falls Short in Creating a Sense of Urgency

According to city coach of Ghent, describing the landscape influences was very useful. It was essential in communicating the urgency of the problem and triggered a lot of discussion, such as on the concepts of 'ecological footprint' and 'climate neutrality'. However, even though the systems analysis has managed to give a synthesis of different facts and perspectives, it falls short in communicating the urgency and motivation to tackle the problems at hand. It is too much of everything; a clear story line is missing (Maas 2011).

In Aberdeen, creating a sense of urgency worked out better. The system analysis revealed the heart of the energy problem rather than justifying precluded solutions (e.g. subsidizing renewable energy). By revealing the strong dependency of the city's economy on the energy supply sector as it is formed at present, a sense of urgency was created.

Lessons for Future Applications of Systems Analysis Within Transition Management in Cities

Overall, the application of systems analysis methodology should be discussed among the participants and tailored to the characteristics of a city. Discussions on *how* to implement the methodology in the context of a specific city in the transition team and with other experts and actors turned out to be essential. This is important, because it is necessary to come to a mutual understanding about the systems analysis methodology as well as of what constitutes the system. Every city is different and thus the composition of stocks and their characteristics within the sustainability domains varies. Energy use in one city may for instance be placed in the environmental domain (as was the case in Ghent), because it is mainly viewed as a source of CO_2 emissions, whereas in another city it may be considered the fuel that drives the industrial economy and therefore it is placed in the economic domain (as was the case in Aberdeen).

The placement of stocks, and their characteristics should be thoroughly discussed to ensure it makes sense for the participants as well as for the specificities of a city. In Ghent, the identification of flows turned out to be difficult for the people filling in the SCENE model. They discussed the difference between flows and relationships between stocks, as well as whether a flow should be a tangible flow of matter, information or emissions, or whether it can also be used to describe a causal relation. In Ghent, it was decided to name interrelations between stocks and give an overview of the stocks and the relations between them in a system model, to cover all of these types of interrelations.

The different lenses that were applied to analyse the data gathered by means of desk research, expert sessions and interviews were not all used in the two cities, or were used in a different manner. In Ghent, most of the lenses have been used, and proved useful. In Ghent, the lack of a strong trend analysis, which is meant to show unsustainable pathways and unsustainability problems for the future, is an explanation for the weak sense of urgency amongst the members of the transition team and the participants of the transition arena.

The systems analysis methodology as it is presented in this chapter/book is not a cookbook; it should be applied flexible and tailored to the local situation. The methodology should be used in a flexible way; it should not be understood as a clear recipe that can be applied everywhere in the same way. Every city, and the participants in the transition process, is different. The form of the systems analysis, why it is used and which goals should be set to accomplish that should be made clear. Before conducting the systems analysis, it is important to voice expectations, make a realistic planning and decide upon the form of presenting the data, in order to present the systems analysis in such a way that it fits with the target audience. Discussions in the transition team were essential to come to a common understanding of the system and a working method for the systems analysis.

Conclusions

The systems analysis methodology enriches the start of a transition management process in an urban setting. The methodology consists of the following four steps that need to be followed in an iterative way: (i) system definition, (ii) system structuring, (iii) data gathering and (iv) data assessment and analysis. Characteristic of this methodology is the use of a set of lenses that reflect a variety of ways of looking at the system from macro, meso, and micro level. Together, these lenses ensure an integrated perspective and thorough analysis of a city, its government organisation and governance. The systems analysis methodology as we propose it involves a variety of stakeholders: the transition team, consisting of policy officers, a city coach and a process manager, who initiate the transition process and organise and facilitate the transition arena workshops. Because integration of objective data and subjective perspectives is crucial, desk research as well as interactive research – interviews with frontrunners, expert meetings and workshops with frontrunners in the transition arena – is needed to collect and analyse the data.

The methodology we propose should not be interpreted as a cookbook. Every city is different with respect to the local context and customs, and every group of participants differs in background, involvement and knowledge. Therefore, the methodology should be applied flexible and tailored to the local situation.

The steps of the systems analysis methodology should be thoroughly discussed in the transition team. It is particularly important to agree on the purpose of the systems analysis, on what constitutes the system and to come to a mutual understanding about the systems analysis methodology. This discussion can reveal interesting and new opportunities for change, such as unexpected connections, possibilities for synergy, and new perspectives on the problem. This is illustrated by the example of 'housing' in Ghent and the problem perception in Aberdeen. Policy officers from Ghent at first instance did not see the connection between 'housing' and climate and energy, whereas later on, in the process 'housing' became a focal point, because of interesting opportunities related to sustainable social housing and social benefits related to housing. In Aberdeen, the problem perception shifted from anchoring the oil industry for the city's prosperity towards seeking possibilities to maintain and expand the city's prosperity by making use of all assets of the city, not only those related to oil and gas. The systems analysis has opened up possibilities for synergy; it shows that costs in one domain can lead to benefits in other domains or sectors. By broadening the solution pathways, the holistic perspective on the system allows for the exploration of opportunities that were before deemed unprofitable.

The systems analysis serves as a catalyst for mutual understanding and learning, and stimulates holistic thinking. Participants involved in the systems analysis learn from each other. The systems analysis also reveals institutional barriers such as short-term thinking and policy fragmentation to sustainable development, but it is not able to overcome these barriers. Obviously, institutional transition requires a long period of time and the systems analysis only constitutes the first phase of transition management. The political and institutional context remains a focus in the subsequent phases of the transition process.

Overall the systems analysis methodology is an effective method to create an overview of the city under study, and stimulates the creation of a common information base and holistic thinking amongst participants in the transition process. It serves as a good start for the transition management process; the participants share a perspective on the system, its properties, challenges and opportunities, and use the systems analysis as a basis to continue working on viable solution pathways to enable their city to reach their ambitious climate and energy goals.

References

Balthazar T (2008) Stad Gent Lokaal Klimaatplan 2008–2020. 105 acties voor een klimaatneutrale stad. Stad Gent, Departement Milieu, Stadsontwikkeling en Wonen, Ghent

Banathy BH (1988) Systems inquiry in education. Syst Pract Action Res 1:193–212

Covenant of Mayors (2011) The Covenant of Mayors. Website of the Covenant of Mayors

Dahl AL (2009) Financial crisis and the green economy. International environment forum. UNEP/ University of Geneva/Graduate Institute Certificate of Advanced Studies in Environmental Diplomacy, Geneva

European Commission (2010) The EU climate and energy package. European Commission, Climate Action, Brussels

Findeisen W, Quade ES (1985) The methodology of systems analysis: an introduction and overview. In: Miser HJ, Quade ES (eds) Handbook of systems analysis: overview of uses, procedures, applications, and practice. Elsevier Science Ltd., New York, pp 117–149

Frantzeskaki N, Loorbach D, Meadowcroft J (2012) Governing societal transitions to sustainability. Int J Sustain Dev 15(1–2):19–36

Geels FW, Schot J (2007) Typology of sociotechnical transition pathways. Res Policy 36(3):399–417

Grosskurth JA (2008) Regional sustainability: tools for integrated governance. International Centre for Integrated Assessment and Sustainable Development, Maastricht University, Maastricht

Jansen K (2009) Implicit sociology, interdisciplinarity and systems theories in agricultural science. Sociol Rural 49:172–188

Loorbach D (2009) Urban transitions and urban transition management. The case of Rotterdam. Workshop on Urban Transitions, Manchester

Loorbach D (2010) Transition management for sustainable development: a prescriptive, complexity based governance framework. Gov Int J Policy Adm Inst 23:161–183

Maas S (2011) Making sense of complex systems. A systems analysis methodology for urban transition management: the case of Ghent and its climate ambitions. Wageningen University, Wageningen

Miser HJ, Quade ES (1985) The context, nature and use of systems analysis. In: Miser HJ, Quade ES (eds) Handbook of systems analysis: overview of uses, procedures, applications, and practice. Elsevier Science Ltd., New York, pp 1–32

MUSIC (2011) Mitigation in urban areas: solutions for innovative cities. The MUSIC Project, Rotterdam

Newman P, Jennings I (2008) Cities as sustainable ecosystems: principles and practices. Island Press, London

Pahl-Wostl C (2007) The implications of complexity for integrated resources management. Environ Model Softw 22(5):561–569

Roorda C (2012) Presentatie: Resultaten enquête transitiearena Gents Klimaatverbond. Drift, Stad Gent, Ghent

Roorda C, Maas S, Frantzeskaki N, Fortuin K (2012) Systems analysis for urban sustainability transitions. A joint understanding of complexity and dynamics. 3rd International Urban Research Symposium, Belo Horizonte

Rotmans J (1998) Methods for IA: the challenges and opportunities ahead. Environ Model Assess 3:155–179

Rotmans J, Loorbach D (2009) Complexity and transition management. J Ind Ecol 13:184–196

Rotmans J, Kemp R, van Asselt M (2001) More evolution than revolution: transition management in public policy. Foresight 3:15–31

Saravanan VS (2008) A systems approach to unravel complex water management institutions. Ecol Complex 5:202–215

Swart R, Raes F (2007) Making integration of adaptation and mitigation work: mainstreaming into sustainable development policies. Clim Pol 7:288–303

van der Brugge R (2009) Transition dynamics in social-ecological systems: the case of Dutch water management. Erasmus Universiteit Rotterdam, Rotterdam

van Sande I (2011) Presentatie Systeemanalyse, Transitie Arena 1. Stad Gent, departement Milieu, Ghent

Transition Management in Urban Neighbourhoods: The Case of Carnisse, Rotterdam, the Netherlands

Julia M. Wittmayer, Frank van Steenbergen, and Matthew Bach

Abstract The case study in this chapter marks a socially disadvantaged community in the South of Rotterdam, the Netherlands: Carnisse. Transition management was applied in Carnisse in the context of decreasing public service provisions, closing of public facilities, high unemployment and crumbling social cohesion to facilitate the self-organisation of inhabitants to address persistent sustainability problems. This chapter illustrates how the process implementation has been influenced by political, institutional and emotional factors related to the place-specific characteristics of urban neighbourhood transitions. While urban neighbourhoods mark opportunities for self-organisation – through for example emotional connectedness – the political playing field can provide support, or restrain it. The transition management application in Carnisse shows several advantages of applying transition management on the scale of an urban community. It provided a new perspective for the development of that community, increased social capital, supported reflexivity on roles and own behaviours and revealed opportunities for self-organisation.

Introduction

The city of Rotterdam has always been a port city and traditionally the port workers from the late 1800s and 1900s settled on the southern banks of the Maas River and closest to the city harbours area. With the gradual automation of shipping activity and the move of the port out of the city centre, the neighbourhoods south of the river began to face increasing hardship. By 2007, the national government listed Rotterdam Old-South, which includes Carnisse, as one of forty neighbourhoods requiring special attention, as they were found to exhibit a range of social, economic, and infrastructure problems (Ministry VROM 2007a, b). Within

J.M. Wittmayer (✉) · F. van Steenbergen · M. Bach
Dutch Research Institute For Transitions (DRIFT), Faculty of Social Sciences, Erasmus University Rotterdam, Rotterdam, The Netherlands
e-mail: wittmayer@drift.eur.nl; vansteenbergen@drift.eur.nl; bach@drift.eur.nl

N. Frantzeskaki et al. (eds.), *Co-creating Sustainable Urban Futures*, Future City 11, https://doi.org/10.1007/978-3-319-69273-9_8

Rotterdam South, the district municipality of Charlois, of which Carnisse is one of the eight districts, has shown an interest in addressing these issues, but has often been hampered in its efforts by government budget cuts resulting from the 2008 financial crisis as well as an expansion of market-thinking to welfare service provision. In Carnisse, specifically, old welfare structures were being dismantled and public facilities, such as community centres, were particularly hard-hit.

Carnisse with its close to 11.000 inhabitants is one of the forty most 'disadvantaged neighbourhoods' or 'special interest neighbourhoods' in the Netherlands (Ministry VROM 2007a, b).[1] While governmental interventions in Rotterdam have mainly focused on physical renewal, with the slinking budgets in the aftermath of the economic crisis in 2008, this has not been an option anymore (Kennisbank Platform31 2011). Slowly, the national neighbourhood approach changed and more emphasis was put on the role of inhabitants and citizens, who should become more active in addressing and solving problems in their living environment (Visitatiecommissie Wijkenaanpak 2011a, b). Illustrative is the following quote from a report on the current neighbourhood approach of the Ministry of Internal Affairs: "*We search for different relationships between governments, institutions and citizens. Attempts to give concrete shape to these, often still in rudimentary form, occur precisely in these neighbourhoods*" (Deetmans et al. 2011: 7). This reconceptualization of the role of inhabitants and citizens was accompanied with a revised role for the Ministry of the Interior and Kingdom Relations in the 40-Neighbourhood-Programme. They describe this change as follows: "*from active financial commitment linked with targets to a more facilitative role, acting on request in relation to what others do*" (Ministerie BZK 2014: 2). This shifting understanding is part of a broader discourse on the changing roles of citizens and governments in what came to be known as 'Big Society' in the UK (Kisby 2010; Ransome 2011) and 'participation society' in the Netherlands (Putters 2014; Tonkens 2014).

It was against this backdrop that a transition management process was started in Carnisse in the year 2011. The process had the overall aim to empower the local community to live more sustainably: due to its focus on communities it was referred to as community arena. This application was financed and carried out through a European research project, InContext,[2] that sought to understand the workings of inner and outer contexts in local transitions processes (Wittmayer et al. 2011a, b). In addition, it was co-financed by a national government-funded project aimed at

---

[1]Carnisse has the single lowest average income to spend per year in Rotterdam (€ 23.300,– in 2014), it has a relatively old and neglected housing stock, there is a lot of mobility and migration streams in the neighborhood (approximately 55% of the people live less than 5 years in the neighborhood). Carnisse also scores low on different municipal indexes (in regards to safety, social cohesion, and housing).

[2]For more information on InContext, please visit www.incontext-fp7.eu or consult the following project documentation: Wittmayer et al. (2011a, b, 2012, 2013a, b, c). The descriptions of the process contained in this chapter build on this project documentation as well as on Wittmayer et al. 2014.

increasing the resilience of the neighbourhood and its communities, Veerkracht Carnisse.[3] In this chapter we outline the transition management process as well as its outcomes in Carnisse. In doing so, we highlight the suitability of such a process for experimentation with new actors, roles and relations as an aspect of addressing persistent problems.

Methodology

The transition management process in Carnisse was implemented through an action-research approach led by the first and second author. Both authors were involved in the neighbourhood through the projects outlined above in the period from 2011 to 2015. During the community arena process outlined in this chapter, they conducted some 80 interviews with different foci, they were present in the neighbourhood both formally at public or other meetings and informally in the streets, and they organized 6 deliberative meetings, a broadening and an evaluation meeting as well as 6 action-oriented meetings. Next to interviews and meetings, we reviewed secondary literature about the neighbourhood and its broader embedding in Dutch neighbourhood policies.

Arena Process

Due to several waves of popularity in neighbourhood policy discourses (Uitermark 2003) there is a long tradition of interventions aimed at addressing living conditions on diverse topics like safety, clean streets, social cohesion projects, cultural diversity projects, poverty-reduction programmes, etc. Therefore, residents in neighbourhoods like Carnisse are rather familiar with these interventions and have increasingly grown weary of them (Van Steenbergen and Wittmayer 2012). A cultivated mistrust towards outsiders (like entrepreneurs, professionals, civil servants, researchers, etc.) is combined with a critique of top-down approaches that stigmatize the neighbourhood and its residents, which have become part of the problem that these approaches actually want to address.

At the same time, there is an increasing trend of self-organized and bottom-up initiatives by local communities and social entrepreneurs in The Netherlands (Hajer 2011). However, this trend is often incorporated in the dominant framing of the ‘participation society’, a withdrawal of the state and dismantling of welfare structures (see introduction). This combination of discourses leads to the following

---

[3]For more information on Veerkracht, please visit www.veerkrachtcarnisse.nl or consult the following project documentation: Wittmayer et al. (2014), Van Steenbergen et al. (2013), Van Steenbergen et al. (2013), and Van Steenbergen and Wittmayer (2015).

challenge: in areas where the level of self-organization among local communities is perceived as the lowest, the demand for self-organization to tackle multiple systemic problems becomes the highest (i.e. the more extensive and complex the challenges, the higher the demand for self-organization and self-resolving capacities). In addressing this challenge, the public sector is increasingly looking for innovative modes of governance. Modes where the formation and maintenance of networks aimed at sustainable futures and resilient communities were fostered. It was in this search that researchers were asked to experiment with transition management in Carnisse. By doing action research and identifying new actors, roles and relations, the suitability of transition management for addressing persistent problems at the scale of an urban neighbourhood was explored.

Setting the Scene for Transition Management

The transition management process in Carnisse was set in a rather interesting institutional setting. As the process was partly financed by an EU-project (InContext 2010–2013), the responsible researchers were quite free in the actual ways of application. Before they started, they developed a methodology for implementing transition management in communities (the community arena), which integrates backcasting and transition management with aspects of social learning (Wittmayer et al. 2011a). The actual application however was also determined by the co-financing, which came through a national policy programme for Rotterdam-South. It was through their administering role for the national funding for the project 'Veerkracht Carnisse'[4] that a closer collaboration with the local government took place. In this collaboration, the open nature of both process and outcome led to a certain amount of scepticism by local policy makers (What is new or different? What will the process deliver in concrete results?). Other barriers in the process of getting the Veerkracht project approved were the bureaucratic accountability relations (e.g. conflicting interests) and the high fluctuation amongst policy makers.

Against the backdrop of such an institutional setting, a long history of policy efforts and institutional participatory processes have, on the whole, left inhabitants weary of 'outside' involvement. Such distrust is also linked to these projects' tendency to portray the neighbourhood as disadvantaged, an image with which many locals are frustrated and in which they do not recognise themselves. Along

[4]The project 'Veerkracht Carnisse' (2011–2015) consisted of a consortium of four partners, all of whom had been involved and worked in their respective domains and with their respective approaches in other neighbourhoods of Rotterdam South. They joined forces based on their shared perspective that neighbourhood development should be more holistic and based on the power of individuals and communities. Due to their non-conventional approaches, it took a negotiation and lobbying process of almost four years before the project was accorded and could start in September 2011. Veerkracht Carnisse translates from Dutch as: Resilience Carnisse.

with the budget cuts, changing responsibilities of government and citizens as well as the erosion of old welfare structures outlined above, this proofed a challenging context to frame and start a community arena.

Therefore, the overall process design was partly done in a participatory and interactive meeting in November 2011 with five inhabitants and professionals in Carnisse who were motivated to think along about the future of the neighbourhood. During this meeting, the participants and researchers defined the overall aim of the transition management process as "*supporting and stimulating inhabitants to shape and take ownership of the future of their neighbourhood and formulate desired (government) activities*" (Wittmayer et al. 2014: 471). The main focus however was on discussing the methodology developed as part of the InContext project (Wittmayer et al. 2011b) and its contextualisation for Carnisse. Based on the meeting, the process design was adjusted (see Fig. 1) to allow for having a deliberative process in parallel with a more action-oriented process. The former was open to invited participants only and focused on developing a shared problem description, futures images and possible pathways ('thinking' – sections "Framing the transition challenge", "Envisioning a sustainable neighbourhood", "Reconnecting short/long-term", "engaging and anchoring"). The latter was open to everyone and focused on experimenting ('doing' section "Getting into action"), mainly the re-opening of a closed-down community centre. Eventually, these two processes were linked by taking the community centre as a symbol of the current and possible future state of Carnisse.

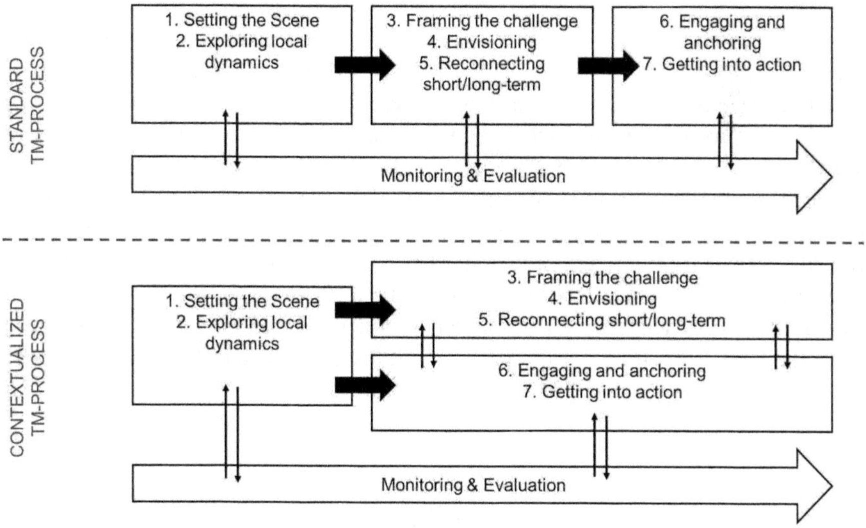

Fig. 1 Adapted process design (Source: Wittmayer et al. 2013b: 19). (The methodology for the community arena did number and name the process steps differently than is done in this book. The differentiation was made between (1) preparation and exploration, (2) problem structuring and envisioning, (3) backcasting and agenda building, (4) experimentation and implementation and (5) monitoring and evaluating)

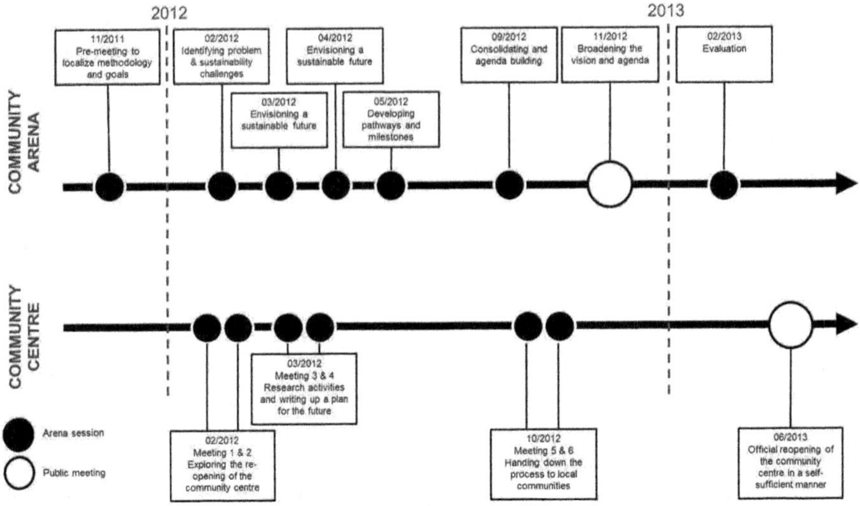

Fig. 2 Adapted overview of community arena process in Carnisse (Source: Wittmayer et al. 2014)

Both processes were translated in several participatory arena sessions with local residents, entrepreneurs and professionals. Over a period of 2 years a total of 13 sessions were organized by the researchers. Also, contributions were made to two public meetings. For a complete overview of both the processes and the sessions see Fig. 2

Both processes were driven by a transition team consisting of two active members, i.e. action researchers (first and second author) plus six more people which were either researchers at a distance, or locally relevant partners of the Veerkracht project. Specific activities were supported also by Master students or student assistants. As such this process can be considered as driven by the action researchers who took on and experimented with several roles in the process, such as process facilitator, change agent, knowledge broker, reflective scientist and self-reflexive scientist (Wittmayer and Schäpke 2014). Not being the traditional reflective scientist only, also led to discussions and a following negotiation of the actual role of the researchers during the above mentioned pre-meeting. The group agreed upon the term ‘activating researcher’ to mark the difference between the activities that the researchers engaged in, in relation to those of other researchers.

Exploring Local Dynamics: The System and Actor Analyses

From August 2011 to February 2012, the transition team worked on the system and actor analyses.

System Analysis

The system analysis was based on desk research, about forty interviews, fieldnotes of several formal meetings with the neighbourhood, as well as numerous informal conversations that often took place simply in the street. The analysis sought to elucidate both macro and micro-level issues, and to identify key topics for Carnisse. It therefore consists of two parts. The first looks at the wider influences affecting the neighbourhood of Carnisse, such as historical factors and the national and local policy environments. The second focuses more closely on the specific characteristics of Carnisse through five themes (as described in Wittmayer et al. 2011a).

(i) The first theme, "Green: population and surroundings", considers both the significant proportion of young persons (45% of the population is below 30) and the relative absence of green public spaces. There is, in fact, only one park, that highlights a lack of recreational spaces for residents. Taken together, these characteristics point to a high potential for development – Carnisse has the potential to become more attractive socially, environmentally and economically.

(ii) The second theme, "Social ties and bonding", revealed a lack of ties between inhabitants and their neighbourhood. This may be linked to the fact that Carnisse is often a starting point for those immigrating to the Netherlands and Rotterdam – they picture their time there as a temporary stop on the way to a better place. The constant flow of people both in and out of the area also makes it more difficult for social cohesion to develop and a sense of rootedness to emerge.

(iii) The third theme, "Housing: quality and stock", shows that on the whole the situation is far from ideal – nearly all apartments are built before the 1960s: they are small and due their thin walls very noisy. The exteriors are neglected, giving the neighbourhood a bleak appearance, and residents are primarily renting from large investors making few investments for the improvement of housing quality. It would seem that this situation is very much a symptom of Carnisse's wider characteristics.

(iv) The fourth theme, "Mobility and stability", highlights the lack of balance between its two components. There is a dual movement of influx by younger and less educated individuals and outflow of those who have become somewhat better-off. At the same time, mobility in social terms is severely restricted within the neighbourhood with few opportunities existing to move up the socio-economic ladder.

(v) The fifth theme, "Power(-less) policy", focuses on the problems faced by local policies and their makers. Many of the existing interventions have not been able to durably improve the situation of the neighbourhood. In fact, policy often appears bogged down in bureaucracy, conflicting interests, and a more general lack of attention to environmental issues. Yet, current policy and politics are dominant when it comes to directing developments in Carnisse:

e.g. historically they control the money (through e.g. subsidies), the dominant discourse and own the public land and facilities.

On the basis of the system analysis and subsequent deliberation in the community arena process, the researchers formulated the transition challenges for the neighbourhood in terms of developments (Van Steenbergen and Wittmayer 2012):

- From short term acting to opening up spaces for future thinking;
- From cramped and nostalgic to resilient and innovative, and
- From controlled risk limitation to learning from alternatives.

Actor Analysis

The interviews that served as a basis for the system analysis also were part of the actor analysis. Interviewees were identified through desk research (press articles, internet, policy documents, etc.), a snowballing method and involving local contacts in Carnisse and adjacent neighbourhoods. Rather than a complete network mapping, the actor analysis was more of a pragmatic nature focusing on identifying suitable interview partners for insights on the neighbourhood and participants for the deliberative arena process. The latter are referred to as 'frontrunners' in the transition management literature. Identifying these so-called frontrunners proved challenging, since not many residents and/or community members fitted the ideal-type profile (e.g. have large and diverse networks, have strategic skills, able to think in abstract ways, have a systems perspective, have access to resources, etc.).[5] Another difficulty was the increased fluctuation of professionals and residents in the neighbourhood (Carnisse is often perceived as an 'arrival district'). For instance, civil servants at the district municipality are changing very quickly, and also the professionals at e.g. welfare organisations change a lot (due to re-organisations, budget cuts and lay-offs), which leads to a certain erosion of institutional networks. A third difficulty was the framing of people's roles. For instance, when is a resident that is involved in a certain community project a resident, a volunteer, a professional or a social entrepreneur? Which role one takes up or gets assigned by others has consequences on how their interests and positions are perceived. The negotiation of these roles was a constant factor during interviews, observations and workshops.

Because of these difficulties, the notion of frontrunner was contextualised to mean inhabitants and professionals who are passionate about and active in the neighbourhood, feel the urgency for change, have new ideas and/or think about or engage in creative actions. Also the term frontrunner was transformed into a verb ('frontrunning'). This opened up the concept and thus making it more inclusive:

---

[5]For more information on the ideal-type profile and selection-criteria on frontrunners see Loorbach (2007) and Rotmans and Loorbach (2010).

everyone is able to 'run in front' on their own terms, in different roles, etc.[6] Based on these adaptions in the methodology, the researchers invited 15 frontrunners to join the deliberative process of the community arena – as a group they came from a diversity of networks and perspectives.

Framing the Transition Challenge

During the first deliberative meeting in February 2012, the researchers presented a first problem framing based on the above outlined system analysis. The community arena group engaged in an intensive discussion to adapt and amend this analysis leading to the following five main topics set the context of the 'rich and turbulent' history of the neighbourhood (see Textbox 1). These topics were closely linked to the developments of the closed-down community centre.

Box 1: Adapted Problem Framing (Source: Wittmayer et al. 2014: 471–472)

1. Struggle for survival: Carnisse is known as a working-class neighborhood, with poverty, low incomes and a small array of shops. The economic crisis and the accompanying government budget cuts left deep marks, and old welfare structures were being dismantled, such as public facilities, the local inhabitant organization, the welfare organization and the district municipality.
2. The individual and the collective: Increasing individualization in Carnisse allows for personal freedom, but also means that greeting each other has become an exception rather than the norm. While everybody seems to 'be busy with their own lives' (Interviewee A, 2011), there is a common longing for more cohesion and a 'shared neighborhood feeling' (Interviewee B, 2011). This goes beyond individuals and includes a longing for more synergy and cross-pollination between institutional actors.
3. Diversity: While the neighborhood is diverse in some regards – hosting about 170 nationalities, many different official churches and a variety of worldviews –, it is less so in terms of housing stock, street scenes, public space and shops. Many inhabitants expressed their frustration with the negative image of a 'deprived neighborhood' and were eager to relativize it by pointing to the many initiatives that were arising from within the community.

(continued)

[6]See Jhagroe and van Steenbergen (2014).

Box 1 (continued)

4. Connectedness: Carnisse shows relatively high degrees of migration. The young, poorly educated and newly arrived immigrants move in, and the relatively better off move out (usually starting families). This constant flow of people hinders bonding between people and the laying down of roots; Carnisse is seen as a transit station toward a better living environment. There is, however, a stable core of people with a nostalgic sense of the past, and places such as schools, churches and community centers that support the establishment of bonds.
5. Public space: The quality of the housing stock in Carnisse is poor, which is related to the high degree of private ownership by large investors. The uncared-for exteriors give the neighborhood a desolate look, and the old interiors exacerbate social and economic problems. The public space is neglected and unappealing (e.g., there are few parks or green spaces), and many complain about it.

Envisioning a Sustainable Neighbourhood

As outlined in the beginning, in parallel to this deliberative process, the researchers were also engaged in a more action-oriented process which focused on the re-opening of a community centre (see section "Getting into Action" for an outline on this). A number of people in the neighbourhood had heard about this action-oriented process and were concerned about the developments of the community centre. Thinking that the researchers would have a meeting about the latter, they were turning up to the following meeting which was initially dedicated to developing a future vision for Carnisse. Therefore, the researchers redesigned the meeting to critically discuss and analyse the role of the community centre for the inhabitants including their emotions, needs and relations with regard to it.

Only the third meeting was then focusing on drawing up a vision for the neighbourhood in 2030 in which the community centre played an important role. The vision was called 'Blossoming Carnisse' and tells the future of the neighbourhood along six visionary images:

1. To living with each other (topic of social cohesion)
2. To a green sustainable oasis (topic of ecology)
3. To diverse housing styles (topic of housing)
4. To a local social economy (topic of economy)
5. To places for everybody (topic of inclusiveness and social infrastructure)
6. To working together for blossoming (topic of democracy and claiming rights).

Fig. 3 Impressions of the illustration of two visionary images

The vision focused much on issues of social cohesion and living together, as well as public and private spaces. Different economic and collaborative practices were enabling factors. While the six images sound rather tame and appealing, they can be considered 'radical' within the context of Carnisse as they countered the dominant policy image of the neighbourhood and put the inhabitants in the driving seat of the development of their neighbourhood – rather than waiting for policy interventions. The vision as such as well as the process thus contributed to creating an image of the inhabitants as change-agents who made their voice heard.

Reconnecting Short/Long-Term

The researchers consolidated the input of the third meeting into text and images of the vision, which were discussed and agreed upon during the fourth meeting in May 2012. During this meeting, the focus was on developing pathways to connect the long-term vision with short-term activities through using a backcasting methodology. The participants split in three groups focusing on two visionary images each to explore possible pathways guided by a facilitator. For each visionary image, they discussed and agreed upon change elements, specific activities and key actors in three points in time reasoning back from 2030 to 2012. This led also to a connection of already existing initiatives and projects with the broader goals. Towards the end of the meeting, the Community Arena group discussed the next steps. The frontrunners/change agents expressed their interest in sharing and discussing the vision and emerging narrative about Carnisse's future with the broader neighbourhood (residents, entrepreneurs and professionals).

After the meeting, the researchers consolidated the input together with the vision into a transition narrative which for each of the six visionary images outlined a short description accompanied by an augmented photograph (see Fig. 3), as well as current related projects and a short fictive narrative or interview with a person in 2030 about how the visionary image would have been realized.

Engaging and Anchoring

An opportunity to organize a neighbourhood conference in collaboration with the district municipality as part of their official community forum emerged in November 2012. As such, the presentation of the vision and the ensuing discussion were just part of the agenda for that evening. The collaborative preparation as well as the evening was however, a first step into confronting and connecting policy and inhabitants' perspectives in Carnisse. There was a special invitation including an announcement of the vision and according to the director of the community organisation, twice as many people showed up than at previous events of this kind.

For the group the goal of the evening was to present the vision – while the researchers framed the overall process, each visionary image was adopted by one frontrunner/change agent and presented from within the audience. This presentation was followed by a rather fierce discussion revolving around the need for necessary short-term action (such as cleaning up pavements) rather than delving into far-away visions. The question of who was to implement activities leading to a realization of the vision was discussed. This led to a second part of the programme, which was to showcase all the activities, initiatives and projects that were already ongoing in Carnisse. This was to show that a lot is already happening within the neighbourhood by different actors to take steps towards the vision – even if not done so intentionally.

The discussion amongst the audience showed that it is difficult to change engrained patterns of behaviour about the expectations and behaviour between government and citizen. While the latter were used to being asked for their opinion – rather than being asked to take an active role, the former was used to citizens 'complaining' rather than taking initiative and showcasing strategic directions for the neighbourhood.

Getting into Action

As outlined earlier, this deliberative process was organised in parallel to a more action-oriented process, which focused on the re-opening of a community centre. The closure of the centre in January 2012 had been a consequence of the bankruptcy of the local welfare organization after having lost a tender for welfare work in Carnisse. The community centre had been run for decades by this welfare organization and offers several different facilities and rooms, such as a coffee house, kitchen, gym, and meeting rooms, which were used by primary schools a kindergarten and a welfare organization.

During the 'Exploring local dynamics' phase, many inhabitants and professionals had indicated their concerns about the closure in their interviews.

The researchers invited them for a meeting in February 2012 to explore the current state of knowledge and activities as well as intentions related to these concerns. The meeting led to the formation of a local action group aimed at the re-opening of the centre under community self-management. Over the course of four initial meetings as well as bilateral informal contact, the group drew up a business plan, collected more than 300 signatures through a petition, came to understand the ownership structure, and searched for contact with or were approached by different representatives of the district municipality, the welfare organization and the city municipality.

In October 2012, after two more broad meetings the researchers withdrew into a reflexive and monitoring role. Not wanting to wait for too long, the group had reopened the community centre 'unofficially' (a tolerated way of squatting). This was also the time when the Municipality of Rotterdam had accepted ownership of the building and had become a party with which to negotiate for a rental agreement.[7] A social entrepreneur had volunteered to support the community members in setting up the daily management and operation of the centre. Four working groups dedicated time to work on financial questions, a management construction, and an inventory of practical daily tasks and of recruiting potential volunteers. After initial enthusiasm about her role, the social entrepreneur was asked to leave, as there were suspicions of her acting out of self-interest. At the beginning of 2013, a foundation was established as a legal entity, which reopened the community centre officially in June 2013. Since then, the foundation has changed once and the board of this second foundation as well due to internal insurmountable disagreements of board members. The board of the foundation, fulfils most of the daily tasks supported by other volunteers and continues the dialogue with the municipalities. The latter mainly revolved around the legitimacy of the foundation as well as the actual rental sum to be paid and continued up to 2016.

During the process, one of the burning issues had been the differentiation between volunteers and 'professionals'. The district municipality was sceptical about working with professionals due to the perception that they only acted out of self-interest. In the same vein, the inhabitants as well as the social entrepreneur were convinced that only community members and no professionals should be involved in the process of setting up the centre. However, one of the driving forces of the local action group had been a professional working at a local school in Carnisse. This led to a constant negotiation of roles and corresponding positions, legitimacy and interests, which in turn led to several (unresolved) conflicts within the foundation, between volunteers and with the district municipality.

[7]At first this was refused due to the ownership structure (which is referred to as 'Rotterdam construction'), where the owner of the building is a different party (in this case the welfare organization) than the owner of the ground (in this case the Municipality of Rotterdam). And to make it more complicated the supervisor of the building was the district municipality Charlois.

Evaluation and Monitoring

In February 2013, an evaluation meeting took place where the participants of both, the arena process and the experiment, evaluated the process and the outcomes and formulated future ambitions. Just before the broadening event in November 2012, monitoring interviews were held with the core group of participants. A total of seven interviews were conducted in person focusing on the process, the group and the individual level as well as the future outlook. In February 2013 the evaluation meeting was held focusing on the current situation and future outlook, as well as aspects of empowerment, sustainability and pro-social behaviour.

During the evaluation meeting, the group expressed their interest in making the vision known to more people within the neighbourhood and they developed a strategy for doing so. They expressed their pride for the vision and felt ownership of it. Different participants aimed to approach different target groups, e.g. youth, churches, entrepreneurs. They also agreed upon organizing a network day in June 2013 where the vision should be central together with the actions that are employed to realize it and the start of a new icon-project. The researchers could support this network day as part of the co-funding that they received for the 'Veerkracht Carnisse'-project, which allowed for involvement for another two years.

Results

To discuss the results of the process, we focus on outlining the more tangible results as well as the outcomes of the deliberative and experimentation processes. When starting the process, the focus was on enhancing the transformative potential of the community and to empower individuals by addressing the question of how life in Carnisse will be in the year 2030. Throughout the deliberative process, from September 2011 to February 2013, around 150 inhabitants, policy officers, politicians and other professionals became involved either through interviews, by taking part in the visioning or experimenting process or through the public event. Tangible outcomes are the vision document 'Blossoming Carnisse' which directly addresses the challenge of dispersed policy action by showing an overall direction and red line for possible developments in Carnisse. A year after the first public presentation, one of the directors of the district municipality acknowledged the strengths of the vision and its importance for the neighbourhood. However, the document did not have policy impact, although it inspired several community arena members to start experimenting with projects and engagements related to the ideas. A second tangible outcome is the re-opened community centre which offers countless activities for children and their parents, for elderly people, marginalized groups (ex-drug addicts, psychiatric patients, etc.) as well as for many different religious and ethnic groupings within Carnisse.

Concerning the deliberative process, its set-up was very different from earlier participation processes known to the inhabitants. The latter ones had focused on informing and consulting the public and were judged critically by residents. However, still local policy makers and inhabitants of Carnisse were sceptical whether the open-ended community arena process could deliver the tangible results they were seeking. While the latter seemed to matter less during the evaluation, participants of the community arena stated that they felt free in terms of putting topics on the agenda rather than having an agenda being imposed on them. This meant that the process could focus on topics they were interested in and felt for. Next to these feelings of empowerment (cf. Hölscher et al. chapter "Empowering Actors in Transition Management in and for Cities", this volume, Schäpke et al. 2017), the participants also indicated that they had learned about the state of the neighbourhood as well as specific skills (e.g. speaking in front of a larger audience, mobilizing others, fundraising). They also felt reinforced in the feeling that a group of inhabitants can make a difference although they do not have 'formal' power and that they could operate more independently from institutions. The latter specifically played a role in relation to the re-opening of the community centre.

While the experimentation culminated in the re-opening of the community centre in June 2013 and is running independent from the action research, the deliberative process stopped with the last officially organized evaluation meeting in February 2013. The participants have not met afterwards in the same constellation while they do continue to meet each other through their engagements for the neighbourhood. However, as outlined above, some of the participants experimented with the ideas discussed during the deliberative meetings and outlined in the vision 'Blossoming Carnisse'. The welfare organization, for example, supported inhabitants to initiate a local swop-shop. Another participant together with a group of professionals and the inhabitant's organization further developed the idea of an internship position through which the neighbourhood would attract new 'social work'-talent.

However, according to most of the participants the arena did not manage to develop sufficient tangible actions for people to make an impact (or that they were too optional/without obligations). Also, it was insufficiently connected to current policy and governmental structures. Most participants felt that the most important decision makers were absent (housing corporations, local municipality, etc.).

Conclusions

The community arena in Carnisse differs on a few aspects regarding the more regular transition management methodologies. First, the contextualization of the process was a notable feature. Playing into the local dynamics and intricacies proved to be challenging due to the specific problems of the neighbourhood such as the erosion of institutional and social networks and the sphere of mistrust and competition which were a result of cutbacks in public and private funding and a history of policy interventions from outside-actors. Another significant adaption was the process design adaptation with experiments running in parallel to the deliberative workshops. And a third adaptation was the opening of the frontrunner-concept as to include local change-makers and engaged residents without presupposing certain elitist standards.

This contextualization opens discussions on the place-based dimension of transitions. It shows us that transition processes do not operate in a vacuum, but that they are entangled in highly political playing fields. Next to this political dimension, there is also an emotional dimension to consider on this micro-scale of neighbourhoods. People have emotional ties to certain places and topics which they express during interviews and workshops. Also, when tied to a certain place the direct consequences and impact of actions or experiments is immediately visible and tangible. Transitions thus come alive on this scale. However, transition challenges are mainly tied to social challenges, rather than the more classical ecological issues of sustainability. In Carnisse the focus was more on social cohesion, equity, community engagement/empowerment, and democratic legitimacy. Sustainability was mainly framed by participants as something that is durable, as a desire for consistency over a longer period of time (in reference to the erosion of networks).

Since change became so tangible in Carnisse it also had consequences for the people involved. Talking and acting on change, often was connected to a reconsideration of one's own and each other's role(s) in the neighbourhood and/or communities. Also, the community arena was seen by the participants as a process that stimulated the notion that people could influence the current and future state of Carnisse. It helped the participants to get an overview of activities in Carnisse and to voice their perspectives on the state of Carnisse. The open agenda of the arena helped in getting these diverse perspectives on the table and out in the open. Transition processes on the micro-scale of neighbourhoods are thus related to discussions on local democracy and legitimacy of which type of change is needed for the system under debate. The reopening of the community centre led to new resources, namely new social capital (e.g. ties and networks of engaged residents and volunteers) and symbolic capital (e.g. the group became a powerful actor in the institutional network of Carnisse).

In sum, the process encouraged local residents and members of communities to see themselves as change-agents and act like them. And while doing so they are actually challenging the engrained expectations and behaviours. For instance, the notion that top-down policies and interventions of the status quo are needed to fix the problems of the neighbourhood. These taken for granted assumptions were intensively discussed and thus the transition management process opened up more fundamental discussions in regard to the type of change Carnisse is confronted with, the type of change that is needed and also whose perspective of change is dominant and/or is imposed. Thus, the transition management process predominantly created awareness on the different scenarios of change and the role(s) one takes in these scenarios and in doing so further explicated the search for new modes of governance at the local scale of urban neighbourhoods.

Study Questions

From local to global: is the example (process and/or content) of Carnisse transferable to other city parts in the Netherlands, European cities or even the world?

In how far is 'the local' (e.g. an urban neighbourhood) fitting for transitions challenges which mostly play out at a larger system level?

How far does one go in making adaptations to a transition management process in order to play into the local context? And how does one tap/play into the local (power) dynamics? In how far does one risk that a process loses its transitional (and radical) roots?

In how far are issues of democracy addressed in transition studies and transition management processes? How does one cope with (the interplay of) hegemonic and marginal voices and positions in transition management processes?

How can the sense of agency be elaborated in transition studies and transition management processes? How does this tap into the discussions in this chapter relating to issues of roles and empowerment?

References

Deetmans W, van der Lans J, Scherpenisse R (2011) Doorzetten en Loslaten. Toekomst van de Wijkenaanpak. Deel 1: Eindrapportage. Den Haag

Hajer M (2011) De energieke samenleving. Op zoek naar een sturingsfilosofie voor een schone economie. Planbureau voor de Leefomgeving, Den Haag

Jhagroe S, van Steenbergen F (2014) Van koploper naar koplopen: een pleidooi voor inclusief leiderschap. DRIFT-essay 2014.04. Erasmus University Rotterdam, Rotterdam

Kennisbank Platform 31 (2011) Problematiek Rotterdam-Zuid wordt nationale opgave. Online at: http://kennisbank.platform31.nl/pages/21215/Nieuws/Problematiek-Rott (accessed 14 May 2015)

Kisby B (2010) The big society: power to the people? Polit Q 81(4):484–491

Loorbach D (2007) Transition management: new mode of governance for sustainable development. Erasmus University Rotterdam, Rotterdam

Ministry VROM (2007a) Actieplan Krachtwijken. Van Aandachtswijk naar Krachtwijk. The Hague, Ministry VROM

Ministry VROM (2007b) Kamerbrief - Wijkenselectie ten behoeve van wijkenaanpak. DGW/S2007033161

Ministerie BZK (2014) Voortgangsbrief Wijkenaanpak 2014. Kenmerk: 2014–0000550800. Den Haag, Ministerie BZK

Putters K (2014) Rijk geschakeerd. Op weg naar de participatiesamenleving. Sociaal en Cultureel Planbureau, Den Haag

Ransome P (2011) 'The Big Society' fact or fiction? - a sociological critique. Sociol Res Online 16 (2):18

Rotmans, J. Loorbach, D. (2010). Research into the governance of transitions: a framework for transition management. In Grin J, Rotmans J,. Schot, J. Transitions to sustainability development. New directions in the study of long term transformative change. 140–160). New York: Routledge

Schäpke N, Omann I, Wittmayer JM, van Steenbergen F, Mock M (2017) Linking transitions to sustainability: a study of the societal effects of transition management. Sustainability 9(5):737

Tonkens E (2014) Vijf misvattingen over de participatiesamenleving. Afscheidsrede. University of Amsterdam, Amsterdam

Uitermark J (2003) Social mixing and the management of disadvantaged neighbourhoods: the Dutch policy of urban restructuring revisited. Urban Stud 40(3):531–549

Van Steenbergen F, Wittmayer J (2012) Carnisse in transitie? Een verkenning van het verleden, het heden en de toekomst van een Rotterdamse wijk. DRIFT-uitgave. E 2012.05

Van Steenbergen F, Wittmayer J (2015) Veerkracht Carnisse, in De stad kennen, de stad maken. Online boek: http://www.kkshetlevendeboek.nl/boek

Van Steenbergen F, Wittmayer JM, Loorbach D van Hoop J (2013) Denken vanuit Doen. Monitoringsrapport Veerkracht 2011–2013. DRIFT, Erasmus University Rotterdam, Rotterdam

Wittmayer JM, Schäpke N (2014) Action, research and participation: roles of researchers in sustainability transitions. Sustain Sci 9(4):483–496. https://doi.org/10.1007/s11625-014-0258-4

Wittmayer JM, van Steenbergen F, Quist J, Loorbach D Hoogland C (2011a) The community arena: a co-creation tool for sustainable behaviour by local communities. Methodological guidelines. Deliverable 4.1, InContext: EU ENV.2010.4.2.3-1 grant agreement n° 265191

Wittmayer JM, van Steenbergen F, Bohunovsky L, Baasch S, Quist J, Loorbach D Hoogland C (2011b) Pilot projects getting started. Year 1 Status Report. Deliverable 4.2, InContext: EU ENV.2010.4.2.3-1 grant agreement n° 265191

Wittmayer JM, van Steenbergen F, Baasch S, Feiner G, Omann I, Quist J, Loorbach D (2012) Pilot projects on a roll. Year 2 pilot specific reports. Deliverable 4.3. InContext: EU ENV.2010.4.2.3-1 grant agreement n° 265191

Wittmayer JM, Schäpke N, Feiner G, Piotrowski R Baasch S (2013a) Action research for sustainability. Reflections on transition management in practice. Research Brief/Deliverable 5.2. InContext: EU ENV.2010.4.2.3-1 grant agreement n° 265191

Wittmayer JM, van Steenbergen F, Baasch S, Feiner G, Mock M, Omann I (2013b) Pilot projects rounding up. Year 3 Pilot-specific report. Deliverable 4.4. InContext: EU ENV.2010.4.2.3–1 grant agreement n° 265191

Wittmayer JM, Mock M, van Steenbergen F, Baasch S, Omann I, Schäpke N (2013c) Taking stock – three years of addressing societal challenges on community level through action research Pilot specific synthesis report. Deliverable 4.5. InContext: EU ENV.2010.4.2.3-1 grant agreement n° 265191

Wittmayer JM, Roorda C, van Steenbergen F (2014) Governing urban sustainability transitions – inspiring examples. DRIFT, Rotterdam, pp 24–29. 10.13140/RG.2.1.2367.7606

A German Experience: The Challenges of Mediating 'Ideal-Type' Transition Management in Ludwigsburg

Katharina Hölscher and Julia M. Wittmayer

Abstract When the city of Ludwigsburg in Germany experimented with the transition management approach, it had already formulated ambitious sustainability strategies and programmes, institutionalised sustainability in a cross-sectional city department and established a variety of participatory processes to collect citizen input for strategic planning. In this chapter, we reflect on the transition management process in Ludwigsburg and its results against the backdrop of how 'ideal-type' transition management got mediated in a context with well-developed sustainability strategies and participatory planning processes. The case of transition management in Ludwigsburg reveals several issues that relate to a questioning of the distinctiveness of transition management as co-creation process, its 'added value' and its fit with existing policies and planning processes. The process outcomes might be an indication of challenges in that regard: the results were treated as 'input' to the city government's plans, were not able to inspire 'transformative' thinking, did not result in concrete actions and, finally, roles and relationships of the city government vis-à-vis the citizens were largely re-confirmed. Nevertheless, the results do include new ideas about the future of Ludwigsburg, build on a systemic perspective and the experience with the new methodology triggered the involved policy officers to reflect on the city government's current inflexible and narrow approaches to participation in city planning.

Introduction

Ludwigsburg is a middle-sized town with about 87,000 inhabitants located in an economically prosperous region close to Stuttgart in the South of Germany. Ludwigsburg's economy is based on automotive supply, engineering, finance, software development and media. As Germany's largest baroque city it is

K. Hölscher (✉) · J.M. Wittmayer
Dutch Research Institute For Transitions (DRIFT), Faculty of Social Sciences, Erasmus University, Rotterdam, The Netherlands
e-mail: holscher@drift.eur.nl; wittmayer@drift.eur.nl

N. Frantzeskaki et al. (eds.), *Co-creating Sustainable Urban Futures*, Future City 11, https://doi.org/10.1007/978-3-319-69273-9_9

characterised by its baroque outlook including three castles. What made Ludwigsburg interesting for applying transition management were the high ambitions and activities of the city government in promoting sustainability and CO_2-reduction and the efforts to incorporate citizen participation into planning processes.

Ludwigsburg has signed the Covenant of Mayors and initiated a Local Agenda 21 process in 2001. In 2006, Ludwigsburg published its Urban Development Concept that identifies principles and strategic objectives for urban development in diverse areas, including attractive living, mobility, co-existence of generations and nations, energy, cultural life and education (City of Ludwigsburg 2006). The strategic planning concept was developed in the course of a long-term participatory process involving 70 interviews and three-yearly Future Conferences to discuss relevant topics, targets and strategies. The city has received several awards for its efforts, such as the German Sustainability Award 2014 for exemplary administration, council and civil society.

An Energy Master Plan was published in 2011 as part of the Urban Development Concept to promote energy reduction, energy efficiency and renewable energy sources (IER 2011). This plan also included the consultation of experts, interest groups and citizens, for instance in form of an Energy Round Table. During the course of setting up this Energy Master Plan, the city government of Ludwigsburg recognised that it could not achieve its goals alone. For instance, the city government is only responsible for about 2% of the total energy use. It therefore searched for ways to involve actors from the city in implementing actions.

These sustainability strategies and participation processes are firmly embedded within the organisational structure of the city government. The Lord Mayor of Ludwigsburg was a key driving force for local sustainability and participation since his first election in 2003. In 2008, the Department for Sustainable Urban Development (DSUD) was established as a cross-sectional unit that is an immediate subordinate to the Lord Mayor. It thus possessed a rather powerful position, yet due to the high profile of the sustainability topic it also faced high work pressure while lacking internal acceptance by and linkages with other city departments. Ludwigsburg enacts its sustainability policies in the wider context of the national nuclear phase out and the German 'Energiewende' (the national energy transition). The state of Baden-Württemberg, where Ludwigsburg is located, is the only state that is governed by a coalition led by the Green Party.

Against the backdrop of this institutional and policy context, Ludwigsburg became part of a European project to decrease its CO_2 footprint: Through participation in the MUSIC-project – a collaboration of five cities and two research institutes – the city aimed to catalyse and mainstream carbon and energy reduction. One strategy was to implement a process methodology based on transition management to mobilise stakeholders to take action towards CO_2-reduction. Transition management has been proposed as a methodology to bring a diversity of actors together to co-create sustainability visions and pathways and to become active in their implementation through transition experiments (Wittmayer et al. chapter "Transition Management: Guiding Principles and Applications", this volume; Wittmayer et al. 2015; Loorbach 2010; Frantzeskaki et al. 2012). As such, it places great emphasis on mutual decisions, ownership, empowerment and innovative solutions and is usually framed as complementary to and distinct from other policy and planning processes. The main 'potentials' of transition management are that it can boost new ideas, facilitates the implementation of new practices and results in a questioning and redefinition of roles and relationships (Wittmayer and Loorbach 2016).

As outlined, sustainability planning in Ludwigsburg already included several participation processes. The goal of such participation was to collect input from citizens to city plans, while transition management builds on strong notions of co-creation and critical reflection. We focus in this chapter on how 'ideal-type' transition management got mediated in a context that shows a high degree of institutionalised participation and well-developed sustainability policies. We show how context-related issues in Ludwigsburg have influenced the framing, focus and design of the transition management implementation as collecting input and new ideas for the city's strategic plans.

In the following, we first briefly outline our methodology before describing the transition management process in Ludwigsburg in terms of how the different process phases have been implemented and what the challenges and results were (section "Transition management in Ludwigsburg: process implementation and results"). This provides the basis for a critical reflection on both process and results in consideration of how 'ideal-type' transition management is mediated when confronted with context-specific conditions (section "Discussion: 'Ideal-Type' transition management in context").

Methodology

The involvement of the authors was co-funded by the MUSIC-project (Mitigation in Urban Areas: Solutions for Innovative Cities, www.themusicproject.eu), an EU-Interreg IVb project running from mid-2010 until mid-2015. The second author was involved in a transdisciplinary engagement as transition researcher in Ludwigsburg from 2011 to 2014. Her tasks were to support policy makers of the city in applying transition thinking and organising a transition management approach fitting the local needs and context. This included explaining transition theory and transition management to the Ludwigsburg transition team, teaching and supporting them in making a system and an actor analysis as well as in conducting related interviews. The first author was involved in an evaluation of the transition management process in Ludwigsburg in 2012/2013. As part of these activities, we conducted 22 interviews with participants of the process and local policy officers, attended six transition team meetings, seven transition arena meetings and three follow up and broadening events. In describing the process, we also have made use of previous project documentation and publications (Hölscher 2013; Roorda and Wittmayer 2014).

Transition Management in Ludwigsburg: Process Implementation and Results

Transition management, applied as part of the MUSIC-project as outlined above, was operationalised into seven process phases that (mainly) take place in different actor settings (Table 1, see also Roorda et al. 2014; Wittmayer et al. chapter "Transition Management: Guiding Principles and Applications", this volume; Hölscher et al. chapter "Empowering Actors in Transition Management in and for Cities", this volume). A particular feature of the transition management process in the MUSIC-project was that the process was initiated and implemented by the city governments. In Ludwigsburg, the earlier mentioned DSUD had the lead role within the transition team. They were responsible for the process framing, embedding the process in the local context, selecting the participants for the transition arena and guiding the implementation of the different process phases. The core transition team consisted of two policy officers and rotating trainees from the DSUD, the head of the department of civic engagement, a transition researcher (i.e. second author) and an external facilitator.

Figure 1 provides an overview of the timeline of the transition management process in Ludwigsburg. Before the first transition arena, the framing of transition management in the Setting the scene-phase was important and time-consuming – lasting from January to November 2011. After a short system sketch and selection and invitation of actors to participate in the arena, a first transition arena kick-off meeting was held in November 2011. Six subsequent arena meetings took place between March and September 2012, involving at least ten core participants with diverse backgrounds. During these meetings, a vision and transition pathways were developed. Several dissemination activities were implemented to communicate the results of the process to the Lord Mayor and city council and within the city government. In addition, a cross-departmental working group on energy was set up. In the winter of 2012/2013, three broadening events took place where the vision and the transition pathways were presented to, discussed and further developed with other city actors. The working group 'local energy transition' was formed, consisting of arena participants and newly interested persons. However, it proved challenging to move into action and implement transition experiments. Also, no transition agenda that presents the vision and transition pathways had materialised.

In the following, we review the implementation and results of the respective process phases. We outline who was involved in each phase, what has been achieved and what were the main challenges and opportunities during each phase. We place particular attention to how the abundance of participatory planning processes and well-developed sustainability strategies in Ludwigsburg influenced the framing, positioning, perception and design of the process and its implementation.

Table 1 Overview of the transition management phases in and for cities

Phase	Activities	(Main) actor setting
1. Setting the scene	Formation of the transition team	Transition team
	Setting up the process plan	
	Creation of space within local government	
2. Exploring local dynamics	System and actor analysis	
3. Framing the transition challenge	Problem definition	Transition arena
4. Envisioning a sustainable city	Drawing up a transition vision	
5. Reconnecting long- and short-term	Developing transition pathways and a transition agenda	
6. Engaging and anchoring	Disseminating transition agenda and engaging new actors and networks	Transition network(s)
7. Getting into action	Implementation of transition experiments	

Adapted from: Roorda et al. (2014: 15)

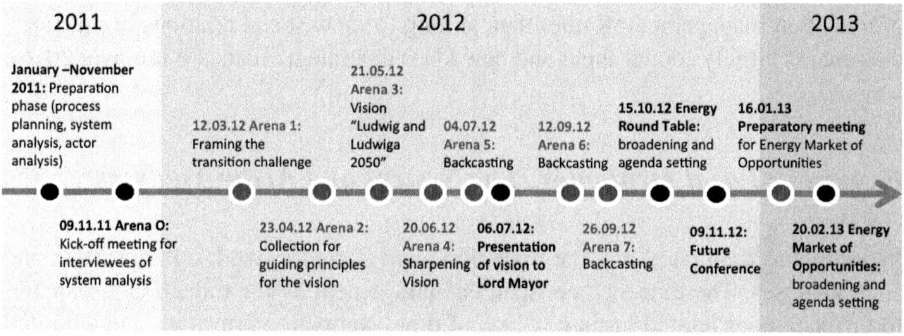

Fig. 1 Timeline of the arena process in Ludwigsburg (Adapted from: Hölscher 2013: 75)

Setting the Scene for Transition Management

The transition management process was started in Ludwigsburg with the formation of the transition team in January 2011. The initial phase of framing and positioning the process took almost a year and turned out to be very challenging for all actors involved. The implementation of the process was located within the DSUD, which faced high working pressure along with competing priorities. This was exacerbated by the fact that no official mandate was sought for the process, which left responsibilities and tasks distributed according to availability of personnel. This constellation of factors contributed to the long duration of the setting the scene-phase and had repercussions on all other phases.

The initial focus of the transition team was on discerning what transition management could mean for Ludwigsburg and how it added value and related to on-going participatory and sustainability-related planning processes. The city government found transition management to closely resemble its participatory planning processes such as the Future Conferences. The goal of these Future Conferences and related participatory planning processes is gaining input from citizens for devising policy and action plans.

The transition researcher supported and coached the involved policy officers in understanding and contextualising the transition approach. However, a structural lack of time within the DSUD along with competing priorities hindered a more profound engagement of the policy officers with the approach and its underpinnings. In addition with an action and task-oriented mind-set this led to a focus on implementing the activities of the process phases rather than the underlying heuristics of transition management. The transition management process was framed to complement existing participatory planning efforts. Namely, as a process in which citizens and professionals could deepen the energy theme in preparation of the next Future Conference in 2012. Its guiding question became: "How can the 'Energiewende' [energy transition] be achieved locally in Ludwigsburg?". As such, the implementation of the process was focused on one of the three potentials of transition management. Rather than aiming for new social relations or practices, the process largely sought input and new ideas to strategic plans (Wittmayer 2016).

Exploring Local Dynamics: The System and Actor Analyses

In parallel to the first phase, the transition team started to work on the system and actor analyses. These needed constant encouragement by the transition researcher. Main challenges related again to a lack of time, competing priorities, and a limited recognition of the relevance of the respective analyses as input for a participatory process. In addition, an institutional focus on solutions rather than problems inhibited the clear framing of transition challenges.

The system analysis was started with a short system sketch based on an internal system analysis workshop with policy officers from the city government, about 30 interviews with local actors (citizens, entrepreneurs and professionals) and input collected during the first arena meeting in November 2011. During the analysis of this input, it was challenging for the involved policy officers to switch from a focus on summarising their achievements towards identifying and acknowledging unsustainability problems. As a middle ground, the system analysis phrased the state of the various system elements in terms of questions (e.g. "How do we use energy?"). The themes that were identified in the analysis include high housing prices, citizen participation, demographic change, traffic and mobility, energy and air and noise pollution. This demonstrates how the system analysis sparked a broadening of the energy theme towards broader sustainability concerns that are interconnected. The system analysis took the form of a power point presentation, hence did not extend beyond a 'sketch' and lacked information on the status of the main system elements as well as their interrelations. For example, it failed to integrate existing reports on the state of energy in Ludwigsburg. This proved to be a detriment to the later process phases. For instance, participants sought to formulate a visionary target for increasing the green in the city by 2050, but did not know the current amount of green. As such, the system analysis mainly served as a good toehold for starting the discussion in the transition arena meetings.

Parallel to the system analysis, the transition team conducted an actor analysis between June and November 2011, also involving other officials from the DSUD. The long-list contained 100 actors and was subsequently analysed by dividing the actors according to background, gender and types of power. This analysis showed that uncommon groups such as craftsmen, priests or youngsters could still be added and that numerous innovative individuals were already known from previous public participation processes. The transition researcher's main task was to continuously sensitise the transition team to the fact that a transition management process is not about including representatives of every possible group, but rather to select frontrunners – i.e. innovative, intrinsically motivated individuals from diverse backgrounds who are dedicated to co-creating a sustainable energy provision in Ludwigsburg (Wittmayer et al. 2012; Wittmayer et al. chapter "Transition Manage ment: Guiding Principles and Applications", this volume; Hölscher et al. chapter "Empowering Actors in Transition Management in and for Cities", this volume). This helped to stimulate thinking beyond the usual suspects and triggered the team to invent ways to reach actors who have not been involved in earlier participation processes. The short-list consisted of 32 actors from civil society (13 actors), business (10 actors), intermediaries (7 actors) and knowledge institutions (2 actors). All were committed to the city and sustainability, and most of them were and still are active in local initiatives in the energy field. There was a diversity in backgrounds: including for example a movie director, a teacher, activists of the city's cycle path initiative and the city's solar initiative. Thirty interviews were conducted also in relation with the data gathering for the system analysis. All interviewees were invited to join the arena meetings; a core of ten participants became motivated to attend all the meetings – others joined sporadically or dropped out. This core group did not include entrepreneurs or young people.

Framing the Transition Challenge

During the kick-off meeting in November 2011, a first version of the system analysis was presented and additional input collected. The system analysis was presented to and discussed by the arena participants at the first official arena meeting in March 2012. The framing of the system analysis in terms of questions had marked a step forward on parts of the city government in that it presented itself somewhat vulnerable and took a step away from their focus on solutions. However, it also led the participants to seek answers to those questions rather than further digging into the problems. As a result, this process phase did not achieve a concrete framing of the transition challenge and did not arrive at a shared understanding of the problems as is envisioned in the 'ideal-type' transition management methodology.

Nevertheless, the framing process succeeded in broadening the problem focus from energy towards a more systemic view that considers, for instance, welfare and liveability, education, mobility and green. This enabled the participants to link their 'pet issues' to the overall energy theme and to identify multiple interdependencies that supported the later development of transition pathways (e.g. the connection between education and energy consumption enriched the view on societal conditions needed for an energy transition). One participant expressed her content that mobility was regarded "an energy topic", because it is "such a hot topic that it was often left out".

Envisioning the Energy Transition in Ludwigsburg

The following three arena meetings (arena 2, 3 and 4) between April and June 2012 were dedicated to visioning. After an initial meeting on the vision, some participants requested to add arena meetings for visioning to make the vision more concrete.

Based on the input of the arena meetings, the transition team proposed guiding principles for addressing the previously defined challenges Ludwigsburg is facing. Based on these principles, the arena participants developed storylines describing "One day in the life of Ludwig and Ludwiga in 2050" in working groups. A graphic designer visualised this plot, which centres on Ludwig showing his cousin Ludwiga, who is visiting from Brazil, how Ludwigsburg has achieved the energy transition locally (Fig. 2). This storyline is accompanied by a more descriptive vision that outlines detailed objectives for identified change topics: attractive Ludwigsburg/housing, energy, mobility, economy and consumption, awareness, culture and education, and green and circular economy.

The main value of the vision is that it nicely illustrates the storyline and the change topics, connecting different objectives for a sustainable future of Ludwigsburg. As such, it facilitates communication of the vision and providing inspiration. However, the vision fails to re-orient towards systemic change. While most arena participants and transition team members regarded the vision as containing interesting ideas, many criticised the focus on technological solutions such as electric cars or smart household appliances or the lack of ambition. The lack of re-orientation towards systemic change might be linked to the local culture that is known for its pragmatism: according to one policy officer from the transition team, it was not easy to get the participants in the right mind-set for the visioning exercise.

Fig. 2 Ludwig and Ludwiga explore Ludwigsburg in 2050: the energy plants in front of the castle (Source: Faber 2012)

Reconnecting Short- and Long-Term

In the subsequent three arena meetings, backcasting was conducted to reconnect the present situation outlined in the system analysis to the long-term vision through transition pathways that describe how the vision was to be achieved. In these meeting, the participants split into different groups around the change topics of the vision and developed a total of 21 pathways within five clusters. The pathways were developed by identifying milestones (activities or events) for 2035, 2025 and 2015 and the accompanying key actors. The pathways focus on the following themes: green in the city and circular economy; energy efficiency, renewable energy generation and energy awareness; civic participation, social interaction and cultural activities; mobility; and sustainable consumption. These relate to and encompass the key themes identified in the vision.

This backcasting process was comparatively long with three sessions spread over 3 months. Moreover, some groups met in addition to the scheduled arena meetings to further elaborate specific pathways. This was largely due to the premise of the policy officers in the transition team that the pathways should be fully based on the input by the participants with minimal processing by the transition team. The reasoning for this approach was to not distort the input (i.e. a framing as a classic civic participation process). This decision might also reflect the time constraints within the DSUD that resulted in limited time for processing inputs and preparing the arena meetings. Because the backcasting took a lot of time, the development of transition pathways at times felt cumbersome to the participants. Additionally, they partially had to work on pathways beyond their topics of interest and expertise. According to one participant, "sometimes one is really a bit floundering and simply makes something up, to put something there, pretty much. But one made an effort.

With some issues it was also indeed well comprehensible [. . .]. With other issues it was simply difficult." Altogether, the backcasting exercise might have contributed to a loss of momentum in the process and exacerbated the getting-into-action phase.

A striking observation from the developed pathways and corresponding actions is that the participants identified the city government as key actor responsible for working towards the vision in most of the pathways. For example, many pathways suggest a high amount of regulation. This is somewhat contradictory to the primal motivation of transition management, namely to stimulate reflection on roles to achieve a transition in a joint effort of all kinds of actors (Hölscher et al. chapter "Empowering Actors in Transition Management in and for Cities", this volume).

Engaging and Anchoring

After the arena meetings, the involved policy officers from the city government and the arena participants engaged in several activities to communicate the developed vision and transition pathways and to engage new actors in further elaborating the pathways and developing and implementing concrete actions. Many of these efforts hinged on the initiative of the city government – specifically the policy officers from the DSUD. Those neither had time nor priority to become a driving force. As a result, to date no official transition agenda has been published that effectively communicates the vision and pathways and could be used for creating common commitment.

A main achievement in anchoring the results of the process in Ludwigsburg was due to its connection to other on-going (participatory) planning processes and the ambition of achieving the energy transition locally. Three broadening events were held in the winter of 2012/2013 to present, discuss and further develop the draft transition agenda with a wider group of city actors: The Energy Round Table, the Future Conference 2012 and the Energy Market of Opportunities. The city government and arena participants invited a wide range of actors to these events. The first event resembled the participatory forums the city government had implemented earlier to collect input from citizens to support planning processes. Through providing input for the second, the Future Conference, the transition management process was connected to an on-going participatory planning process resulting in the Urban Development Concept. The Urban Development Concept is connected to formal decision-making processes, and ensured to some extent that (some of) the results would get anchored. The third, the Energy Market of Opportunities, established the working group 'local energy transition' and was specifically designed to identify concrete actions as well as actors willing to implement them. This working group consisted of some arena participants and participants from the broadening events. The group still meets to discuss ideas for concrete initiatives to facilitate the energy transition in Ludwigsburg. In a further effort to engage societal actors in the future of the city, the city government launched the website 'MeinLb.de'. On the website, actors can directly get into contact with the Mayor and the city government, suggest ideas and connect with other actors that are interested in developing and implementing projects. Also the working group 'local energy transition' is using this portal to coordinate and link efforts to the broader sustainability agenda of the city. Overall, the transition management process was not able to spur a self-organising network of actors. Rather, extensive coordination and engagement from the city government was needed.

The vision and two preliminary pathways were presented by some arena participants to the Lord Mayor as early as July 2012 (amid the arena meetings and the backcasting phase), and the city council had been informed about the process. This established linkages between the arena process and official decision-making processes and put the developed transition agenda on the agenda of the city – albeit with no clear consequences/outcomes. The involved policy officers also engaged in some communication efforts within the city government and invited colleagues from different departments and from the city council to join the Energy Round Table in October 2012, which some of them did. A major achievement in terms of structural change within the city government was the setting-up of a cross-departmental working group energy.

Getting into Action

The arena process led to the identification of several activities to pursue the vision for Ludwigsburg, such as a sustainability shopping basket pointing to the fact that biological and fair trade products are not necessarily more expensive than conventional ones, e-bike- and car-sharing initiatives, a book tree for exchanging used books and an energy contest to raise awareness for energy saving. As the arena meetings centred largely on developing a vision and transition pathways, the main input for activities came from the three broadening events and the work in the group 'local energy transition'.

The process of actually getting into action was cumbersome, however. The working group 'local energy transition' found it challenging to implement the diverse project ideas. For example, project ideas were put on hold because arena participants who initiated them moved away. Much hinged on the involvement and engagement of the city government, which was challenged to find actors willing to take the ideas up. The first follow-up activity was a sustainability exhibition in autumn 2013. This exhibition would have taken place also without the arena process, but the involved communications agency actively sought linkages to the established energy working-group. The city government pushed the connection of such 'external' projects and initiatives under the overall umbrella of the energy transition. A 'sustainable celebrity dinner' was planned to advertise regional, ecological and fair food.

Overall, the transition management process did not seem to succeed in sparking concrete additional actions of the involved arena participants. Rather, the city government was the driving force and facilitated project elaboration and connections with on-going initiatives. This might be due to the initial framing and perception of the process as 'just another' participation process to provide input to city planning. As a result, the participants did not seek activities additional to their previous engagement such as for example in the cycling initiative to realise the vision for Ludwigsburg. Relatedly, they regarded the city government as main actor in implementing actions and in pushing the follow-up process and supporting initiatives by other city actors. However, within the city government there was no time, priority or space to do so. The loss of momentum after the backcasting exercise might have created a gap between defining visions and strategies and bringing them into action.

Discussion: 'Ideal-Type' Transition Management in Context

The review of the transition management process and its results in Ludwigsburg reveals how context-specific conditions affect the way transition management gets implemented and, consequently, the efficacy of the suggested 'ideal-type' methodology. In this section, we discern some of the main issues that emerge in mediating 'ideal-type' transition management (as e.g. outlined by Wittmayer et al., chapter "Transition Management: Guiding Principles and Applications", this volume) in a context that shows a high degree of institutionalised participation, time pressure and well-developed sustainability policies. We discuss how these issues mediate the implementation of the transition management process and the intended results.

Allocating Time and Prioritising

Transition management requires time for and commitment to the implementation of the process phases and principles (Roorda et al. 2014) as well as a basic understanding of the transition approach. The individual process phases ask for thorough preparation, such as the positioning of the process in the respective context, the system analysis and the actor selection. Next to attention to these operational tasks, to engage with the principles of transition management – including creating space for co-creation, flexibility, long-term and radical thinking and learning – openness and attention is required. In the first place this concerns the transition team that designs and prepares the process.

In Ludwigsburg, the DSUD that was responsible for the implementation of the transition management process suffered from a structural lack of time, competing priorities and a lacking official mandate that would have clearly assigned responsibilities and time investments to specific individual officers. The transition team members very much focused on 'executing' the MUSIC-project, for which the city government received funding from the EU and of which the transition management process was part. As a result, only limited priority was given to the transition management process, its operational implementation and a thorough consideration of the underlying transition approach and process principles. As there was neither time nor priority to go in depth into the transition approach, the positioning and contribution of the transition management process in relation to the city's existing strategies, participatory planning processes and projects proved difficult. The process planning and the system analysis remained superficial and required constant encouragement. The unfinished system analysis could also not provide enough information (e.g. on existing amount of green in the city), which hindered the later process phases.

Throughout the implementation, the involved policy officers could only invest minimal time in preparing arena meetings, processing the results and following-up on the process outcomes. For example, an official finalised transition agenda document that communicates a shared direction and connects efforts has yet to materialise. Similarly, the lack of processing of the transition pathways prolonged the backcasting exercise and might have contributed to a loss of momentum before moving into action. However, the involved policy officers nevertheless succeeded in creating linkages with other processes and initiatives in the city, connections with political processes and agendas and in establishing new working groups. This demonstrates the efforts and achievements of the involved officials despite a lack of time and space.

Other experiences with transition management substantiate the need for time and priority to the process preparation and implementation. A review of Local Agenda 21 and transition management processes shows that participatory governance often faces constraints due to economic, human, time and knowledge resources (Wittmayer et al. 2015). In comparison, a transition management process in Ghent (Belgium, also as part of the MUSIC-project) faced another kind of time pressure: The process gained much attention and priority from the city government because it was linked to the local elections. On the one hand, this provided the involved policy officers with considerable time and space to engage in the system analysis and processing. On the other hand, the processes needed to finish within a certain time frame and was pressured to yield results, hence losing its openness (Hölscher et al. 2016a, b; Nevens and Roorda 2013; See Maas et al. chapter "Starting Up Transition Management: A Closer View on the Systems Analysis and How It Initiated Transformative Thinking in Ghent and Aberdeen Cities", this volume).

Differentiating the 'Transition Approach'

The operationalisation of transition management in phases is similar to existing participatory processes (Wittmayer et al. 2015). It is the underlying transition approach, which is based on systems thinking, that demands a different mind-set in the implementation. The ideal-type transition management methodology seeks to create space for the involved actors to enter a shared searching, learning and experimentation process (Loorbach 2010; Frantzeskaki et al. 2012; Wittmayer et al. chapter "Transition Management: Guiding Principles and Applications", this volume). It suggests several process phases to develop a common understanding of a sustainable future, give an impulse for local change through innovative strategies and concrete initiatives and to motivate actors to tackle challenges and seize opportunities for a sustainable city (Roorda et al. 2014). Transition management is nevertheless an intrinsically flexible approach, and its principles, process phases and instruments need to be contextualised and adapted to the specific context it is applied to (ibid.). Hence, it is critical to understand and discern the distinctiveness of transition management and what it means in the concrete context in which it is implemented.

At the start of the MUSIC-project in Ludwigsburg, there seemed little need for an alternative governance approach to formulate strategies and implement projects. The extensive participation processes to complement city planning activities appreciate the role that society at large (rather than policy alone) can play in achieving a more sustainable world. However, these processes largely resembled public consultation (i.e. giving input) rather than co-creation. The difference between these approaches and an ideal-type transition management process (e.g. the focus on redefining roles and developing innovative solutions) had not crystallised sufficiently through the implementation to neither the involved policy officers in the city government nor the arena participants.

Due to a lack of time and priority as outlined above, the involved policy officers did not sufficiently engage with the transition approach. They considered transition management as yet another participatory process to support the city government's planning efforts and were mainly interested in the different phases and activities. The process was framed as "the city needs your help" instead of "we are all in the same boat and want to co-create our future together". Over the course of the process implementation, however, the experience with the new methodology triggered the involved policy officers to reflect on its approaches, appreciate the openness of transition management and to take more systemic perspectives in strategies and projects.

It seems that the (initial) process framing also confirmed the expectation of the arena participants that they would embark on a 'conventional' participatory process. The division between the transition team being composed of policy officers and the transition arena of actors from other backgrounds might have additionally confirmed this expectation. As a result, the participants provided input to the city government's strategic plans. The formulated strategies assign most implementation responsibility to the city government or other public actors. While they testify ownership and a feeling of responsibility for the future of the city, this did not result in additional actions or responsibilities being taken by the arena participants.

Differentiating transition management essentially refers to a profound engagement with and contextualisation of the methodology's process-specific elements and principles. This builds on an appreciation of transition management's 'promises' and 'philosophy'. Other cases of transition management reflect similar challenges in positioning the process and conveying how it is different, therefore providing an added value. As indicated above, the case of Ghent was marked by a lack of experience with participatory processes, a culture of control, a need to yield quickly visible results and a high level of priority on the results of the transition management process. This restrained the openness of the involved policy officers towards co-creation and open-ended results (Hölscher et al. 2016a, b). Avelino (2009) describes the difficulty to adapt transition management to on-going projects, because project leaders were reluctant of applying the methodology and struggled with administrative hassle.

Embracing Complexity

'Systems thinking' is one of transition management's basic principles that underlie the above-described transition approach. The approach builds on a conceptualisation of complex adaptive systems and as such acknowledges complexity, feedbacks, persistence and radical change (Frantzeskaki et al. 2016, chapter "Introducing Sustainability Transitions' Thinking in Urban Contexts", this volume; Loorbach 2010). The design of the transition management process phases are derived from an understanding of how systems change (e.g. the criteria to invite frontrunners, Loorbach 2010; Hölscher et al. chapter "Empowering Actors in Transition Management in and for Cities", this volume). To deal with complexity, transition management as governance approach emphasises the need for flexibility and continuous learning (Loorbach et al. 2011).

Additionally, the process seeks to inspire a systemic perspective on the particular transition challenge that a transition management process addresses as well as the developed strategies and solutions. A systemic perspective enables to address the interrelated causes of a problem rather than its symptoms and to create synergies between different goals and needs and to align diverse sustainability initiatives in a city towards a common agenda. The individual phases and methods, such as the system analysis or the visioning exercise, support the development of a shared and systemic perspective of a problem and its solutions while being mindful of the complexity of problems and achieving sustainability (Franteskaki et al. 2012; Wittmayer et al. chapter "Transition Management: Guiding Principles and Applica tions", this volume).

In Ludwigsburg, there was already some systems thinking on the strategic city governmental levels, i.e. strategizing about sustainability in a broader sense that connects sustainability to different topics. For example, sustainability was (and still is) a priority of the Mayor who established the DSUD as a cross-sectional unit that connects (sustainability) to different departments and agendas. However, this approach was not translated into the operational level of project implementation, where the transition management process had to operate. On this level, project goals were identified from a rather narrow perspective. The question of the energy transition was framed solely from the perspective on energy rather than considering how it connects to, for example, consumer behaviour, mobility and food.

The system analysis and problem definition of the transition management process succeeded in opening up the initial energy theme towards social and environmental elements (e.g. culture, mobility, education, housing). They opened problem perspectives and aligned different backgrounds and interests under the theme of the energy transition. This systemic approach to the energy transition is reflected in the vision and variety of transition pathways. The draft transition agenda provides an umbrella for diverse strategies, goals and initiatives that promote systemic sustainability. As such, it enabled actors to link their activities to the overall framing of 'achieving the energy transition in Ludwigsburg together'. For example, a participant who was active in the bike paths initiative in Ludwigsburg before the process could link her engagement to the envisioned sustainability transition. Similarly, the involved policy officers connected the process and its outcomes to a diversity of existing strategies, projects and initiatives. However, the agenda was never made publicly available. This limited the extent to which it could act as a connector of different actors and initiatives in Ludwigsburg.

This systemic perspective was also taken up on a more institutional level of the city administration. The cross-departmental working group energy resonates the need to connect different, otherwise isolated city departments that work on different themes to create synergies between their goals and initiatives.

Other cases show similar experiences with regards to both the challenges and successes of promoting a systemic perspective. Some report on the positive value of especially a system analysis and heterogeneous group composition for broadening up the problem scope and creating awareness of complex interdependencies (Hölscher et al. 2016a, b; Maas et al. chapter "Starting Up Transition Management: A Closer View on the Systems Analysis and How It Initiated Transformative Thinking in Ghent and Aberdeen Cities", this volume). The recognition of complexity in both problems and (the search for) possible solutions can be perceived as a major contribution of transition management that is different from other participatory approaches.

Overcoming Entrenched Role Relationships

Transition management seeks to support and boost self-organising actor networks that co-create a sustainability transition by building collaborative transitions networks (Wittmayer and Loorbach 2016; Nevens and Roorda 2013). The arena participants are stimulated to reflect on their roles and identify how they could contribute to a sustainability transition. This implies that existing relationships and roles of actors are challenged: Who should be responsible for investing time and energy into organising service provisions and creating and disseminating sustainable practices? As such, ideal-typical transition management opens a way to question and experiment with alternative social relations, such as between local governments and citizens, or between citizens and businesses (Wittmayer and

Loorbach 2016; Wittmayer 2016). This builds on an open and co-creative process setting during the process implementation, in which decisions on process and content are taken following mutual deliberation and agreement, that promotes ownership and a feeling of shared responsibility (Hölscher et al. 2017; Avelino 2011: 64–66). The transition team who initiates and guides the transition management process implementation fulfils a central role in this regard. It decides who is invited, chooses facilitation techniques and sets the boundaries in which the interaction takes place (Hölscher et al. chapter "Empowering Actors in Transition Management in and for Cities", this volume; Wittmayer and Schäpke 2014; Wittmayer et al. 2012).

In Ludwigsburg, there was no attention for the possibility of the process to support the re-defining of social relations. The process was framed in a way that it sought new ideas and initiatives from the arena participants – as such approaching them as in other participatory planning processes and not as partners in the quest for a more sustainable city. As a result, the process reconfirmed rather than challenged deeply entrenched role relationships: The city government was an established leader in developing and implementing sustainability strategies to which citizens provide input through participatory processes. The rationale of transition management to engage in a joint searching and learning process and to identify and take up shared responsibilities did not receive attention and therefore did not sufficiently materialise.

The framing of the process as 'seeking input from citizens' did not convey the rationale of transition management to develop and take up a common direction. The developed system change perspective therefore did not reflect a shared commitment towards promoting sustainability in Ludwigsburg. Additionally, the process framing did not approach the arena participants as 'frontrunners', i.e. they were not given the feeling of being appreciated for their ideas and activities and of being able to carry change. They were also not significantly involved in the planning of the process or given the opportunity to provide input to it – nor did they show interest in this. Hence, to many participants the process did not represent an exclusive forum that would stimulate them to engage in 'something new', and many were already engaged in other forums that related to their specific topic of interest. As a result, roles and responsibilities remained unquestioned and the input from the arena process re-confirmed the city government as the main actor responsible for fostering sustainability in Ludwigsburg. A striking observation is that the level of trust between arena participants and involved policy officers was high from the beginning, which contributed to a positive atmosphere during the arena meetings. The participants appreciated being in a group of like-minded people and valued the engagement and motivation of the involved policy officers.

Experiences with other transition management processes highlight that the shift towards a co-creative process setting is demanding (Avelino 2009; Hölscher et al. 2016a; Hölscher et al. chapter "Empowering Actors in Transition Management in and for Cities", this volume, 2016b, 2017). It requires everybody involved in the process to "learn new skills, new attitudes, new behaviours, new ways of relating, and even a new language" (Randolph 2000:97; cf. Avelino 2011). This highlights again, the need for time and space to engage with the transition approach and the underlying principles. The implementation of transition management in Montreuil (France, also in the context of the MUSIC-project) stands out: The openness of the process and philosophical attitudes of the arena participants stimulated discussions on what the discerned challenges and the developed vision mean for different forms of governance and a (re-)distribution of roles and responsibilities between the government and citizens (Krauz 2016; Hölscher et al. 2017). This contributed to a shared commitment, in which the city government and citizens agreed to work together towards a better future for the city – e.g. in form of an institutionalised transition arena.

Conclusions

The case of transition management in Ludwigsburg reveals several issues through which an 'ideal-type' transition management process is mediated when being realised in a particular context. In Ludwigsburg, transition management was confronted with a context that already showed a high level of institutionalised participation processes to support city planning and well-developed sustainability strategies. Hence, the city government viewed itself – and rightfully so – as already implementing a successful approach to addressing sustainability. It was not looking for new inspirations in that regard. In combination with a high workload within the DSUD, which was responsible for the transition management process, these context conditions challenged the adaptability of transition management and the positioning and contribution of the methodology in that specific context. The outcomes might be an indication of this – though the experience was still recognised as valuable. The process nevertheless resulted in a systemic perspective in the transition agenda that provided input to the city's strategic sustainability plans, the creation of a network of likeminded people and cross-departmental connections.

In Ludwigsburg, the main question was about how to position transition management as complementary to other participatory processes that support the city's planning processes. The contribution of 'ideal-type' transition management lies in its emphasis of co-creation, re-defining of roles and responsibilities in governance processes such as transition management but also in the longer-term pursuit of sustainability transitions, and innovative solutions and approaches.

The analysis showed how the phases of transition management are mediated by context conditions such as expectations and preconceptions as well as lacking time and priority. The methodology then becomes disconnected from its basic assumptions and guiding principles that emphasise flexibility, learning, co-creation and long-term thinking. Such an understanding of the transition management process in Ludwigsburg can support the positioning and framing of the process and its results. In Ludwigsburg, the main rationale for the process was derived from the other participatory processes, i.e. to collect new ideas, while other 'potentials' of transition management, such as new initiatives and roles and relationships (cf. Wittmayer and Loorbach 2016), were not taken into consideration. In light of this context setting we might critically question whether transition management was actually the most suitable approach, or whether other approaches or adaptations might have been better suited to address the challenges and ambitions in Ludwigsburg.

Ultimately, the analysis raises the question about the extent transition management is actually different from 'conventional' participatory approaches – especially considering that it shares similar individual process phases (e.g. visioning, backcasting) – and how this difference can be (and needs to be) mediated in the specific context it is applied to. For example, though the transition management process in Ludwigsburg was implemented so as to collect new input, the engagement of a limited number of local frontrunners and the system analysis have been new process design features. These have contributed to an opening up of the problem focus and to an explicit consideration of radical change. Overall, the experience with the new methodology triggered the involved policy officers to reflect on their prior participatory approaches and they came to appreciate the openness of transition management.

References

Avelino F (2009) Empowerment and the challenge of applying transition management to ongoing projects. Policy Sci 42:369–390

Avelino F (2011) Power in transition. Empowering discourses on sustainability transitions. PhD thesis. Erasmus University of Rotterdam

City of Ludwigsburg (2006) City development concept. Shaping the future of our town together. Ludwigsburg

Faber J (2012) Ludwig und Ludwiga: Visionen und Wege. Wie schaffen die Energiewende in Ludwigsburg. Ein Besuch im Jahr 2050, presentation to the Lord Mayor, June 1, 2012, Ludwigsburg

Frantzeskaki N, Loorbach D, Meadowcroft J (2012) Governing societal transitions to sustainability. Int J Sustain Dev 15(1):19–36

Frantzeskaki N, Hölscher K, Bach M, Avelino F (2016) Introduction to sustainability transitions thinking. In: Frantzeskaki N, Hölscher K, Bach M, Avelino F (eds) Co-creating sustainable urban futures. A primer on applying transition management in cities. Springer, Tokyo

Hölscher K (2013) Seeds of change? Evaluating the transition arena for urban climate governance in Gent and Ludwigsburg. Master thesis. Utrecht University

Hölscher K, Avelino F, Wittmayer J (2016a) Empowering actors in transition management in and for cities. Understanding different roles in urban sustainability transitions. In: Frantzeskaki N, Hölscher K, Bach M, Avelino F (eds) Co-creating sustainable urban futures. A primer on applying transition management in cities. Springer, Tokyo

Hölscher K, Roorda C, Nevens F (2016b) Ghent: fostering a climate for transition. In: Loorbach D, Wittmayer JM, Shiroyama H, Fujino J, Mizuguchi S (eds) Governance of urban sustainability transitions. European and Asian experiences. Springer, Tokyo, pp 91–112

Hölscher K, Wittmayer J, Avelino F (2017) Opening up the transition arena: an analysis of (dis) empowerment of civil society actors in transition management in cities. Technol Forecast Soc Chang. Corrected Proof. https://doi.org/10.1016/j.techfore.2017.05.004

IER, Institut für Energiewirtschaft und Rationelle Energieanwendung (2011) Integriertes Klimaschutz- und Energiekonzept für Ludwigsburg. Ludwigsburg

Krauz A (2016) Transition management in Montreuil: towards perspectives of hybridisation between 'top-down' and 'bottom-up' transitions. In: Loorbach D, Wittmayer J, Shiroyama H, Fujino J, Mizuguchi S (eds) Governance of urban sustainability transitions. European and Asian experiences. Springer, Tokyo, pp 133–152

Loorbach D (2010) Transition management for sustainable development: a prescriptive, complexity-based governance framework. Governance 23(1):161–183

Loorbach D, Frantzeskaki N, Thissen WH (2011) A transition research perspective on governance for sustainability. In: Jaeger CC, Tàbara JD, Jäger J (eds) European research on sustainable development, volume 1: transformative science approaches for sustainability. Springer, Berlin, pp 73–90

Nevens F, Roorda C (2013) A climate of change: a transition approach for climate neutrality in the city of Ghent (Belgium). Sustain Cities Soc 10:112–121

Randolph WA (2000) Terms of empowerment/exemplars of prevention: toward a theory for community psychology. Am J Community Psychol 15(2):121–148

Roorda C, Wittmayer J (2014) Transition management in five European cities – an evaluation, DRIFT, Erasmus University Rotterdam, Rotterdam

Roorda C, Wittmayer J, Henneman P, Van Steenbergen F, Frantzeskaki N, Loorbach D (2014) Transition management in the urban context. Guidance manual. Drift, Erasmus University Rotterdam, Rotterdam

Wittmayer JM (2016) Insights and lessons for the governance of urban sustainability transitions. In: Loorbach D, Wittmayer J, Shiroyama H, Fujino J, Mizuguchi S (eds) Governance of urban sustainability transitions. European and Asian experiences. Springer, Tokyo, pp 157–173

Wittmayer JM (2016) Transition management, action research and actor roles: understanding local sustainability transitions. PhD thesis. Erasmus University Rotterdam, The Netherlands

Wittmayer JM, Loorbach D (2016) Governing transitions in cities: fostering alternative ideas, practices, and social relations through transition management. In: Loorbach D, Wittmayer J, Shiroyama H, Fujino J, Mizuguchi S (eds) Governance of urban sustainability transitions. European and Asian experiences. Springer, Tokyo, pp 13–32

Wittmayer JM, Schäpke N (2014) Action, research and participation: roles of researchers in sustainability transitions. Sustain Sci 9(4):483–496

Wittmayer J, Frantzeskaki N, van Steenbergen F, Roorda C, Henneman P (2012) Introducing a profile for urban change agents: experiences from five European cities, Drift Working Paper. Erasmus University Rotterdam, Creative Commons, Rotterdam

Wittmayer J, Roorda C, Van Steenbergen F (eds) (2014) Governing urban sustainability transitions – inspiring examples. DRIFT, Rotterdam. 10.13140/RG.2.1.2367.7606

Wittmayer JM, van Steenbergen F, Rok A, Roorda C (2015) Governing sustainability: a dialogue between local agenda 21 and transition management. Local Environ: Int J Justice Sustain. https://doi.org/10.1080/13549839.2015.1050658

Wittmayer J, van Steenbergen F, Frantzeskaki N, Bach M (2016) Transition management in urban context. In: Frantzeskaki N, Hölscher K, Bach M, Avelino F (eds) Co-creating sustainable urban futures. A primer on applying transition management in cities. Springer, Tokyo

Transition Management for Local Sustainability: A Case Study from La Botija Protected Area, San Marcos de Colón, Honduras

Giorgia Silvestri and Niki Frantzeskaki

Abstract In this chapter we present two methods that we employed to facilitate the exploration for environmental stewardship in La Botija protected area: transition management was the transdisciplinary methodology that connected a co-created sustainability vision with the co-development and mobilization of local action. The ecosystem services framework was employed to systematize knowledge about resource use and ecosystem degradation and identify links between malpractices and ecosystems' deterioration. The transition management process activated participants to realise the co-created transition agenda by forging new partnerships, enabling self-organisation in community groups and empowering existing community groups. A highlighted learning outcome is the recognised need to develop own knowledge and community capacity to overcome local challenges, deviating from the dependence on external aid.

Introduction

In this Chapter, we reflect on the adaptation and application of the transition management process in rural communities of 'La Botija' protected area in San Marcos de Colón, Honduras. The context of the investigated case is characterized by wicked problems, unsustainable lock-ins and a persistent maladaptive situation. Sustainability transitions studies propose that problems persisting over time require a fundamental change in the ways of organising (structures), ways of thinking (culture) and in the ways of doing (practices) in order to allow for sustainable solutions to be considered and take over (Frantzeskaki and Haan 2009; Grin et al. 2010). With this understanding, it is important not only to recognise the drivers of persistence – drivers that sustain a maladaptive persistent state or, as addressed in social-ecological research, a trap (Carpenter and Brock 2008) – but also to find

G. Silvestri (✉) · N. Frantzeskaki
Dutch Research Institute For Transitions (DRIFT), Faculty of Social Sciences, Erasmus University Rotterdam, Rotterdam, The Netherlands
e-mail: silvestri@drift.eur.nl; n.frantzeskaki@drift.eur.nl

N. Frantzeskaki et al. (eds.), *Co-creating Sustainable Urban Futures*, Future City 11, https://doi.org/10.1007/978-3-319-69273-9_10

ways for fundamentally reorganising the social and technological sub-systems as well as the human-ecosystem relations to steer clear from traps.

Transition Management represents an alternative governance approach for activating, strategizing and mobilising action to sustainability which seeks to support and condition the transformative change of socio-political landscapes, socio-technical practices while fostering social and policy learning (Frantzeskaki et al. 2012; Loorbach et al. 2015; Loorbach 2007; Loorbach and Rotmans 2010; Markard et al. 2012). With the aim to contribute to transformations towards a more sustainable use of resources as well as to increase collaborations and trust among local actors and institutions, an action research project has been developed by three research institutes and a NGO in La Botija protected area in San Marcos de Colón, Honduras. In this research project, transition management was deployed as a conceptual and operational framework to set up the transdisciplinary research architecture (Frantzeskaki et al. 2012).

The case of application of transition management in 'La Botija' area illustrates multiple challenges for adapting and implementing transition management in contexts characterised by persistent problems and by a 'gap' between civil society and governmental institutions. Particularly, different power relations between actors belonging to different societal levels represent an important factor influencing transition management process. For this reason, adaptations of transition management process were considered.

In the following chapter, we first present the case context. In section "Transition management process in La Botija protected area", we describe the adapted transition management methodology applied in La Botija area. Section "Discussion" includes the results of the developed transition management process considering the transition pathways and the transition agenda developed through the participatory process. Section "Conclusions" includes the conclusions.

Case Context

The Republic of Honduras is, with approximately 7.7 million inhabitants, the second most populated country and the second poorest country in Central America (Andréu 2012). Honduras is a lower middle-income country characterized by persistent poverty and income inequalities. Despite the favorable economic outlook, the country faces the highest level of economic inequality in Latin America and the country is yet not on a track of welfare improvement nor of poverty alleviation across urban and rural areas (IFAD 2013). Honduras was declared one of the heavily indebted countries by the World Bank and the International Monetary Fund, and was made eligible for debt relief in 2005 (IMF 2013). Poverty in Honduras affects 60% of the population, 36% of which live in extreme poverty. In rural areas, these percentages increase to 63% and 50%, respectively (IFAD 2013). Poverty affects the way priorities are assigned, creating an array of vulnerabilities across service sectors and related impacts such as: lack of and deteriorating infrastructures, degraded ecosystems and mismanaged natural resources, low access to lands, low agricultural productivity resulting overall to high rates of unemployment. The agricultural lands cover approximately 28% of its territory and contribute to 39% of total employment. With agriculture being an employment provider sector, forest areas are under stress in becoming agricultural land with the inhabitants employing fires as a land-clearing strategy (FAO 2003).

The case study area is La Botija protected area, located in the municipality of San Marcos de Colón, in the southeast part of Honduras, at the border with Nicaragua. La Botija was declared a protected area in 2005 by Decree No. 385-2005 delivered by the National Congress of Honduras. La Botija is a multiple-use protected area (Area de Usos Multiple La Botija'(AUMLB)) where limited logging and agricultural activities are allowed and ecosystems conservation and ecotourism should be promoted. The protected area has a surface area of 19,079.81 ha (191 km2) which represents 34% of the surface area of the Municipality of San Marcos de Colón that holds 562,9 Km2 at an altitude of between 500 and 1700 m (Flores Velásquez et al. 2008; Sánchez Villa 2013). The main economy of the investigated area is subsistence agriculture with small-scale commercial farms. The agricultural production of La Botija area includes coffee, corn, bananas, sugar cane and beans.

The area consists of 22 rural communities and faces multiple social, environmental and economic problems such as environmental pollution, low economic resources, and limited access to services. The area is important to Hondurans and Nicaraguans as a water basin with the headwaters of the Río Comalí that provide the 70% of the water supply for the municipality of San Marcos de Colón on the north of the area (Macknick et al. 2012).

The protected area is managed by an assemblage of NGOs and institutes including ODESA, Enrich the World and INADES NGOs, APROBOSQUE local association, Municipality of San Marcos de Colón and the Institute for Forest Conservation (ICF – Istitudo por la Conservación Forestal in Spanish). The management of the area follows an Area Management Plan called ‘Plan de Manejo del Area de Usos Multiples Montaña La Botija’ approved by ICF in 2009. With the majority of lands being private, a common plan for the management and conservation of the protected area is difficult to be established between land owners and land users (mainly farmers) also weighted by the fact that the majority of large landowners do not reside in the area. Adding to this situation, Hurricane Mitch in 1998 had a devastating impact in the already aging infrastructures in the San Marcos de Colon Municipality, followed by a very slow-pacing infrastructure repairing phase that is still ongoing (ECLAC 1999).

Methodology

The participatory action research methodology adopted for the development of strategic agenda builds on the transition management approach as an overarching framework to consider how transitional change towards a sustainable future can be enabled (Loorbach 2010; Rotmans et al. 2001). There is a limited academic literature analysing interventions and/or possible strategies to enable transformative change in developing countries with a focus on co-creation, knowledge co-production and sustainability transitions (Binz and Truffer 2008; Poustie et al. 2016; Simon and Leck 2015). To address this knowledge gap, in this chapter we show how applying transition management as a transdisciplinary framework organises a co-creation of: 1) transformative narratives in the form of visions and 2) strategies in the form of transition pathways.

The project of application of Transition Management in La Botija protected area presented in this Chapter was co-organised between Rey Juan Carlos University in Madrid (Spain), University of Pisa in Italy, Enrich the World NGO in Honduras and DRIFT (Dutch Research Institute For Transitions) with Erasmus University Rotterdam in the Netherlands. The first author was the initiator of the project and was involved as action researcher in the local area in 2014. The first author conducted the context analysis, designed and organized transition arenas and she analysed the results of the process. The second author continuously supported the first author in the adaptation and application of transition management in the investigated area.

Participatory observations were regularly conducted in the different rural communities of the protected area during the fieldwork period. The first author lived with a local family in El Jocote community. This involvement in the rural life of both a family and the communities allowed observing the diversity of perspectives, interactions, activities, and views of the local communities.

The context analysis is based on information obtained from different, complementary data collection methods: (i) literature review of reports and data related to local infrastructures, government policies, legislation and previous activities of NGOs and local associations; (ii) 43 in-person semi-structured interviews and 14 focus group interviews; (iii) participation in 14 meeting and public events organized by different local institutions and organizations, (iv) participatory observation of rural life of 13 communities of La Botija protected area and (v) 14 workshops as part of TM workshop series#1. All semi-structured interviews were recorded, reviewed, transcribed and translated by the first author from Spanish to English. Information obtained from participant observations were also considered. All the workshops were recorded. During the workshops, the researcher kept notes of information and descriptions provided by the participants. All the recorded and written information were reviewed and transcribed in Spanish. A report in Spanish was developed and provided to each community. In total twelve reports were developed. The reports were translated from Spanish to English, and further analysed.

The transition management process included a total of forty-nine workshops (divided into workshop series #1, #2 and #3) at community level and three crosscutting workshops inviting participants from different communities. Throughout the workshops, the researcher acted as facilitator of the transdisciplinary process with the collaboration and support of local teachers. During the explanation of the problems' interconnectedness and the importance of ecosystem services, the facilitator focused on stimulating critical thinking about the current local problems and their practices. During the development of transition pathways, the facilitator motivated collective thinking between participants and asked questions for indicating support for possible future collaborations.

Transition Management Process in La Botija Protected Area

In the following part, we present the phases of implementation of the adapted transition management process in 'La Botija' protected area. The transition management process is divided in the following phases: (1) process preparation; (2) exploration of the local context; (3) adaptation of the process to the context and design of transition management workshop series and (4) transition management workshop series and results. The transition experiments that developed as an outcome of the transition management workshop series are presented in section "Discussion".

Process Preparation and Setting

The first phase of the process (i.e. process preparation and setting) consisted in (1) desk research related to the investigated context, (2) Skype interviews with local actors and (3) first development of a process plan.

Exploration of the Local Context

The context analysis represents a baseline assessment of the current local conditions and dynamics of the context. It includes three parts: the problem analysis, the analysis of opportunities and the actor analysis.

The context analysis permitted to identify the social, economic and environmental dimensions and included the current institutional setting (e.g. politics, organizations, capacities, connections between organizations, policies and strategic plans). The opportunities analysis seeks to outline the existing but yet non-capitalised local opportunities for improving local sustainability. The actor analysis intends to clarify the relationships between the different stakeholders that operate in the local area to understand institutional arrangements (both formal and informal) and represented the information base for selecting transition arena's participants. The actor analysis considered actors from local community, actors engaged in the non-profit sector and actors from governmental institutions.

Adaptation of the Process to the Context and Design of Transition Management Workshop Series

Based on the results of the context analysis the transition management process methodology was modified to adapt it to the local context of the investigated area. These adaptations were taken into account for the design and organization of the transition arenas. The adaptations included:

- The stakeholders' selection and engagement strategy was adapted to the local context by revisiting the selection criteria of 'frontrunners' and by considering power dynamics. In La Botija case, local frontrunners included pioneering committees, community groups and individuals not only on service delivery (e.g. food production) but also on establishing ecosystem stewardship relations. Due to the strong power tensions between actors from local communities and from the governmental institutions, the transition arenas did not include actors from the governmental institutions.
- The ecosystem services framework was used during the problem structuring and strategy development phases as an analytical tool to dissect and examine the problem and how it is reproduced, as well as to structure the discussions on strategy development by pointing out the human-ecosystem relations and how they are established in the area.
- Different numbers of workshops were organized in each community considering the cultural context of each rural community and the involved stakeholders. Specifically, in some of the rural communities it took more time for the researcher and facilitator of the process to build trust with participants who have not been confident in sharing their opinions and views.
- Capacity building activities were developed in parallel to the transition arenas in collaboration with local residents and local organisations such as NGOs and community based initiatives. The capacity building activities included workshops and farming on multiple topics such as sustainable farming, water and sanitation issues, health and disease prevention and environmental protection.

The frontrunner characteristics taken into account for the identification of arenas participants were the following: (a) system thinking; (b) motivation to act upon achieving local sustainability; (c) resourceful actors with access to knowledge and local networks; (d) ability to mobilize existing underutilized resources; (e) collaborative attitude in establishing networks and coalitions. Based on these criteria, we selected the following participants: teachers, residents, coordinators of local committees that deal with improving living conditions (i.e. water committee, health committee, committee for the environment protection), representatives of the Women's Network ('red de mujeres' (RED) in Spanish), and actors from local and national branches of NGO's. The coordinators of local associations and NGOs and the teachers of the local schools helped in identifying residents of local communities with frontrunner characteristics. Over time, the transition management workshops 'opened up' to engage in the discussions local residents interested in the deliberations (Roorda et al. 2014).

Transition Management Workshop Series and Results

In total 49 workshops were realised in the rural communities of the investigated area. The transition management process included three workshop series organised in every community: TM Workshop Series #1 aimed at understanding the context and what creates persistence. TM Workshop Series#2 aimed at co-creating a vision and guiding principles. TM Workshop Series #3 aimed at formulating transition pathways and a future agenda. In each community were organized at least three workshop series.

Following the participants' needs and interests in multiple communities, additional follow-up workshops (4–9) were realised to summarize findings and outcomes as well as to identify roadmaps for local action. Additionally, cross-cutting workshops were organized in order to connect frontrunners from multiple communities. In total three cross-cutting workshops were organized with participants from eleven rural communities. The cross-cutting workshops contributed to sharing of information and ideas, fostering dialogue between participants, building trust and networks amongst local frontrunners. Every transition arena had between 9 and 20 practitioners in attendance.

Workshop Series #1: Understanding the Context and What Creates Persistence

TM Workshop Series #1 aimed at understanding the context and what creates persistence, and included sessions for (i) problem structuring, identifying domains of change and local opportunities, (ii) description of the local context dynamics, (iii) connecting environmental and social problems in La Botija area.

Additionally, in this section we describe the main results from the context analysis collected from both the system analysis and workshop series#1 that were divided into the following themes: (a) Local economy; (b) Under-developed and under-invested infrastructures; (c) Geophysical conditions; (d) Environmental problems; (e) Social problems and (f) Institutional problems.

Problem Structuring, Identifying Domains of Change and Local Opportunities

Participants were encouraged to express their needs and describe persistent problems, domains of change and opportunities of their community and of La Botija area. Participants were facilitated to underline causes of the problems and to identify domains that will require changes for these persistent problems to be overcome in the near future. Additionally, the opportunities of each community and of La Botija area were identified. During this phase the following questions were formulated: 'Which are the problems of your communities?', 'Which are the needs of you, your family and your community?', 'What are the opportunities of the area and your community?' The root causes of the problems were identified by asking: 'Why do these problems persist?', 'Which are the areas of the community that require a change?'. The inputs obtained during the first workshop series were used to enrich the context analysis previously described.

Description of the Local Context Dynamics

Following the expression of the problems and challenges, the participants were asked to describe: (1) the key organizations and stakeholders (e.g. community organizations like water and health committee, women's group, NGOs, and local authorities) including outcomes of their activities; (2) how these actors relate and (when applicable) collaborate; and, (3) additional opportunities (e.g. favorable geophysical conditions, existing skills and knowledge).

Connecting Environmental and Social Problems in La Botija Area

In this session, the multiple environmental, social and economic problems affecting La Botija protected area were described and articulated. Participants were facilitated to develop a more holistic and systemic thinking of the interconnections between social and ecological problems using illustrations and maps. The main topics regarded water and soil pollution, soil erosion, nutritional and health problems, deforestation and fire risks, as well as the low collaboration, planning and governmental support. In this session of the workshop, we employed a simplified and adapted version of the framework of ecosystem services to systematize the discussion on interconnections between social and ecological problems, or simply between social practices and their ecological impacts (TEEB 2010, 2011; MA 2005; Gómez-Baggethun and Barton 2013). Participants were asked to describe goods and benefits received from local and regional ecosystems and the facilitator used images of the ecosystem elements located in La Botija protected area (e.g. waterfalls, different forest types of the area, species threatened with extinction) and of the children of the area to stimulate the discussion and the reflection about ecosystem services and their importance for future generations.

Context Analysis

Local Economy

La Botija protected area is characterized by a low financial and human capital. Job opportunities are almost non-existent and the majority of families live on subsistence farming or on the wages of temporary employment of coffee and corn harvests. Subsistence farming is the main economic activity in the area, with farmers being able to only support their families. The existing assets in the area, including social and natural capitals, are unsustainably managed. Accessibility of lands for cultivation is limited due to the privatization. In La Botija, a small number of large landowners owns most of the land and a vast number of small owners and tenants cultivate small plots of land. The large landowners, in some cases, abandon the lands and do not allow small landowners to use their lands. The low access to land causes discontinued agricultural production that, in turn, inhibits production for commercialisation and burdens start-up of local enterprises. This is further constrained by limited management knowledge of farmers in setting up new businesses. All the problems described above contribute to fuel migration to cities.

Regarding the opportunities, the area is relatively safe and characterized by a low level of criminality and low risk of malaria. A cooperative, called COCASAM, exports organic coffee (especially to USA and Germany), enables the improvement of organic coffee production and trains local farmers about organic coffee production methods. ODESA NGO promotes the local economy with the establishment of two microenterprises of cheese production in two communities while two rural banks provide loans to local farmers. Another opportunity is the local production of different produces (e.g. beans, bananas, corn, sugar, coffee, etc.) with traditional methods that is seen as a business opportunity. For example, in La Mesas de Cacamuya and Las Flores communities, residents produce organic coffee following a traditional method. In La Laguna and Los Chaguites communities, active women groups collaborate and organize regular meetings in which they openly discuss and address present challenges: a form of building social capital. Developing sustainable tourism is recognised by the residents as an economic opportunity. This however requires coordination between local authorities for the promotion of the area, for training tourist guides, for developing information material and for improving the accessibility of the area. Aprobosque is a local non-profit organisation representing the interests of landowners aiming to protect forests, to increase agricultural productivity and to create local opportunities. In order to achieve this purpose, Aprobosque actively seeks support from the government, NGOs, and other international organizations and lobbies for the local government to improve landowner rights and conditions.

Under-Developed and Under-Invested Infrastructures

In every community, the residents expressed the need to improve and establish new water and sanitation infrastructures for improving drinking water quality. The irrigation system is underdeveloped with some residents using rainwater tanks for the recollection of water, while the majority sources surface water for irrigation. Lack of environmental awareness of overexploitation of surface water and inability to build irrigation canals underlies and worsens this situation. The under-resourced and underequipped healthcare system was discussed both by residents and healthcare employees. La Botija area has two health centres located in Las Trementinas and Duyusupo communities. Health centres operate with limited human resources, medical equipment, technical instruments and medical supplies. Many residents prefer to go to Nicaragua for medical assistance and emergencies, especially the communities located closer to the Nicaraguan border and the more isolated ones such as El Sarzal, Las Trementinas and Las Delicias. Many houses need to be restored, especially in the poorest communities of El Mal Paso, Los Chaguites, El Sarzal and Cacamuya. Mobility infrastructure is discontinued and underdeveloped, with communities of El Sarzal, Los Chaguites, Las Trementinas and El Guajiniquil being isolated and completely disconnected and inaccessible from the municipality of San Marcos de Colón. The other communities are connected to the municipality with infrequent and unsafe public transport. The majority of the communities are also digitally isolated with neither telephone nor internet. Centralised waste management is non-existent due to the inability (both financial and organisational) of local government to invest in such infrastructure. The inaccessibility of the area and the concentration of low economic income families in the area make local authorities reluctant to invest in waste management services. In the majority of the cases, local residents illegally dispose or bury their household garbage or burn them by using open incineration systems.

The majority of the communities explained that the quality of the educational services has increased over the past decades and several schools were built through international cooperation programmes. Nevertheless, there are still several challenges such as the low quality of the schooling facilities, the scarcity of educational materials and the low quality of water and sanitation services in schools (e.g. the school of Cacamuya community has no drinking water). Education is appreciated by the communities and in El Mal Paso and Las Mesas de Cacamuya communities, residents expressed interest to acquire skills for health care in order to provide health care services to their communities. In some communities, residents are mobilised for improving infrastructure services and they set up community committees. In El Mal Paso community, for example, active citizens set up a health committee that works on improving health services and increasing skills and knowledge of healthcare practitioners. The health committee collaborates with the health centre of San Marcos de Colón that provided trainings and materials. Another example of community self-organisation comes from the Portillo Grande community in which a parents group works for improving the school facilities. NORDEX is developing a wind park with 27 wind turbines in the territory of San Marcos de Colón covering also the South area of La Botija area. The project is an opportunity for renewable energy production and local employability. The international programme for Sustainable Rural Development (Emprende Sur) by IFAD aims to increase the income, employment and food security of small-scale agricultural producers and micro-entrepreneurs in the southern region of Honduras. Particularly in La Botija area Emprende Sur restores streets connecting coffee producers with the municipality of San Marcos de Colón.

Geophysical Conditions

Some lands of La Botija area are characterized by a forest vocation, meaning that agricultural production is either too difficult or too expensive. The rural communities located in these areas (e.g. El Mal Paso, Los Chaguites, Cacamuya) have dry and infertile soil that is costly and difficult to cultivate. The area's soil fertility and ecosystem quality suffer from the spread of various plant pests and diseases. One of which, is the coffee plants' pest called 'Roya', or 'coffee rust', a fast-spreading fungus that infects the leaves of the coffee plants and caused a loss of approximately 80% of coffee production in 2013.

The studied area is characterized by a high risk of floods and hurricanes during the humid season. This risk increased in the past decades due to forest destruction, soil erosion and climate change. Hurricane Mitch caused destruction of water and sanitation systems, schools, and houses that need to be repaired. Next to these, communities located on mountains and hills (such as Los Ranchos, El Mal Paso, El Sarzal) are exposed to higher risks of landslides. Looking at opportunities related to geophysical conditions, La Botija area has rich ecosystems, high biodiversity and threatened species providing environmental benefits and representing potential touristic attractions. The investigated area is strategically located at the border with Nicaragua and it is rich in water, supplying water to several communities both in Honduras and Nicaragua. Additionally La Botija is one of the few areas in Honduras without malaria (outbreaks).

Environmental Problems

The ecosystem services framework allowed to connect underlying challenges with explicit environmental degradation impacts.

Unsuitable wastewater infrastructures, lack of waste management, excessive use of chemical fertilizers, and unsuitable livestock management deteriorate provisioning ecosystem services of water supply and water infiltration. The environmental contamination in La Botija protected area is the result of unsuitable wastewater infrastructures, lack of waste management, excessive use of chemical fertilizers, unsuitable livestock management, and forest destruction through deforestation and fires. The lack of awareness and limited knowledge about the severity of impacts of unsustainable practices such as excessive use of fertilizers are the main drivers of these persistent problems. Drinking water sources are contaminated by wastewater, agriculture runoffs and ineffective livestock management. Consumption of contaminated water has frequently caused disease outbreaks in the area (diarrhea incidents in adults and children).

Forest fires deteriorate habitat creation and water retention ecosystem services and deforestation deteriorates habitat creation, biodiversity and water retention ecosystem services while provide material extraction ecosystem service. Almost all the forest fires in La Botija area are human-initiated. Farmers deploy fire as a means for soil mineralization, grass growth and for creating new lands for livestock. Fires also occur in the south and south west areas of La Botija (Las Trementinas, Las Delicias, Guajinijil, El Sarzal) that are closer to the Nicaraguan border and even more difficult to access for fire mitigation. Deforestation occurs due to financial incentives of selling wood. In this context, ICF appears to directly support deforestation by providing deforestation permits to large-scale farmers without monitoring of their practices motivated by financial incentives outside the law. The destruction of forests through fires and deforestation leads to destruction of food crops, soil erosion and loss of biodiversity.

Even though in the majority of the communities these problems prevail, Las Mesas de Cacamuya and Las Flores communities have maintained organic coffee production. Farmers and residents from these communities talked about fertile and productive soil contrasting the reality in the neighbouring communities. Residents are slowly becoming more aware about how unsustainable practices and environmental contamination are related and how health diseases and low crop productivity are connected. Therefore, in the communities of Las Delicias, Las Mesas de Cacamuya, El Mal Paso and Portillo Grande, locals work in environmental committees to address the environmental contamination ('comite' por el medio ambiente' in Spanish). The main activities of these environmental committees include environmental awareness meetings and events as well as forest fire prevention activities organised in Portillo Grande and El Mal Paso communities. Other local organisations and initiatives such as Aprobosque, RDM, COCASAM and ODESA, as explained previously, contribute with developing local activities for environmental protection.

Social Problems

The fieldwork unravelled the following social problems: (a) low level of motivation for self-organisation and taking action, (b) low self-confidence, (c) limited collaboration amongst communities and (d) low awareness about socio-ecological sustainable practices. Residents are rarely motivated to take actions and start new projects and there are limited collaborations amongst residents. Local frontrunners described the difficulties to find active collaborators. Indicatively, the failure of a project for the development of a micro-enterprise in the community of El Mal Paso showcases the causes of social challenges present in the area: the low collaboration, organisational capacities and trust amongst participants. Even the involved NGO failed in building trust between participants and in developing skills that respond to their needs (e.g. team work and collaboration, funding management) for maintaining the project over time given the low economic turnover of the micro-enterprise in its initial period.

Malnutrition is the result of limited financial access to a variety of foods and a low nutritional knowledge. This further results in cognitive functioning deficits and learning difficulties, especially in children. Malnutrition is present in the most isolated communities such as El Sarzal and Cacamuya. The existence in the area of local active groups, committees and associations represents an opportunity for a sustainable transition of the area. The RED (Network of women in English), for example, aims at increasing awareness about social and environmental issues, collaborations, women employability and empowerment, education and development of skills in the communities of La Botija and in San Marcos de Colón. Additionally, in the majority of the local communities, residents set up community committees with the aim to solve local problems, organise new partnerships and seek new opportunities for their communities (i.e. Health, Environmental, Education, Water and Forest conservation committees).

Institutional Problems

Governmental departments are characterized by poor organization and coordination of activities concerning the area's problems, low collaboration for sharing of environmental and social information and limited communication with other institutes and organizations resulting in unclear roles and responsibilities.

The low support and the overall failing institutions are visible in low quality of infrastructures, limited or often non-existent environmental protection and provision of poor public services. The environmental protection agencies at local, regional and national level are characterized by low communication and collaboration. These institutions poorly collaborate and share information with local organisations and initiatives of the area.

There is a consensus that the main problem with environmental legislation in Honduras is how it is implemented and monitored (USAID 2004). Notwithstanding, the causes of environmental problems are linked to under-enforced environmental laws such as the forestry laws. According to the Honduran forestry laws, the forests are considered property of the forest landowners that should ensure the protection and safeguard of forests (e.g. activities for the fire protection and combating deforestation and forest degradation) and be responsible for emergency actions. The lack of environmental monitoring and enforcement of the forest law to its full extent, i.e. also including penalties of malpractice, results in forest landowners employing fires as a means of deforestation and land cleaning. A commonly voiced driver of such 'blind spots of legislative enforcement' is corruption. In view of this malpractice, self-organised forest conservation groups (e.g. aprobosque, forest conservation committee and committee for environmental protection) do not get legitimised and are left to operate in the shadows of the law to protect the forests.

Next to this there are neither incentives nor support from governmental institutions for development of local businesses that can ignite employment in the communities. The low support by the government is considered as a factor increasing the uncertainty of the longevity of small enterprises.

Different forms of collaboration between the municipality of San Marcos de Colón and civil society (e.g. ODESA, RDM, COCASAM, community committees, etc.) represent opportunities for the area. Additionally, some of the teachers working in the schools of La Botija area represent an important source of information, awareness, and knowledge on social-ecological issues of the communities.

Workshop Series #2: Co-creating a Vision and Guiding Principles

During the envisioning phase of the workshops, participants were asked to express and to share their aspirations for the future of their community and La Botija area. This session aimed to stimulate thinking about long-term aspirations rather than quick fixes to system's problems. Visioning is considered a key method in sustainability research and problem solving and according to the literature 'positive visions about our societies' future are an influential, if not indispensable, stimulus for change' (Wiek and Iwaniec 2014). Indicatively, the questions asked were the following: 'What are your dreams for you and your children for the future?', 'What do you dream for your community and La Botija area?'. Participants were invited to explain the reasons of their visions and to formulate principles describing the desired future.

The guiding principles and visions expressed by the participants during the multiple workshops organized in the different communities are divided into seven themes: (1) health and wellbeing; (2) environmental quality and protection; (3) clean and safe drinking water; (4) prosperous and sustainable economy; (5) infrastructures' quality; (6) collaboration, trust and motivation; and (7) education. The improvement of the quality of infrastructures was considered among the participants one of the most important visions to achieve in the future. Participants highlighted their dream to restore and build water services in their communities and to have access to better quality of drinking water. Other aspirations about infrastructures regard the renovation and construction of sanitation services. They understood that their health is connected with the quality of water they consume.

Another vision theme concerns the economic development and the creation of job opportunities. People dream about stopping to live on subsistence agriculture and instead setting up local small-scale businesses. In many communities, participants expressed their dream to improve the quality of education and in this way create an 'opportunity to change' and a 'fundamental tool to build a new generation and to reconstruct the future' of their communities. Another recurring theme during the envisioning phase was the desire to improve collaboration, motivation and willingness to change within communities.

Workshop Series #3: Formulating Transition Pathways and a Strategic Transition Agenda

TM Workshop Series #3 aimed at formulating transition pathways and a future agenda and included sessions for (i) formulating strategic transition pathways and (ii) creating a transition agenda with short, medium and long-term actions. Each transition pathway built on the guiding principles and visions developed during the workshop series 2 and summarised in a poster that included images of healthy, liveable and sustainable communities. Next to presenting the visions, a 'businesses as usual' scenario was shown in which the different risks and vulnerabilities were presented in case transformative actions are not taken. These included health diseases, fires, flooding and landslides. By presenting these two contrasting contexts (vision versus 'business as usual') we aimed to support the development of transition pathways by creating a sense of urgency for thinking of action and to escape the 'business as usual' situation. The development of transition pathways as courses of action aiming to progressively improve the current situation and aid in achieving the vision, also contribute to building awareness amongst participants about the expected outcomes of identified actions.

At a later stage, workshops focused on how transition pathways may be operationalized in the specific context by the development of short, medium and long-term actions. The participants were asked to identify a set of strategies to achieve the visions. During this stage, strategic thinking was supported and stimulated for generating different actions. Specifically, the objective was to identify a broad range of actions at individual and community levels that could overcome current problems and to reach future visions.

Participants were asked which strategic actions could enable specific changes. When participants explained the possible actions to develop in their territory, discussions and sharing of opinions were strongly supported. Participants were asked to describe each defined short, medium and long-term actions by asking questions such as 'what needs to be done now?', 'what needs to be done in the near future?', 'What needs to be done in the far future?' and 'when should each strategy be implemented?'.

Examples of identified actions include the organization of educational activities related to multiple issues (e.g. sustainability, health care, nutrition, microenterprises development, etc.), development of programs for supporting the setting up of microenterprises in the area, the creation of collaborations among frontrunners within and between communities and among existing organisations and institutions, and institutional support for improvement and provision of infrastructures.

Back-casting was the method employed for the formulation of strategic actions that were then structured into transition pathways. Backcasting is a decision support method that allows for first designing a desired future and then identifying how this desired future could be achieved (Robinson 2003; Robinson et al. 2011) (Table 1).

Table 1 Transition athways and examples of strategic actions identified for the sustainable development of La Botija protected area, Honduras

	Time period	Examples of short, medium and long term strategic actions (leading agency)
Pathway A: Improve health quality and wellbeing		
Improve health services	Short term	Employ human resources already trained on health care by the Health Center of San Marcos de Colón
		Educate residents and children on good practices of hygiene and toilet use
	Medium term	Educate and train community residents (e.g. health committees members) as health care professionals
		Provide equipment, medicines and materials to health centers
	Long term	Renovate and restore health centers
		Implement new technologies and advanced medical equipment
Improve green and public infrastructures	Short term	Promote residents' collaboration for improving and maintaining community goods (e.g. school gardens, schools, public spaces and buildings, etc.)
		Support planting of trees in the communities
	Medium term	Develop activities and projects to raise awareness about sustainability, social responsibility and ethics
	Long term	Improve and maintain the quality of public green spaces involving the community (e.g. set up community gardens and parks, school gardens, etc.)
Improve nutrition and enrich diet	Short term	Educate and train communities on healthy nutrition and diets (e.g. nutritional lessons in the schools
	Medium term	Set up community gardens and promote fruit trees cultivation
	Long term	Promote sale of healthy food in local small grocery stores

Pathway B: Improve environmental quality and protection		
Promote waste management	Short term	Organise constantly community 'clean-up' days and cleaning activities
		Install prohibition signs against unlawful deposit of waste along roads and into the communities
		Educate community about good practices of waste management, recycling management and reuse of materials
	Medium term	Collaborate with the Municipality and other institutions for organising recovery and rehabilitation of contaminated areas by waste (i.e. illegal dumps)
		Demand the establishment of waste collection services in La Botija area by local institutions
		Set up waste recycling activities
		Lobby for the establishment of controls and prohibitions by the institutions (e.g. penalties)
Promote forest protection and reforestation	Short term	Develop fire prevention measures (awareness-raising campaigns and special trainings)
		Support community actions on forest protection
		Promote trees planting in the communities
		Develop a proposal for requesting the increase of support to the community committees by institutions (ICF/UMA/municipality) with materials, equipment and economic resources
		Enhance knowledge and skills of members of environmental protection committees
		Monitor and control forest protection activities and measures
		Demand the organisation and increase of controls, prohibitions and penalties by institutions on illegal fires
		Support plants nurseries businesses and forestry business development providing economic incentives and training
	Medium term	Learn about the current forestry laws and develop strategies to change them

(continued)

Table 1 (continued)

	Time period	Examples of short, medium and long term strategic actions (leading agency)
		Improve the organization and forestry management at governmental level
	Long term	Organise campaigns to stop corruption and illegal traffic of wood
		Set up a system of payments for ecosystem services (PES) to local farmers and landowners
Increase organic practices and sustainable use of soil	Short term	Collaborate with COCASAM cooperative for educating communities residents on sustainable farming and use of organic fertilizers (theoretical and practical activities)
	Medium term	Set up organic community gardens, school gardens and family gardens
	Long term	Employ community frontrunners with experience on organic agriculture practices and technics to teach and train other people of the community (train the trainers)
Increase awareness about environmental issues	Short term	Provide training to make residents aware about the importance of ecosystem services, sustainable exploitation of natural resources and territory values
	Medium term	Develop an environmental education plan for the schools of the communities
		Organise community awareness campaigns on different environmental topics
	Long term	Train stakeholders (belonging to different governmental institutions) on environmental issues
Support sustainable livestock management	Short term	Educate farmers on sustainable livestock management to avoid water contamination and to permit reforestation

Prevent flooding, landslides and plant diseases	Short term	Promote water management and rivers and streams protection activities by communities residents (e.g. maintenance of rivers and streams banks)
	Medium term	Collaborate with COCASAM cooperative to train farmers on plant diseases preventative actions and activities
	Long term	Support local committees for the environmental protection with provision of materials, equipment and trainings
		Prohibit deforestation and support reforestation
		Employ innovative infrastructures technologies
Pathway C: Sustainable water management and quality of water		
Improve water infrastructures, sustainable water management and quality of water	Short term	Carry on a first research for analysing water hidrogeology and water quality and quantity
		Support the use and provide water filters
		Improve current water infrastructures efficiency and capacity
		Support water committees (e.g. provision of materials and equipment)
		Monitor contaminants levels in water
	Medium term	Evaluate the state and condition of the existing water infrastructures
		Analyze water resources distribution in the territory (e.g. streams, water sources, ground waters)
		Develop an initial scientific and technical assessment for the creation of water projects
	Long term	Develop water projects in the priority intervention areas (e.g. digging of water wells and supply of purifying systems)
		Use abandoned drinking-water sources
		Improve the water quality through innovative technologies

(continued)

Table 1 (continued)

	Time period	Examples of short, medium and long term strategic actions (leading agency)
Support water management education	Short term	Educate residents on water treatments (e.g. disinfection)
		Train residents on water values, practices for water protection and prevention of its contamination
	Medium term	Train water committees members to become community water management professionals
	Long term	Train governmental departments on water management
Pathway D: Improve and create infrastructures		
Improve sanitation services	Short term	Educate inhabitants on the suitable ways of latrines building (e.g. location of latrines far from streams and rivers)
		Evaluate the sanitation services in the different communities
	Medium term	Restore the existing latrines in a poor condition
		Build new latrines and sanitation services
	Long term	Develop sanitation services projects
		Improve quality of health care services providing equipment, human resources and materials
Improve communication channels quality	Short term	Improve roads quality
	Medium term	Improve the bus service
		Offer wireless internet access
	Long term	Offer newspapers and book sale in the communities
Improve housing quality	Short term	Promote the restoration of damaged houses structures
	Medium term	Provide energy services to new houses (i.e. solar panels)
	Long term	Increase electricity services
Improve school quality	Short term	Improve infrastructure quality of damaged schools (the school with priority of intervention is 'El Jocote' community school)
	Medium term	Provide water and sanitation services to each school
	Long term	Improve quality of transports for schools

Create and improve irrigation systems	Short term	Support the development of family rainwater harvesting systems
	Medium term	Develop irrigation systems for agriculture production
	Long term	Promote the use of innovative irrigation systems
Pathway E: Promote local economy and job opportunities		
Promote business activities, enterprise and projects development	Short term	Organize family and community allotment gardens
		Improve collaboration and engagement of residents into community committees and local associations
		Support the collaboration between large landowners and small farmers
		Improve existing projects (e.g. projects of ODESA, CARE, Enrich the World)
		Support and empower community committees and associations
		Maintain projects and activities over the time
		Involve women in business and enterprise development
		Plan and create microenterprises for production of local products
	Medium term	Promote handcraft activities
		Develop plant and trees nursery business
		Develop laying hens projects at community or family level
	Long term	Develop a project for organic coffee cycle production: coffee nursery, growing and final processing through a community micro-enterprise (Las Mesas de Cacamuya)
Support sustainable tourism	Short term	Train residents for the development of tourism activities (e.g. tourist guides, etc.)
		Set up footpaths for trekking in the different touristic attractions of the area
		Organize public events
	Medium term	Increase the promotion of the touristic attractions of the area
		Ask for nstitutional support and better tourism management
	Long term	Improve infrastructures and communications channels (see pathway D)

(continued)

Table 1 (continued)

	Time period	Examples of short, medium and long term strategic actions (leading agency)
Rational, efficient and sustainable exploitation of available resources	Short term	Use abandoned lands for agriculture of for the development of other activities
	Medium term	Develop forestry businesses
		Develop biomass businesses
	Long term	Develop land use management plans by the governmental institutions (i.e. municipality and ICF)
		Develop a technical assessment of the available resources of the area
		Develop a business plan for sustainable exploitation of available resources of the area
		Change the projects aims to be implemented in the areas (e.g. development of biomass projects, trees nurseries, forestry business, etc.)
		Develop in-depth researches on lands vocation for ensuring the best land uses
Pathway G: Improve cooperation, participation, empowerment, trust and motivation		
Promote cooperation, trust union and aggregation into the community	Short term	Support community involvement and participation of people organizing practical activities, community meetings and events
		Improve networks building and openness to new opportunity and more interested in new projects and opportunities (more open-minded)
	Medium term	Promote motivational training to take actions and to increase people self-esteem
	Long term	Promote communication, collaboration and synergy between active residents (frontrunners) and between them and community committees members creating exchanges and interactions opportunities (e.g. meetings, events, courses, etc.)

Support and empower community committees and associations	Short term	Organize practical activities with economic effects, especially in the case of RDM (e.g. microenterprises, handcraft production, community gardens, community laying hens, etc.)
		Provide economic funding, provision of material and equipment
		Train, educate and motivate committees members
	Medium term	Strengthen collaboration between the members' groups of different committees
		Increase information about committees activities
		Support participation and engagement of residents in committees activities
	Long term	Creation of networks between committees of different communities
		Encourage residents and committees members searching for projects and opportunities and creating new connections
		Improve network building and collaboration between community committees and external associations and projects
Improve cooperation and collaboration between communities and governmental institutions	Short term	Improve communication and sharing of information between communities and governmental institutions
		Organize meetings and workshops with the participation of communities residents and governmental institutions officers
	Medium term	Support network building, sharing of information and collaboration between different associations and projects and between these and community committees
		Change the community management considering the involvement of citizens
	Long term	Support the maintenance of projects and activities over the time

(continued)

Table 1 (continued)

	Time period	Examples of short, medium and long term strategic actions (leading agency)
Pathway F: Improve education		
Increase knowledge, skills and values	Short term	Different types of team-teaching work and trainings and new forms of motivation for learners (e.g. organic agriculture, waste management, water management, fire prevention and forest conservation, health care system, community cooperation, ethic values and civic duty, etc.)
	Medium term	Organize training that support future visions creation
	Long term	Organize training to support collaboration and trust among the students and develop mutual understanding, solidarity and ethical values
Provide opportunities to continue to study after primary school	Medium term	Scholarships by private and public funds (e.g. travel reimbursement to go in San Marcos de Colón, provision of school materials, etc.)
	Long term	Build a secondary school in the south-east area of La Botija protected area
Prevention of criminality and alcohol and drugs consumption among the adolescents	Short term	Develop educational, recreational and cultural activities for children and youth (e.g. sport, theater, practical lessons and workshops, etc.)
	Medium term	Open an educational community center
	Long term	
Develop professional skills and expertise	Short term	Develop effective and efficient professional trainings courses for specific jobs for community residents
	Medium term	Provide trainings on entrepreneurial thinking and strategies
	Long term	Incorporate professional training in the secondary schools of 'El Jocote' community (from seventh to ninth school grades)

Cross-Cutting Workshops

Three cross-cutting workshops were organised in La Botija protected area with the aim to support collaborations and sharing of knowledge, practices and expertise amongst frontrunners from different communities. These workshops were realised after the completion of each workshop series and connected the different communities.

Discussion

The adaptation and of the transition management process reveals the importance of carefully studying a context and its conditions prior to implementing transition arenas, especially in contexts characterized by persistent problems and power tensions among societal actors. In this section we discuss the outcomes of the implementation of the adapted transition management process in La Botija protected area.

Transition arena workshops increased individual and collective motivation to take actions towards transformative change. The common understanding of the current local challenges and opportunities, together with the reflection on the value of the ecosystem services, the envisioning phase and the identification of possible pathways, have enabled the development of multiple actions. The transition arenas participants understood the urgency to take individual and collective actions in order to address community's challenges by themselves, without waiting for the intervention of local authorities or NGOs.

At individual level, transition arenas played an important role in enabling participants to recognise the impacts of their own unsustainable practices (the relations between social-ecological dimensions and ecosystem services) and promoted thinking of change of their daily practices. At collective level, the transition arenas fostered the creation of new collaborations and the reinforcement of existing ones in order to take transformative actions towards more sustainable communities.

Understanding and Reflecting on the Interconnectedness of Problems and the Value of Ecosystem Services

The identification of challenges and opportunities, in combination with the reflection on the way social practices impact the provisioning of ecosystem services, resulted in a better understanding of the complex problems to be addressed in La Botija area and the ways 'traps' are reinforced in the area. The envisioning phase played an important role in motivating the participants to unlock their minds and creativity from current persistent problems into thinking about actions to deal with them and to start a transformative process towards a different future.

Giving Voice to Individual and Collective Desires and Ambitions

Transition management represents aparticipatory method promoting the expression of individual and collective ideas, desires, interests and ambitions of local stakeholders. The transition arenas' participants had the possibility to express their opinions, inner desires and visions for themselves and their communities. Thus, transition management differs from other capacity building processes through which local stakeholders are trained to develop skills and knowledge without taking into consideration their specific interests, willingness and visions.

Identification of Knowledge and Learning Needs to Solve Local Problems

A main learning outcome of the transition management process in La Botija is the realization (by the participants) of the need to develop their own capacity and expertise in order to take actions to overcome local challenges. Expressing this willingness to create their own-capacity in order to take transformative actions by themselves, local practitioners deviate from the expectation of aid provisions. Due to the transition management workshops and the learning achieved by participating in them, participants self-organized in collaborative groups and requested to the facilitator to organize additional meetings, lessons and trainings in order to enhance their knowledge and expertise related to specific topics (e.g. organic agricultural technics, forest conservation, organic composting, prevention of plant diseases, development of entrepreneur's skills, nutrition and health). The participants realized the need to acquire specific skills, and knowledge for developing specific projects and activities defined in the transition agenda.

In Las Mesas de Cacamuya community, where the tradition of the production of organic coffee is strong, the members of the 'fuerza activa' asked to be supported for the development of a collaborative project involving the production, storage and process of the organic coffee. A group of women of Los Chaguites voiced their interest to develop a bakery's microenterprise and their desire to learn different tools and methods related to microenterprise management, team work, cooking skills, etc. Additionally, they voiced their interest to learn about organic agricultural methods such as organic composting for the development of a community garden. In Las Lagunas community, a women's group expressed their interest to learn handcrafted technics. In Las Mesas de Cacamuya community, a group of people expressed their willingness to be trained as nurses for contributing to the health services in the area. In Las Trementinas and Las Delicias communities, men requested support to develop trainings for forest conservation, protection and maintenance. In Cacamuya community, the participants asked the acquisition of knowledge for sustainable farming.

The willingness to acquire new knowledge and skills in every community confirms the impact of transition management process in inspiring and motivating participants to take up individual and collective actions at the aftermath of the transition agenda creation. These activities stem from their interests and desires while they consider the challenges, opportunities and the visions expressed during the transition management workshops.

Creating New Collaborations and Transition Initiatives

The transition management process promoted the building of trust, social synergies and new collaborations. Overall it enabled and enriched social capital building for transformative action realisation. The transition management workshops activated the establishment of new transition initiatives, as well as empowered and reinforced existing ones. An example is the setting up of La Botija Education Network ('Red educativa de La Botija' in Spanish) and new community groups in Los Ranchos and Cacamuya communities. The empowerment of existing collaborative groups is manifested by the activities of community groups in Las Mesas de Cacamuya, Los Chaguites and Mal Paso communities.

Some months post the transition management workshops, the teachers of the primary schools, most of which participated in the transition arenas, created an organization with the aim to enhance the sustainability education and management of the area. In July 2015 they organised the 'Fiera del Reciclaje' ('Recycling fair') event where they promoted environmental awareness by selling products made by waste materials and by organizing several activities on related issues.

The transition arenas contributed in the empowerment of existing initiatives and in fostering new partnerships. The transition management process therefore provided the opportunity to create new spaces for developing new values (e.g. collaboration) and knowledge, promoting a joint understanding and shared responsibility of the current unsustainable practices and the urgency to take collaborative actions towards transformations, as well as fostering autonomous self-organisation towards initiating those actions.

Realisation of Transition Agenda Post-Workshops and Cross-Cutting Workshops

The transition arena process empowered workshop participants to take actions for solving local problems and challenges in multiple rural communities. Overall, the participation and engagement in the transition management workshops/process had an activation effect on participants to take action to realise the co-created transition agenda.

In Las Mesas de Cacamuya community, after the third transition arena, the 'fuerza activa' (FA) group organized a 'clean up' day of the rural community mobilizing the inhabitants of the community for the cleaning activities. Next to this, they collaborated with four teachers of the primary school and the nursery in involving the children of the community during the 'clean up' day. As explained by one of the members of FA, they aimed at improving the value of their community and, at the same time, at promoting the awareness about the importance of maintaining a clean and uncontaminated living space. The members of FA decided to organize an additional meeting for developing collaborative activities. During the meeting they discussed multiple topics such as possible strategies to reuse an abandoned medical center that they were currently using as community meeting space and that could be converted into a community building for developing collaborative activities and projects. During the meeting, they shared information about multiple topics such as specific organic agricultural techniques, agricultural protection measures and treatment of plant diseases (e.g. for coffee plantations).

During one of the transition arenas, the members of FA expressed their desire to influence behavioural change related to many practices (e.g. use of organic agriculture techniques, forest protection, fire prevention, sustainable water and waste management) as well as improving community collaboration and cooperation. In order to achieve this objective, they asked the facilitator to organize further meetings including an additional broader transition arena open to all community residents including the children for promoting sustainability awareness and additional workshops with the aim to develop a community project. The facilitator accepted the requests and co-organized with the FA a broadened transition arena opened to all local citizens and three meetings for the development of a collaborative community project for the production, storage and processing of organic coffee.

In the case of El Mal Paso community, the active group of the community, consisting of people active in different community committees (e.g. water committee, health committee, etc.), requested the support of the facilitator to clean an illegal landfill provoking the contamination of the water source and the soil of their community. The facilitator supported the frontrunners in connecting with the Municipality's Environmental Department (Unidad Municipal Ambiental in Spanish) and the Health Center of San Marcos de Colón for co-organising a cleaning activity of the landfill area. In Las Trementinas community, the local environmental committee Comitè de Naturaleza organized monthly meetings to plan strategic actions for increasing environmental awareness. In Los Chaguites community, a women's group, after the organization of the transition arenas, requested to learn about developing a bakery microenterprise. They asked information about the most cost-effective products to produce and sell in the rural area and in the municipality of San Marcos de Colón. In this case, the facilitator connected the women's group with ODESA, a local NGO working in several rural villages for the creation and maintenance of microenterprises (e.g. microenterprise of milk products in Guajinijil).

Many teachers participating in the transition management workshops expressed their interest in introducing environmental lessons in their courses. The mobilisation of the teachers' group included educational actions like raising awareness about sustainable use of resources, setting up school and community gardens, reflecting on ethical values, taking actions to support local economy, enable continuous engagement across the communities and foster new collaborations.

The organization of cross-cutting workshops enabled the transfer of knowledge, ideas and best practices across communities. Particularly, representatives from more active communities were a source of inspiration for other frontrunners coming from less organized and cohesive communities. Through these workshops participants realized that they could benefit from developing and strengthening collaborations between active members from different communities.

Conclusions

This Chapter presented an empirical case of adaptation and application of transition management in La Botija protected area, a rural area located in the municipality of San Marcos de Colón, Choluteca, Honduras. The investigated area is affected by a multitude of complex social, environmental and economic problems such as environmental degradation, lack of infrastructures, low institutional attention and economic resources, low cooperation and motivation among local stakeholders.

The application of the adapted transition management process in La Botija area, in one hand, validate the method as a tool for promoting, stimulating and motivating participants to take up actions towards transformative change and, at the same time, reveal the need of further research for adapting and designing transition management to contexts characterized by power tensions and lock-ins.

Regarding the outcomes of the application of transition management to communities of La Botija area: first, the multiple actions co-developed and realised in multiple communities validate the method as a tool for promoting, stimulating and motivating participants to take up actions towards transformative change. The in-depth reflections and understandings of local challenges and opportunities, the appreciation of ecosystem services, together with the envisioning phase and the sharing of opinions, aspirations, knowledge, and expertise during the transition management workshops promoted the development of transformative actions at community and regional levels. Second, the process allowed the different communities to reflect upon their socio-ecological practices and to recognise their local core values while appreciating ecosystem services. It is therefore important to develop processes that move towards improving the communities' capacity to understand and apply sustainability principles through engagement, recognition of their own abilities and the decision of taking responsibilities for present and future actions contributing to local sustainability (Flint 2013). Third, the co-creation of the transition pathways motivated participants to seek and learn new skills for addressing local challenges and implementing specific actions. Fourth, the process enabled new collaborations and community initiatives, as well as empowered and activated existing ones. Fifth, the process allowed the expression of personal and collective desires, interests and visions of local residents that had a motivational and transformational impact at personal and collective levels.

Considering the outcomes of the implemented process we can infer that transition management is suitable for empowering low scale communities if adapted to the specific context conditions and local needs and can be considered as a strategic orientation method for the co-development of projects by local authorities, local communities (and community frontrunners), NGOs and international cooperation agencies. Nevertheless, additional research is needed in order to further improve the method to better fit with contexts characterised by power discrepancies between societal actors. For instance, it is important to design the transition management process considering also the inclusion of actors from institutions in order to support institutional and governmental change.

References

Andréu C (2012) Rural poverty in Honduras: despite progress, an ongoing challenge. Glob Majority E-J 3(2):100–109

Binz C, Truffer B (2008) Leapfrogging in infrastructure-identifying transition trajectories towards decentralized urban water management systems in China, DRUID conference paper, Copenhagen, pp 1–26

Carpenter SR, Brock WA (2008) Adaptive capacity and traps. Ecol Soc 13(2):40

ECLAC – United Nations Economic Commission for Latin America and the Caribbean (1999) Honduras: assessment of the damage caused by hurricane Mitch, 1998. Implications for economic and social development and for the environment, pp 1–20

FAO (2003) The state of food insecurity in the world, Rome, pp 1–37

Flint RW (2013) Practice of sustainable community development, a participatory framework for change. Springer, Berlin

Flores Velásquez P, Martínez de Anguita P, Hsiao E (2008) La conservación en las fronteras: el ciclo de proyectos aplicado a la creación del Parque Binacional 'Padre Fabretto', Fundación Fabretto. Rey Juan Carlos University, Madrid, pp 1–374

Frantzeskaki N, De Haan H (2009) Transitions: two steps from theory to policy. Futures 41:593–606

Frantzeskaki N, Ferguson BC, Skinner R, Brown R (2012) Guidance manual: key steps for implementing a strategic planning process for transformative change. Dutch Research Institute for Transitions, Erasmus University Rotterdam, The Netherlands. Monash Water for Liveability, Monash University, Melbourne, Australia, pp 1–31

Gómez-Baggethun E, Barton DN (2013) Classifying and valuing ecosystem services for urban planning. Ecol Econ 86:235–245

Grin J, Rotmans J, Schot J, Geels FW, Loorbach D (2010) Transitions to sustainable development: new directions in the study of long term transformative change. Routledge, New York

IFAD (2013) Project for competitiveness and sustainable development in the South-Western Border Region (PRO-LENCA), President's report, pp 1–27

IMF (2013) International monetary fund annual report 2013, promoting a more secure and stable global economy, Washington, DC, pp 1–80

Loorbach D (2007) Transition management. New mode of governance for sustainable development, PhD thesis. International Books, Utrecht

Loorbach D (2010) Transition management for sustainable development: a prescriptive, complexity-based governance framework. Governance 23(1):161–183

Loorbach D, Rotmans J (2010) The practice of transition management: examples and lessons from four distinct cases. Futures 42(3):237–246

Loorbach D, Frantzeskaki N, Huffenreuter LR (2015) Transition management: taking stock from governance experimentation. J Corp Citizenship 58:48–66

MA – Millennium Ecosystem Assessment (2005) Ecosystems and human well-being. A framework for assessment. Available at: http://millenniumassessment.org/en/Framework.html

Macknick J Enders S, Escalante J, Martínez de Anguita P, Robinson N (2011) Bi-national water management for economic development in Nicaragua and Honduras, Report of Yale School of Forestry and Environmental Studies, New Haven, pp 1–3

Markard J, Raven R, Truffer B (2012) Sustainability transitions: an emerging field of research and its prospects. Res Policy 41(6):955–967

Poustie M, Frantzeskaki N, Brown R (2016) A transition scenario for leapfrogging to a sustainable urban water future in Port Vila, Vanuatu. Technol Forecast Soc Chang 105:129–139

Robinson J (2003) Future subjunctive: backcasting as social learning. Futures 35:839–856

Robinson J, Burch S, Talwar S, O'Shea, Walsh M (2011) Envisioning sustainability: recent progress in the use of participatory backcasting approaches for sustainability research. Technol Forecast Soc Chang 78:756–768

Roorda C, Wittmayer J, Henneman P, van Steenbergen F, Frantzeskaki N, Loorbach D (2014) Transition management in the urban context: guidance manual. DRIFT, Erasmus University Rotterdam, Rotterdam, pp 1–49

Rotmans J, Kemp R, Van Asselt M (2001) More evolution than revolution: transition management in public policy. Foresight 3:15–31

Sánchez Villa Á (2013) Plan de Estrategia de Ecoturismo Area Protegida La Botija (PEEAPLB), Project proposal, Universidad Rey Juan Carlos, Madrid (España), pp 1–45

Simon D, Leck H (2015) Understanding climate adaptation and transformation challenges in African cities. Curr Opin Environ Sustain 13:109–116

TEEB – The Economics of Ecosystems and Biodiversity (2010) TEEB: mainstreaming the economics of nature: a synthesis of the approach, conclusions and recommendations of TEEB

TEEB – The Economics of Ecosystems and Biodiversity (2011) TEEB manual for cities: ecosystem services in urban management. Available at: www.teebweb.org

USAID – United States Agency for International Development (2004) Environmental policy analysis: report on Honduran environmental laws and their real or potential impact on the intermediate result "improved management and conservation of critical watershed" and the Central America Free Trade Agreement (CAFTA) related activities, pp 1–102

Wiek A, Iwaniec D (2014) Quality criteria for visions and visioning in sustainability science. Sustain Sci 9:497–512

Part III
What Transition Management Could Offer in Different Contexts

Chapter 11: Walsh – Translating transitions thinking and transition management into the city planning world. This chapter needs to be read jointly with Chap. 5 – Wolfram, this volume, so as to contrast the theoretical take on transition management versus strategic urban planning with the in-practice integration of transition management with urban planning in Australia that Chap. 11 offers. What is unique in this chapter is the practitioner's view and experience in how to translate the principles of transition management and transitions' thinking into more operational 'forms' of planning. Specifically, this chapter raises the question of whether, and how, transition management ideas might more directly enter and influence this "turbulent" reform space. Sydney provides the case-setting for the introduction of transition management, and related conceptions, to a recent major planning system reform project for that city. Working from a theoretical meshing of planning, relational-institutional and transitions studies, a new process model for planning system reform projects is introduced: the Planning System Transition (PST) framework. This framework's development and testing, through empirical research, is explained.

Chapter 12: Mai – Transition management in urban China: Co-creation of new homes or a monopoly game? This chapter is an interpretative and analytical take on what transition management applied as a heuristic can offer to urban planning of metropolitan coastal cities in China, including Shenzhen, Guangzhou, Ningbo, Beijing, and Tianjini. This chapter places empirical applications and review of established Transition Management principles in the metropolitan areas in coastal China. Based on content analysis and semi-structured interviews, an overview of state envisioning processes of urban sustainability is first presented as a conflicting context for the disequilibria observed in the environmental governance system across the metropolitan areas. The interactions of observed disequilibria grounded locally potentially weaken the state-led envisioning processes. With the presence of challenges in state-led transitions, the chapter proceeds to case studies in Shenzhen, the city that is practicing the state-led sustainability visions, to analyse how local dynamics of resource and organisational landscape are integrated in the transition process.

Chapter 13: Schwartz – A transition management approach for shrinking cities in the United States. This chapter applies transition management approach as a heuristic to identify areas of possible transformative change and opportunities for new types of governance using the transition management tenets and phases. It comes with the motivation that transition management can be a valuable approach to shrinking cities in the United States. Specifically, the principles of Transition Management may prove valuable to planning and policymaking efforts in the US, particularly in the shrinking cities of the Rustbelt region. In these cities, population loss and deindustrialization pose daunting challenges, but also offer robust and multi-faceted opportunities to pursue more sustainable urban futures. Established planning practices in the US are primarily geared toward managing growth. These practices are often ineffective in cities that are not growing. New, more flexible approaches to planning and community engagement are needed. The principles of Transition Management may prove especially relevant for shrinking cities in the US because the overarching idea that transition management is not about adapting or correcting current conditions, but rather a focus on transformative change.

Chapter 14: Lachman et al. – Context-driven transition management as a necessary vehicle for sustainable urban futures in Suriname. This chapter presents how transition management can be applied to inform sustainable urban development in a fast urbanizing city, Suriname while considering climate change and housing challenges simultaneously. The analysis of the urbanization context and its challenges and the insight view of the existing planning processes posit this chapter as a special contribution to our book. Readers should consider this chapter as an analysts' view on possibilities for methodological advancement of transition management to fit to a very challenging context. Authors summarise their contribution as follows: The capital of Suriname has experienced enormous growth over the past decennia. The decline of industries in other urban settlement have resulted in migration to the capital along with population growth and ongoing country immigration. With the absence of sound urban and spatial planning, the result is urban sprawl where increasing population and traffic density create ongoing specialization of services in the capital with continuously decreasing action radius; there is a tendency where people opt to live further from the city core but which makes the utilization of services in the city's core cost-prohibitive. The answer to this is the creation of new urban settlements which build on new industries with increasing returns. However, this implies a break from existing paradigms with respect to strategizing, policy, planning, tourism, logistics, urbanism, etc. This requires a transition for which management has been broken down in several steps in literature. However, transitions tend to differ along geographical parameters; for instance, regimes tend to be relatively stronger in developing countries. Thus, transition management needs to fit the context which requires assessment tools such as Quinn's culture model and the STEEPL and OPTIMISM frameworks to anticipate pitfalls and opportunities during the transition.

Chapter 15: Jhagroe - Urban Transition Management as a democratic practice? The case of Rotterdam's waterfront regeneration. This chapter examines the democratic aspects and challenges associated with urban transition management. It does so at a conceptual level and critically reflecting on the application of transition management in the context Rotterdam's sustainable waterfront regeneration. As such, it aims to shed a more reflexive light on the increasing academic and public interest in transition management.

Translating Transitions Thinking and Transition Management into the City Planning World

Peter Walsh

Abstract City planning draws out relationships between land use patterns, housing form and density, transport, energy provision and other infrastructures. While commonly struggling with the task, a city's planning system *could* be one of the key integrating forces for tackling the sustainability challenge. As it happens, these particular systems are quite regularly subject to coordinated and relatively well resourced reform efforts, usually *not* generated from a sustainability viewpoint. This chapter raises the question of whether, and how, transition management ideas might more directly enter and influence this "turbulent" reform space. Sydney provides the case-setting for the introduction of transition management, and related conceptions, to a recent major planning system reform project for that city. Working from a theoretical meshing of planning, relational-institutional and transitions studies, a new process model for planning system reform projects is introduced: the Planning System Transition (PST) framework. This framework's development and testing, through empirical research, is explained. This chapter suggests positive prospects for transition management in city planning reform, but also highlights the critical requirement for contextualisation with practical and scholarly planning insights, as exemplified in the PST framework.

Introduction

Ambitions to shift urban configurations towards more sustainable spatial development patterns disturb established interests and therefore run into fierce resistance. In Australia's major cities the systems which underpin urban planning structures and processes are so deeply embedded in a short term, mostly property-related argument ("pro" versus "anti" development/urban change), that there is little room for policy, or even discourse, on the longer term challenges before cities

---

P. Walsh (✉)
Institute for Public Policy and Governance, University of Technology Sydney, Ultimo, NSW, Australia
e-mail: peter.walsh@uts.edu.au

N. Frantzeskaki et al. (eds.), *Co-creating Sustainable Urban Futures*,
Future City 11, https://doi.org/10.1007/978-3-319-69273-9_11

(Gleeson et al. 2012; Ruming and Gurran 2014). This general kind of problem setting is familiar to transitions scholarship, with its interest in sustainability-oriented change at the urban system level. Likewise, for transition management (the point of attention in this book), given its concern with the pragmatic steering of urban system transition towards that goal.

Planning systems are regularly under a reform spotlight, often at the behest of the pro and anti-development protagonists *themselves* (both, and at the same time). A conspicuous theoretical and practical question for those interested in sustainability transitions is what would happen if the regularly occurring planning system reform episodes (usually mandated from within an established *regime*) were *themselves* founded on a transitions approach. This chapter is concerned with probing this question. From 2011 to 2014 a major ("once in a generation") government-mandated effort at the reform of Sydney's planning system was undertaken. This case provided a conducive experimental setting for the question's examination.

The chapter starts with an outline of the connections between city planning systems and the challenge of sustainable development. It then steps into the case study to initially evidence the problems with translating transitions thinking into the city-planning world. It moves on, inductively, to develop a logic which involves the linking of contextualised planning insights to transitions theorisations as a means of helping make transitions ideas more resonant to planning system actors. The chapter then explains how a more detailed "meshing" of planning and transitions ideas was developed into a new process model (the "Planning Systems Transitions" or PST framework). Later in the chapter, the PST framework is applied, deductively, as a critical evaluation tool for the recently concluded planning system reform exercise for Sydney. This results in the identification of a set of insights for planning system reform project conception and design, especially how blockers of transformative change might be better avoided.

Given its centrality to the chapter content, the PST framework can be introduced briefly here as comprising three progressive innovation "horizons", which an organised planning system reform project should attend to if it is to have prospects of achieving transformative change. The original "3-Horizons" model of Baghai et al. (2000) was concerned with *mapping out* transformational potential; to see the deeper patterns which underlie events, and to have an awareness of interdependencies involved in effecting change (Sharpe 2013). The PST framework has the same concern. Its first horizon (H1), poses the question of *intent*. Systems thinking is used to explain why aiming for wider societal benefits (for planning this is captured by the idea of sustainable urban development) can open up opportunities for change which tend to be closed off in business-as-usual (BAU) approaches. Regular approaches to planning reform, which commonly adopt a narrower, property-centric intent, are more inclined to run into blockages because of political competition among sectoral interests. The second horizon (H2) poses the question of what different *understandings* might be required to conceive a reform project underpinned by this intent. The challenge is a complex one and the essence of this horizon is that there be engagement with this complexity, rather than avoidance or an over-focus on overcoming it. On the one hand this concern is with understanding planning's *spatial* intricacies; what is important, and what is less so (and why), in the bringing about of planning-related societal improvements (associated with H1). But also the *institutional* complexities of change in urban systems, discussed earlier. Finally, the third horizon (H3) poses the question: how might *action* be framed mindful of the particular reform intention and the problem understanding. While transition thinking is meshed through the whole framework, transition management comes to the fore here in H3. This is through the guidance it provides on the action phase: agenda setting, participation processes (including alternative democratic processes), and what types of project outputs should be targeted (over others), if transformative planning system change is the goal. Figure 1 provides a graphic overview of the PST framework.

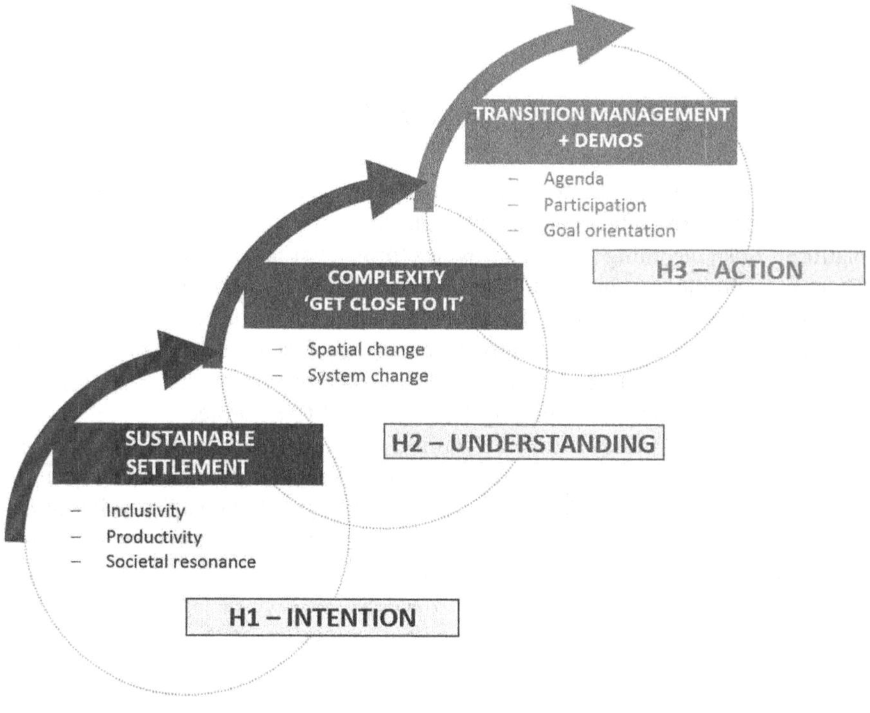

Fig. 1 Planning system transition (PST) framework: a graphic outline (Walsh 2016, p 127)

Planning Systems and Reform

City Planning and Urban Sustainability Governance

The planning system can be thought of as the set of arrangements which underpin the preparation of land use and infrastructure strategies, zoning plans, and related development regulation, and the various participatory activities and decision processes involved in each. For cities, planning systems can be an important integrating force for many other urban system elements central to the sustainability challenge. These include: housing, transport, energy, public health, infrastructure, environmental management and social justice systems.

Like transitions thinking, planning at its foundation is values-based, concerned with managing the spatial aspects of urban change to achieve wider social, economic and environmental goals (the "good city" Campbell and Fainstein 2003, p 1). City planning also, by definition, take us beyond the "here and now" and thus, like transitions, in word if not deed, sees itself as the steering of change over time to a sustainable future. While it may be commonly something very different to this, at its best *perhaps* city planning *is* transition management.

The Sydney Planning System Reform Project

A major project aimed at reforming the planning system for Sydney provides the case which underpins this chapter. Planning reform was one of the key election platforms of an incoming New South Wales (provincial-level) government in 2011. There were political ambitions to *both* reduce wasteful red tape which was seen to be holding back the economy, *and* empower local communities dissatisfied with planning decision making. The central locale of attention for the reform project was the greater Sydney metropolitan area. For the sake of this chapter, the project is being called the Sydney Planning System Reform (SPSR) project.

Over a 3-year period, the SPSR project comprised: extensive participatory processes (with all tiers of government, major stakeholder groups and the general public), an array of technical analyses and documents (including a White Paper and draft Bills), and much (typical) lobbying and political horse-trading about the particulars of change. Ultimately, the project was unsuccessful, on the government's own account, due to it failing to achieve passage of intended legislative changes through parliament.

Transitions Thinking, Planning and the SPSR Project

Translation Problems in Initial Attempts

For this writer, a planning practitioner, involved for over 25 years in advising government on planning system reform, the recent "uncovering" of sustainability transitions (ST) theory provided a new perspective. By the time of the commencement of the SPSR project I was already exploring the prospects for ST theory in the planning system domain. Given my past work in the field, I thought there were good prospects of convincing the "powers that be" of the benefits of including at least some experimental work related to transition management during the SPSR project.

But sustainability transitions is a notably complex and abstract concept, and in my early efforts at engaging in the topic with government officials I was mostly met with blank looks. This early contact with senior colleagues involved in the SPSR effort identified an initial (and continuing) challenge of translating transitions ideas into the city planning world. It was apparent that for transitions ideas to have *impact*, there was a need for better planning contextualisation of the transition ideas, and a better situating of them into local planning settings. The research turned to planning's own scholarship for help in this task.

Mainstream Planning Scholarship and System Reform

My research within mainstream planning scholarship itself, up until then, had exposed little thinking of relevance to the topic of planning *system* reform. Planning scholarship included a number of core paradigms regarding the quest for sustainable urban development. As a point of origin, planning's humanist intentions are foundational to the literature (Friedmann 2008). Then, at its best, the rational technical planning paradigm provides an important capacity to provide images of the sustainable city. It works on detailed associations and the particulars of relations between, on the one hand, technology innovations for cities (such as that related to energy, water and transport), and on the other, integrated spatial patterning and urban design innovation to deliver them (for both infill and greenfield arrangements) (Newton 2008). A key criticism of rational technical planning is whether it can be responsive enough to the pace of change in cities today, and how it might account for unexpected relationships and innovations over time (Searle 2012).

The popular communicative planning paradigm adds a required socio-political dimension with its acknowledgment of the need to step beyond negotiation (with established interests) towards public discourse and collective deliberation (e.g. Dryzek 1982; Hajer 1995; Forester 1999; Healey 2003; Innes 2010; Hendriks

2011). But there is also much in the way of criticism for this paradigm. For communicative planning, it includes its prioritisation to (often unrepresentative) *dialogue processes* over the *substantive*, such as actually *achieving* normative societal goals (Fainstein 2000; Flyvbjerg 2002; Roy 2010).

The most applicable paradigm to planning system reform today is that of market liberalization, or neoliberalism. The importance of market efficiency as a means of encouraging innovation and competition needs to be acknowledged, including for the challenge of sustainable development. But it is the *dominance* of the neoliberal agenda in the reform "project" that is of concern. In practice, it has proven a frustration in the quest for sustainable development. Business and indeed government have been much more ready and able to deliver a polemic on the effect of planning systems on current business and jobs (Westacott 2015), than other voices have been in support of planning as a means to help deliver sustainable social and economic models for our cities of the future.

Beyond the Mainstream: Relational-Institutional Planning Thought

In the search for better understandings and tools to attend to the challenges involved in spatial change, planning researchers have been drawn to *institutional studies*, an area also familiar to transitions scholars. This line of research suggests it is "institutions" (its meaning here is as in its sociological sense, as "social norms" or the "rules of the game", etc. – not just organisations) that give structure to key planning relationships and guide the behaviour of involved actors (Hall and Taylor 1996; Bell 2002; Healey 2007a; Powell 2007; Verma 2007). These institutions encourage certain (strategic) behaviours and discourage others. Under this thinking, institutions are socially embedded and will tend to favour those already privileged. In planning reform endeavours this might include property development interests, local resident groups, along with the bureaucracy or politicians themselves, as election cycles have their effect. The concern is the tendency for concessions and trade-offs among interests to frustrate wider and long-term ambitions. So in a planning reform effort, big system modifications may come about (new planning laws, city governance structures, new metropolitan strategies, etc.) but in their particulars they will be ineffective in bringing about transformative change due to institutional resistance.

Where *relational* thinking joins in, is with its attention to opportunities to moderate institutions (Healey 2006, 2007a). The idea from social-institutionalist thinking is that there is a dualism which is operating here (aligned to structuration conceptions Giddens 1984). On the one hand there is the structuring force of institutions and on the other the possibility to moderate them, though relational interaction which is encouraging of agency. The term "relational-institutional" (my own terminology) can then be thought of as involving the conscious linking of the two dynamics, as a means of opening up change opportunities in planning reform efforts. That is to shift the gaze away from "big system modifications", which might be illusory, towards the importance of stimulating opportunity in reformist efforts for social (or relational) action with this intent (i.e. to directly challenge institutional blockers). It places the "spaces between" the start of a reform process and any final policy shifts (say new laws or governance structures), as the central unit of analysis in a reform effort, rather than the nominal content of the policy shifts themselves (Bradbury and Lichtenstein 2000, p 551).

What's Still Missing: And How Transition Management Can Help

If more sustainable urban settlement and growth management is the aim, then city planning systems need to take on a different and more influential role. The insights above explain planning's critical *potential* but also the institutional complexity with delivering on it. What transition thinking offers, in addition, is a more detailed appreciation of the mechanisms of change in an institutionalised urban planning system. This is through ideas like: (1) the multi-level perspective (how niche activity and/or landscape pressures can re-align dominant planning regimes), (2) the multi-phase concept (importance of timing and thus how *system reform projects* have a special status), and (3) transition management (as a governance orientation and action guideline).

As explained in chapter "Introducing Sustainability Transitions' Thinking in Urban Contexts" (this volume) in particular, the notion is that there are specific patterns, dynamics and mechanisms that drive or block system change and, provided these are understood, the relational-institutional complexity becomes less of an obstacle, and more a practical means of securing some leverage for reform. For a planning system reform project transition management seems to have particular pertinence, with the potential for direct application of heuristics aimed at transformative social innovation, like: the transition management cycle, transition arenas, transition agendas and transition experiments (Grin et al. 2010).

Translating Transitions Ideas into the City Planning World

Conceptualising a Means of Translation

There was a need for a logical means of connecting-up planning and transitions thinking, which provided a concrete spatial and institutional context to the more abstract constructions in transition studies. The approach which was adopted involved a simple bridging of the literatures. This *planning system transition*

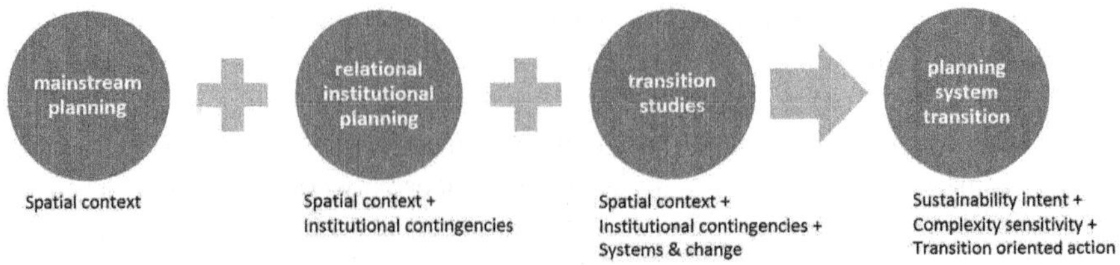

Fig. 2 Planning system transition (PST) logic includes three theoretical pillars

(PST) logic relies on a set of three theoretical pillars: (1) mainstream planning – especially its focus on spatial context, (2) relational-institutional planning – which draws a deeper attention to the idea of reform and the importance of institutional (and relational) contingencies to it, and (3) transitions studies – concerned exclusively with systems and their transformation and especially transition management as the action component. This conceptual connection is suggested graphically in Fig. 2.

Taking the Logic into the World of Practice with Interviews

The PST logic was interrogated in a set of interviews with very senior and/or experienced planning system actors (n = 28), who represented a range of backgrounds and perspectives (government officials, developers, community activists, environmental interests, academics, professional interests, politicians). It helped here that I have had a long-term career in the Sydney planning domain. This provided considerable access opportunities, but also opened up opportunity for critical and candid discussions on shared experiences within the interview. Each of the interviewees had either direct involvement in, or a sound appreciation of, the SPSR project. There is insufficient space to explain the details of the interview process here. But a modified phronetic research approach (Flyvbjerg 2004) sat behind the interviews and set the structure for the sequential approach shown in Fig. 3.

The interview started with reflections on whether planning in Sydney was on the right track (Questions 1 and 2 in Fig. 3). But the question was not just about planning outcomes, it was also about the institutions driving these outcomes. Question 3's inquiry on "what's to be done" was centred on planning system change. The approach was to first ask interviewees why they thought system problems were so hard to shift. This was followed by the interviewer positing some suggestions. First the relational-institutional ideas were put up for discussion as an understanding of how things get stuck. Then transition ideas were offered as a means of steering change, including outlines of the multi-level perspective, and the transition management cycle (outlined at chapter "Introducing Sustainability Tran sitions' Thinking in Urban Contexts", this volume), highlighting points of difference with BAU reform approaches.

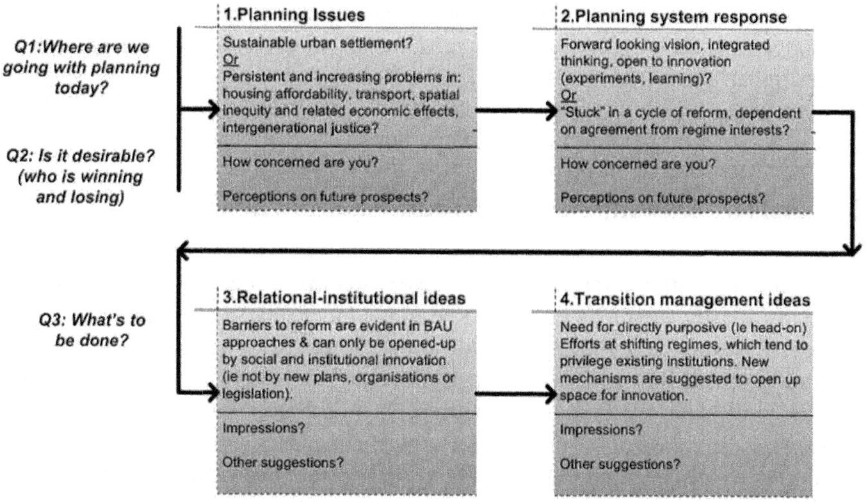

Fig. 3 Outline of structured and sequential approach used in interviews (Walsh 2016, p 150)

Based on the interview approach's adoption of this combination of: (1) spatial and institutional context, and (2) steering of discussion towards relational-institutional and transitions ideas as a means of introducing new thinking; a large majority of interviewees, across the sectors of interest, were able to engage with the concepts (including transition management) and saw them as providing important insights. The change from my initial dialogue with system actors armed with transitions ideas alone (outlined in section "Translation problems in initial attempts") was marked. Some descriptive quotes from interviewees are provided below:

> *A highly useful discussion. This work needs to be picked up in the education of planning practitioners and politicians.*
> (Consultant)

> *... a really interesting discussion... I think it's a good process because you're taking us further than we go now*
> (Academic)

> *...I really like your approach. What I like the best about it is bit by bit, little by little, education and change...*
> (Local Community Advocate)

The position was summed up in questions put at the end of the interview, which sought perceptions on: (1) "how important" the relational-institutional and transitions concepts covered in interviews were, in regard to planning system reform, and then (2) existing levels of "understanding" of the concepts on the part of actors in the planning reform domain. All responding interviewees ($n = 21$) indicated the interview subject matter as *highly* important but levels of existing understanding of the concepts as relatively *low*. This feedback from planning system elites was suggesting an important institutional knowledge gap.

Towards a Planning System "Transition" Framework

One of the suggestions from interviewees was that these non-mainstream planning and transitions conceptions needed to be made more tangible and practical. This viewpoint was aligned with my overall research goal of developing a more pragmatic conceptual tool for planning system reform episodes themselves. A more critical synthesis of the literatures was needed if such a tool was to have practical value in making sense of the processes and mechanisms involved in reform. Each of the literatures are conceptually rich and have significant individual offerings, which can be seen as a (sometimes overlapping) progression, but with some areas of theory foregrounding in places where others are less strong. This suggested meshing of literatures is shown graphically in Fig. 4.

The next step in the research project took this conceptual logic into a framework for the design/implementation, or evaluation, of a planning system reform project in the world of practice. The new practice model, the Planning System Transition (PST) framework and its three innovation horizons of *intention*, *understanding* and *action*,

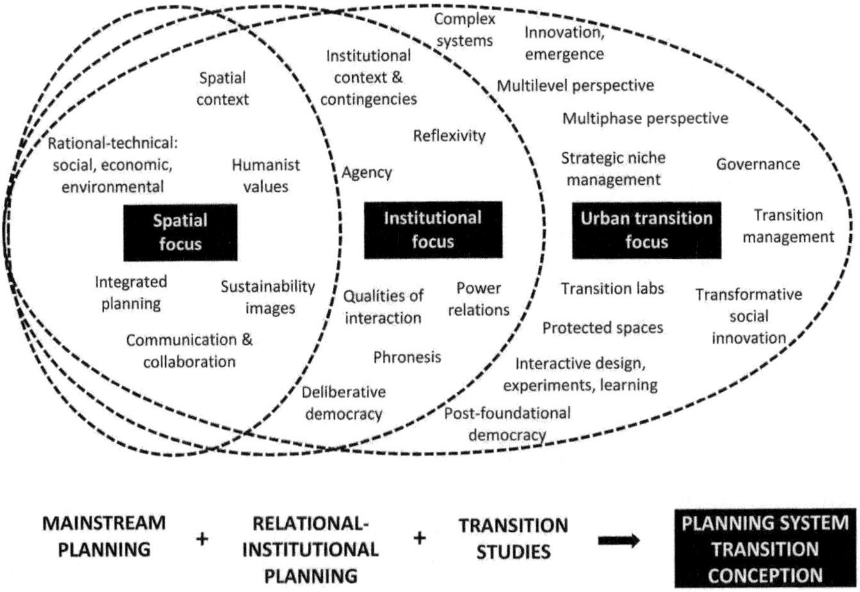

Fig. 4 Deeper synthesis of the literatures behind the PST logic (Based on Walsh 2016, p 124)

aimed at helping steer transformative change, has already been introduced in section "Introduction". The next section applies the PST framework to the SPSR project.

PST Framework Meets the Sydney Planning System Reform Project

This section works through each of the (three) PST horizons in turn. Each subsection starts with a brief outline of the conceptual construction behind the horizon. Then a comparison is drawn with the parallel aspect of the SPSR project. It is in this comparison that more of the particulars of the horizon are able to be drawn out (see Walsh 2016 for a more complete explanation). While it is H3, the action horizon, which relies most heavily on transition management, it will be seen that transition management concepts thread through the entire framework, as do conceptions from planning and relational-institutional scholarship. With this book's attention to transition management, it is worth emphasizing that this (theoretically-founded) "meshing" and "threading-through" is addressing two separate problems.

On the one hand, it provides the means of *translation* of the more complex transitions concepts into the world of regular planning system actors. But it also provides the response to the problem of *regime*-capture. As discussed in chapter "Introducing Sustainability Transitions' Thinking in Urban Contexts" (this volume), transition arenas are normally operating at the edge of the regime involving a select group of transition-minded elites. But here we have a regime-*mandated* reformist project, and a different set of actors are in favour. This framework is suggesting that it is the embedding of transitions conceptions (meshed with the relational-institutional) within the early stages of project conception (H1 and H2), which positions the action stages of the reform project for an atypical approach (in transition management terms) to the dialogue and action planning elements of the reform project.

Horizon 1: Intention

"Intention" Under the PST Framework: Sustainable Urban Settlement

Under the PST framework, establishing *intention* is a first decisive, direction-setting, horizon for a reform effort. There are three aspects: (1) delivering inclusive societal benefits, rather than only selective or exclusive benefits to empowered interests, (2) improving wider city economic prospects, rather than only on property-related economic interests, and (3) systems thinking, and that reform is more likely where "attractors" are part of the package (to act as a magnetic forces for change in path-dependent systems, see Grin et al. 2010, p 116–7).

"Intention" Driving the SPSR Project: Comparative Analysis

Within 3 months of coming into power in April 2011 the O'Farrell Liberal government announced its plans to "overhaul" the planning system. A review of ministerial media statements directly concerned with the planning system reform project and released between the announcement of the reform project and the White Paper's release indicates serious ambitions on the part of government. It was to be a "monumental overhaul of the ... system" (NSW Government 2011)". Each of the nine media statements released during this (pre-White Paper) period highlighted transformative *process* intentions including in regard to: civic engagement, integrity, transparency, certainty and simplifying the overly "complicated" system. The most commonly cited *substantive* (or outcome related) objective of the reforms coming through in the media statements was that of providing "new housing" to address the "current housing shortfall".

By the time of the release of the White Paper in April 2013, the government's intentions had been further clarified. The introduction section to the White Paper, concerned itself with establishing the case for reform. It specified the following as "objectives" of the intended new planning system (NSW Government 2013, p 15):

> *The main purpose of the planning system is to promote economic growth and development in NSW for the benefit of the entire community, while protecting the environment and enhancing people's way of life.*

Inclusive Social Progress or Exclusive Societal Benefits? The language in the White Paper and draft Bills were conspicuous in their foregrounding of the encouragement of economic growth and (housing) development as the core intention of the planning reforms. The backgrounding of social and environmental priorities is also borne out in other aspects of the intended reforms. These include:

- A name change from the key planning legislation from today's "Environmental Planning and Assessment Act" to "Planning Act".
- Weakening of the specific legislative provisions relating to sustainable development. For example, "intergenerational equity" and the "precautionary principle" would be removed from statutory definitions under the exhibited Planning Bill.
- Codifying, within the intended legislation, a set of principles to guide the preparation of strategic plans which clearly privilege short term growth over other environmental concerns (for example, see Principle 10 of the draft Planning Bill's strategic planning principles (NSW Parliament 2013): "Local plans should facilitate development that is consistent with agreed strategic planning outcomes and should not contain overly complex or onerous controls that may adversely impact on the financial viability of proposed development".

Wider Economic Intentions or Property-Centric? The PST framework extends reformist economic intentions beyond the property-centric (i.e. improving prospects for private sector investment in property) to wider questions of city productivity (e.g. congestion-effects, agglomeration opportunities, relationships between housing costs, spatial disadvantage and productivity Piketty and

Goldhammer 2014). The documentation accompanying the intended reforms were limited in their drawing of such wider relationships, falling back on the view regularly expressed by development lobby groups suggesting the planning system as the "chief cause of housing market problems", a fact not supported by the evidence (Gurran and Phibbs 2013). But more importantly here, the focus on this rather limited aspect of the planning system's effect on the economic prospects for the city is seen as a missed opportunity. The situation was available to open up the planning reform conversation to what's actually at stake in regard to wider economic realities for cities like Sydney (Treasury Australia 2010; Newton and Doherty 2014). This could help significantly in explaining why some, otherwise difficult to accept, urban change may be necessary.

Intention as an "Attractor" to Help Shift Institutional Behaviours? There was evidence of enthusiasm during the early stages of the reform project. The suggestion from the Minister was that: "putting community participation at the forefront of planning decisions about how our cities and towns will grow and develop" (Hazzard 2012); would be a revolutionary shift away from the suggested bad old days of "deals-for-dollars culture and red tape that has been strangling investment in NSW" (ibid). But the enthusiasm shifted over time to cynicism. The shift was captured in two sequential independent surveys on attitudes to the proposed reforms (Manion and Marketinfo 2013). A key catalyst was the White Paper's suggested increase to the proportion of development applications (DAs) about which residents would not have a say, that is "code assessment" development (the White Paper target was for 80% of DAs to meet that definition and as such not be affected by comments from neighbours). This became a touchstone for backlash against the entire reform project. Residents' association activists were pointing out that any increased community powers at the more strategic (i.e. conceptual) planning stages would be replaced by dilution of consultation requirements at the DA stage. By the end of 2012 a new group called the Better Planning Network (BPN) had formed with the ambition of overturning the reforms. Corinne Fisher (2012), the convenor of the group, captured the essence of its concerns in an interview, reproduced in part below:

> *These reforms have an overall emphasis on fast tracking development at all costs and that is not what is going to be good for our communities and for this country... We need to have a visionary planning system, integrate fundamental principles: mitigating for climate change, ecologically sustainable development, protecting our heritage, getting good building design that enhances the wellbeing of our community.*

As this conflict around the intended trade-off over community consultation (more strategic and less at the DA stage) built-up a momentum at the community level, there was hardly any discourse from advocates on how the reforms might help shape the future trajectory of the city for community good (beyond more housing supply).

Table 1 summarises the key points of comparison between: (1) the PST ideas related to intention in planning system reform efforts, and (2) the intention brought to the SPSR project as apparent from this research.

Table 1 Comparing SPSR project with PST framework's "intention" (Walsh 2016, p 181)

PST Framework	Characteristics	Planning System Reform Project alignment
1. Intention ***Sustainable urban settlement***	***Inclusive social progress*** *Attention to inclusive societal benefits over exclusive.*	– Little mention of wider societal challenge. Goals centred on growth and "making NSW number one again". – Virtually no mention of social justice concerns, low carbon future or intergenerational interests.
	Improving city productivity *Wide interpretation of economic implications, rather than property-centric.*	– Relatively narrow aspirations for planning system to influence society's wider economic questions. – Economic intentions mostly property related.
	Shifting embedded governance *Intention is framed purposively and ambitiously to inspire system shift, not just respond to interests*	– Idea of increasing the power of people in planning decisions acted initially as a new attractor. – The associated transformative power was weakened as reform particulars entered the debate.

Horizon 2: Understanding

"Understanding" Under the PST Framework: Alertness to Complexity

This second horizon of the PST framework acknowledges the deep challenges involved in steering change in city planning to be more supportive of sustainable development. H2 is aligned with Rotmans (2005) suggestion that "… between dream and reality lies complexity". The *understandings* required to appreciate the planning system challenge under the PST framework operate in two domains. One is empirical and is concerned with spatial planning, development processes and outcomes on the ground. This domain is particularly linked to mainstream planning paradigms. The other is more conceptual, and is concerned with understanding the challenge of system change. This domain draws on relational-institutional and transitions thinking.

"Understanding" Driving the SPSR Project: Comparative Analysis

Empirical Understandings Driving the Reform Project There were just two new pieces of independent expert research which were commissioned by

government (or at least released and promoted) during the project. The first piece links to the PST idea of evaluating planning benefits, and comprised a cost-benefit analysis of the proposed regulatory changes. This research (CIE 2013) indicates the reforms as proposed in the White Paper will bring savings of between $848 m and $1482 m per year if fully implemented. Much of this was associated with the reduced risks associated with community objections to development, and negative government decisions. But there is little regard to the serious risks ahead for cities and wider society due to factors like climate change. There are points of difference in how cities: (1) contribute to such problems and (2) are resilient or otherwise to them. The spatial form of cities is central in many ways to both. Under PST thinking, planning system reform should be closely concerned with city change which improves resilience and reduces risks and impacts for future generations. Among the factors missing today are economic models that factor in sustainability (Doherty 2010). Overall, this commissioned work, or, perhaps its brief in the context of the SPSR project itself, should be seen as a missed opportunity to introduce the wider empirical questions of the economics of a sustainable future and the place of city planning within it.
The second piece of empirical research was a "review of international best practice in planning law" (Stein 2012) which covered three themes: (1) "mediating relationships between state and local government", (2) "the form and substance of the plan", and (3) "development control" (p 3). Whereas the "colour" and detail of Stein's research report certainly *does* show an awareness of: (1) sustainability concerns, (2) growing planning complexity and (3) required planning law responses; this component of the advice is not well represented in the White Paper or Bills, which on the topic of sustainability rarely go beyond the first order idea of avoiding negative impacts. This is evidenced by the Stein report's citation of several particular statutory exemplars which have regard to the core intention of the PST framework. For example, he cites a particular provision on the question of short term vs long term thinking:

> *Ensure that plans are drawn up over appropriate time scales, and do not focus on the short term or ignore longer term impacts and the needs of communities in the future. Planning authorities should consider both whether policies have short term benefits which may have long term costs, but also whether short term detriments (which are capable of being mitigated) may be offset by longer term benefits which are realistically achievable.* (Clause (ii) of UK Planning Policy Statement No.1, as cited in Stein 2012, p 18)

This exemplar is almost completely opposite in intent to the "principles to guide the preparation of strategic plans" as proposed in the SPSR project's exhibited Planning Bill (NSW Parliament 2013).

Conceptual Understandings The PST framework suggests that changes to "hard" system elements like laws, regulations and plans may be illusory without attention to the "softer" challenges of institutionalised norms (cultures, structures, practices). However, the SPSR project was highly focused on changes to these fixed system elements. This is evidenced by the focus on achieving passage of legislative changes through parliament within a tight timeframe sensitive to the 2015 government elections, and the way the rest of the reform program seemed, in the public eye at least, to be dropped by government when the legislative changes failed. However, this did not stop many initiatives arising in the SPSR project which were departures from regular thinking and which confronted "softer" questions of institutional change. Below a set of "transformative changes" nominated in the White Paper are synthesised to present their potential as devices to address institutional blockers.

Strategic Planning and "Growth Infrastructure" Provisioning A complaint about the current system has been the lack of coherence between planning decisions (such as zoning changes to promote growth in infill areas) and planning and provisioning commitments for hard and soft infrastructure. The reforms proposed a "major shift" to a whole-of-government approach for the first time including cabinet-level commitments to ensure linked up land use decisions and infrastructure provisioning (Jackson 2013).

Community Participation Already discussed is the nominated intention to "(place) community participation at the forefront of the new planning system" (Hazzard 2012). While there were complaints about decreased participation at DA levels, clearly it is at the strategic planning stage that the more critical decisions about sustainable urban settlement are made. The White Paper included a new focus on more democratic participation (NSW Government 2013, p 51) and towards the tail end of the SPSR project the government initiated a set of randomly selected citizens' juries or deliberative panels to consider some aspects of the reforms, as means of offsetting the voices of the usual suspects (pro and anti-development) which had been quite dominant.

Governance Shifts A new concept of "Planning Boards" was to be introduced comprising representation from elected government as well as independent participants (appointed by government but not representing a particular jurisdiction). Planning boards would be established at the "sub-regional" level (roughly 500,000 people). The boards were intended to bring a new capacity to influence whole-of-government planning and infrastructure provisioning for this spatial scale, and were charged with effecting the intended upscaling of civic engagement. With appropriate terms of reference, the board arrangements could open up opportunity for more deliberative and democratic approaches, and for the planning narrative to prioritise achievement of the larger and longer term planning goals, than do current very fragmented local government arrangements (at present some 41 local councils have control over planning decisions within the Sydney metropolitan region).

Culture Change A further stated intention of the reforms was to establish a "delivery culture" for successful implementation of the new "world class" planning system (NSW Government 2013, p 34). PST thinking links the need for culture change with new governance arrangements, that is through "mutually consistent and reinforcing changes" to structures and associated practices (Grin 2011, p 4). But the suggested "delivery culture" seems to be requesting more "pragmatic" (NSW Government 2013, p 6), and perhaps therefore *compliant* council planners and CEOs. It presents as a likely candidate for the regular criticisms of "cultural

Table 2 Comparing SPSR project with PST framework's "understandings" (Walsh 2016: 190)

PST Framework	Characteristics	Planning System Reform Project alignment
2. Understanding *Engage with the full complexity of the system reform challenge*	*Empirical understanding* *What's wrong now in planning?*	– Focus was on planning as a brake on the economy, without accounting for contrary empirical evidence. – Little accounting for past reform experiences. – Some incidental openings for discourse on sustainability and long term urban economics in commissioned research. – Relatively low levels of reflexivity and only incidental alignment with PST.
	Conceptual understanding *Planning reform as complex societal change.*	– Some of the reform elements showed reflexivity and struck at key institutional issues. – But the naïve or avoidant focus on legislative change (especially code assessment goals) tended to dominate.

change" in public sector reform programs which suggest they are mostly about changing the "identity" of public servants to be more encouraging of pro-market ideologies over wider public interests (Inch 2010, 2012). Table 2 summarises the key points of comparison between the PST idea of foregrounding an understanding of planning reform complexity, and the understandings brought to the SPSR project.

Rather than bring new understandings, the SPSR project seemed to find itself facilitating familiar aggregative politics (on the one side pro-growth and development, and on the other "save our suburbs") which in the end brought down the intended reforms, including those representing considerable initiative. In the interviews stage of this research, one of the State's most experienced planning consultants put it thus:

> *They couldn't help themselves ... leading with the rhetoric "NSW Number One", the emphasis on growth and development ... when there was enough content in that White Paper to have a completely different narrative. They could easily have focused on consultation and sustainable development but missed that opportunity.*

Horizon 3: Action

"Action" Under the PST Framework: "Transition Management" but in Context

The final horizon of the PST framework is focused on the design of the specific reform project itself. It is concerned with *actioning* the reform episode. Transition management theory, the main subject of this book, underpins suggested *action*, due to its specific attention (i.e. beyond intentions and understandings and) to the steering of actual system change. A particular point of interest here are some baseline incongruities between the transition management approach, which would by definition happen outside of the "regime", and the fact that the PST framework is intended to direct a regime-mandated reform exercise. For example, the opening institutional position for a regime-mandated planning reform project would be expected to be inclined towards:

1. Solving system problems which have the most resonance for politically powerful stakeholders – *whereas* transition management is about searching, learning, cooperative design and strategic management of niche activity to help achieve societal innovation and sustainable development;
2. Self-selected participation in reform arenas (i.e. by already vocal and/or politically connected interests) – *whereas* policy arenas in transitions thinking are more selectively looking for participation by diverse innovators and even outsiders;
3. Shorter term political action (e.g. driven by election cycles) – *whereas* transition management brings a longer-term orientation.

So there are questions as to the prospects of transition approaches associated with the "closeness" of the regime to the reform actions. But this is offset by positive effects which a major government-led reform project brings, such as: (1) the resources allocated by many parties to the work at hand (especially, but not only, government), (2) the attention-creating platform which is achieved (not so easy to otherwise secure given the socio-political competition for attention in a busy world), and (3) the consequential system instability (and thus creative opportunity) which would follow a government-mandated reform announcement. So the challenge for the *action* component of the PST framework is to capture the best of both worlds. Given the focus of this book it is worth spending a little more time in explaining the characteristics of this aspect of the framework and its three points of attention (agenda development, participatory approaches and the orientation of its outputs or results).

Agenda Development In transition management the *agenda* has a specific role in providing a "compass" for the "search and learning process" ahead (Grin et al. 2010, p 158). Under the PST approach the compass has been set by the overall *intention* which has been adopted under H1 (i.e. a wider agenda of sustainable development indicated) and the search and learning process is underpinned by the *understandings* indicated in H2 (i.e. appreciation of the complexity involved). The agenda for a planning system reform project would then take up the opportunity provided by the special (i.e. rare) platform for policy selection which opens up in a major system reform project. It would factor in both the substantive (say city form and infrastructure) challenges as well as the institutional impediments to change. Here the agenda is playing a role *itself* in "seducing" frustrated system actors (politicians, stakeholders and citizens) to the idea that there are trajectories which can be followed which can take the city along these pathways over time, and that their ideas and actions might be important in getting there (i.e. the agenda playing a role as an *attractor* in complex systems thinking). In that sense, the agenda is concerned with positioning for the mobilisation of the creative power of individuals and groups in the next action phase.

Participatory Approaches Given the powerful interests and different values involved in planning system reform, the design of participatory approaches is crucial. That is because it is here that "gamechanging" social rationalities and relationships are either opened up to change or not (Servillo and Van Den Broeck 2012). Structurally, the PST framework follows transition management's conception of the "double level" to participatory approaches (Lissandrello and Grin 2011, p 244). The first level is that of actors from different backgrounds sharing rationalities, deliberating and collaborating on new images on spatial and institutional planning problems. The second level is that which *relates* such images to existing (problematic) system attributes (cultures, structures and practices), including those which guide the behaviour of these participating actors themselves. While the opportunity to include outsiders and creative minds must be taken-up, it is a pre-requisite of a government-mandated reform project that institutional stakeholders (the "usual suspects") be deeply involved. This places an onus of the designing in of high quality ("relational") interactions which move beyond the idea of trade-offs and compromises, into the terrain of new interpretive frames which encourage the creative re-configurations of institutions and networks. The "relational-institutional" idea here is that if there is to be an encouragement of reflexivity and a sense of reciprocal behaviour, as a lead to agency, then less room should be given in participatory processes to those arguing a narrow agenda (e.g. either pro or anti development). Instead participatory process would be centrally concerned with pushing institutional actors (e.g. development interests, NIMBY groups, bureaucrats), to consider wider questions. Joining influential outsiders to the process is then one of the devices to assist.

An additional *participatory* dimension is concerned with *democratic* considerations. There is the already established baseline requirement of ensuring such considerations are designed into the reform episode itself (Hendriks 2009). But because, and unlike some policy domains, there is already a particular expectation of public involvement in planning issues, there is special attention to the potential for democratic approaches to bring legitimacy to reform and *assist* in institutional breakthrough in the city planning domain.

Outputs/Results For complex system settings like city planning, with their many ambiguities and contradictions, the concern would be away from moving too early in locking-in new structures and processes that may have unintended consequences and spark backlash. Instead the expected outputs would be concerned with ongoing facilitation of new perspectives and agency, and "guided variation and selection" in structure and process shifts. But at the political level, the expectation can be inclined towards concrete headline-attracting changes to structure and process.

This to create an appearance, at least, of immediate improvements to appease vocal interests and justify the resource allocation involved. Reform project outputs under the PST logic would seek to reconcile these short term expectations in light of the acknowledged failures and unintended consequences of previous BAU reform efforts over the years, and the counter-arguments presented in support of a *transitions* rationale as evidenced in the rest of the PST logic.

"Action" Driving the SPSR Project: Comparative Analysis

There *were* progressive agenda items within the SPSR project which had potential to bring about institutional shifts. They are introduced above and included: increased funding and commitment to community participation at the key planning stages, new decision-making structures (sub-regional planning boards) to reduce myopic decision-making, a new high level cross agency approach to strategic planning and infrastructure provisioning and certain ideas on "streamlining" planning approvals aimed at reducing the economic drag of what would be reasonably seen as an often unreasonably burdensome process. This *nominal* agenda (based on White Paper content) was "drowned out" by the pro vs anti-development argument (here centred almost completely on the extent of "code assessment" for DAs), and trumped any serious discourse about institutional change.

The participatory approaches, and they were extensive, were unable to provide a circuit breaker for this problem. The participation requirements instituted in the reform exercise were more inclined to the direct facilitation of sectoral interest advocacy than in having institutional actors ask critical questions of themselves. The encouragement of this kind of personal and institutional reflexivity is no easy task. But that only points to the importance of an appreciation of, and engagement with, the complexity of the reform challenge, as introduced with H2. There *were* progressive ideas employed in the participatory aspects of the SPSR exercise. One example is explained briefly below. But overall, there was not a convincing (enough) narrative coming from those advocating the system reforms to counter the argument, employed ultimately in a highly organised effort by what became a large consortium of community activists, that the reforms were more favouring of property development interests than any wider community interest.

The most interesting participatory component of the reforms, from a transition management perspective, was an experiment in deliberative democracy. The government had experts from the University of Technology Sydney (UTS) design and deliver three deliberative panels (2×1.5 day and 1×1 day duration) and eight focus groups (of about 2 h duration) to consider the reforms. Rather than "the usual suspects", participants in the panels and focus groups were "randomly" selected citizens, sourced from a randomly generated telephone list ($n = 3000$). After an initial sifting process, final participants were selected directly by UTS (following short telephone interviews) aiming for representativeness on factors like demography, residence location, personality types, community values and interests.

The design of the sessions was very similar to the adapted "phronetic" approach (Flyvbjerg 2004) adopted in the interviews with elite planning system actors described above. The three questions were again able to be put to this time with regular community members. Question 1 (where are we going with planning?) allowed a presentation, and a citizen level exchange, on the wider spatial and infrastructure shifts and prospects for the future. Question 2 (are such shifts desirable?) allowed factual information to be presented and a discussion on winners and losers and future concerns. Question 3 (what should be done?) provided citizen feedback on various aspects of the White Paper proposals, but also on more substantive issues about city planning and its system.

While there is insufficient room to go to the details here, it was very obvious to myself as an observer for most the sessions that the individuals involved in the process did develop a fundamentally more critical insight into both Sydney planning problems and their own preconceptions. Importantly, the consequence of this was an openness to reform which was hardly seen elsewhere in the whole of the SPSR project.

Turning to the final element of H3, the *results orientation* of the SPSR project was clearly at odds with the PST framework. The reform project saw major new planning legislation as crucial to implementing the nominated agenda suggested above. The effecting or otherwise of legislative change became the absolute centre of attention. A battleground of lobbying and counter-lobbying by interest groups, especially with the minor parties who held the balance of power within the Upper House eventuated. The scenario was one of trade-offs and brinksmanship, rather than deliberation and reflexivity, and in the end no SPSR-related legislative changes passed through the parliament.

Table 3 summarises the key points of comparison between: (1) the PST ideas on a suitable framework for action within a planning reform episode, and (2) the action stages of the SPSR project.

Concluding Remarks

To date, transition management has mostly focused on energy systems, transport technologies, infrastructures and the like (Markard et al. 2012; Markard 2014). There is an increasing interest in the possibilities and limits for its transferability into wider fields of urban change. This chapter attempts to address this interest by its effort at translating transitions thinking into city planning at a systems level. Some problems were found with translating transition thinking into the planning world, especially associated with its complexity and abstraction. But this case study evidences that these can be offset by an effort at contextualization. An approach which meshed transitions and planning theorisations was promising in interviews with experienced planning system actors. The planning system transition (PST)

Table 3 Comparing SPSR project with PST's "action" framework (Walsh 2016 p 206)

PST Framework	Characteristics	Planning System Reform Project alignment
1. Agenda	*New visions/images/objectives in the face of embedded governance deficits.*	– Some creative new imaginings in the nominal agenda, genuinely take on serious governance deficits.
	Translate these new understandings to opportunities for institutional action centred on agency.	– A shadow agenda became more prominent (either pro or anti-development) and became a blocker for agency. – Intentions and understandings were not well suited to counter this dynamic.
2. Participatory approaches	*Arenas created to challenge dominant ideas.*	– Arenas were not generally reflexive. The most powerful were those formed to resist the suggested reforms.
	Learning, new images, new networks emerge.	– No visionary images of what might change for the city with the reforms. – Most powerful networking was strategising to resist the reforms (eg Better Planning Network)
	Design in participation by "frontrunners" with diverse interests.	– Some inclusion of thought leaders. But most participatory work more inclined to support regular sectoral interests.
	Design in democratic procedures.	– Late decision to try deliberative democracy experiments showed latent power.
3. Outputs / results	*Less "one shot" structural change, more facilitating new perspectives and agency*	– Reform success or failure centred on achieving legislative change.
	Reforms position for breakthrough projects to evolve and upscale over time.	– The reform process inclined to have resulted in reform fatigue and burn-out.

framework was introduced as a device for the design or evaluation of organised planning system reform efforts.

When used for the latter purpose, with respect to a large-scale planning system reform effort for Sydney, a series of insights emerged. In particular, it demonstrated how potentially transformative initiatives were either blocked, or not effectively followed up or integrated into the mainstream of the project. The framework should prove useful for those interested in designing future planning system reform episodes with transformative intent.

For transition theorists, this work asks some questions about whether and how transition arenas might be more routinely placed within *regime-based* reform exercises. One of the chief criticisms of transition management as a methodology, finds transition arenas as just another group of elites (e.g. Smith and Stirling 2008). But as far as PST thinking is concerned a combination of factors see the transition arena concept as misplaced in a planning reform project anyway. This is because: (1) the first horizon of the PST framework already establishes the essence of a sustainability vision/images/objectives, (2) a major planning system reform project already has an inbuilt mechanism for the "building up" of a network of actors interested in reform, and (3) the second horizon of the PST framework already requires *reflexive* design of the reform project. This latter aspect is no easy task. It requires exacting attention to the problems of self-interest and institutional blockers throughout a reform endeavor (with the exposure of past institutional failings suggested as a potent tool for this purpose Healey 2007b). The approach adopted in the PST framework therefore stresses the importance of transition management *concepts*, while avoiding the suggested democratic legitimacy problems of a ("TM-compliant") transition arena.

References

Baghai M, Coley S, White D (2000) The alchemy of growth: practical insights for building the enduring enterprise. Perseus Books Group, New York

Bell S (2002) Institutionalism: old and new. In: Summers J (ed) Government, politics, power and policy in Australia, 7th edn. Pearson Education Australia, NSW Australia, pp 363–380

Bradbury H, Lichtenstein B (2000) Relationality in organizational research: exploring the space between. Organ Sci 11(5):551–564

Campbell S, Fainstein SS (2003) Readings in planning theory. Blackwell, Oxford

CIE (Centre for International Economics) (2013) Reform of the NSW planning system – better regulation statement. Prepared for NSW Planning and Infrastructure. Sydney. http://www.planning.nsw.gov.au/Portals/0/planningsystem/CIE_Better_Regulation_Statement.pdf. Accessed 11 Feb 2015

Doherty P (2010) What do we owe future generations. In: Sykes H (ed) Future justice. Future Leaders Publications, Albert Park

Dryzek J (1982) Policy analysis as a hermeneutic activity. Policy Sci 14(4):309–329

Fainstein SS (2000) New directions in planning theory. Urban Aff Rev 35(4):451–478

Fisher, C. (2012) Better planning network – protect the places you love. Network, B. P. Sydney, Youtube. https://www.youtube.com/watch?v=dJfrSqxpmYY#t=13. Accessed 13 Jan 2015

Flyvbjerg B (2002) Planning and foucault – in search of the dark side of planning theory. In: Allmendinger P, Tewdwr-Jones M (eds) Planning futures: new directions for planning theory. Routledge, London, pp 44–62

Flyvbjerg B (2004) Phronetic planning research: theoretical and methodological reflections. Plann Theory Pract 5(3):283–306

Forester J (1999) The deliberative practitioner: encouraging participatory planning processes. MIT Press, Cambridge, MA

Friedmann J (2008) The uses of planning theory: a bibliographic essay. J Plan Educ Res 28 (2):247–257
Giddens A (1984) The constitution of society: outline of the theory of structuration. Polity Press, Cambridge
Gleeson B, Dodson J, Spiller M (2012) Governance, metropolitan planning and city-building: the case for reform. In: Tomlinson R (ed) Australia's unintended cities: the impact of housing on urban development. CSIRO Publishing, Melbourne
Grin J (2011) Theorizing energy transitions. What we know, and what we don't know. ESRC Transition Seminar. University of Liverpool. http://sustainabilitytransitions.info/wp-content/uploads/2011/07/ESRC_TransitionSeminar_Liverpool-Grin.pdf Accessed 24 Jan 2015
Grin J, Rotmans J, Schot J (2010) Transitions to sustainable development : new directions in the study of long term transformative change. Routledge, New York
Gurran N, Phibbs P (2013) Housing supply and urban planning reform: the recent Australian experience, 2003–2012. Int J Hous Policy 13(4):381–407
Hajer MA (1995) The politics of environmental discourse: ecological modernization and the policy process. Clarendon Press, New York
Hall PA, Taylor RCR (1996) Political science and the three new institutionalisms*. Polit Stud 44 (5):936–957
Hazzard, B. (2012). Bold new planning system for NSW unveiled. Sydney, NSW Government. https://members.nsw.liberal.org.au/news/state-news/bold-new-planning-system-new-south-wales-unveiled. Accessed 5 May 2016
Healey P (2003) Collaborative planning in perspective. Plann Theory 2(2):101–123
Healey P (2006) Transforming governance: challenges of institutional adaptation and a new politics of space1. Eur Plan Stud 14(3):299–320
Healey P (2007a) Urban complexity and spatial strategies: towards a relational planning for our times. Routledge, Abingdon
Healey P (2007b) The new institutionalism and the transformative goals of planning. In: Verma N (ed) Institutions and planning. Elsevier, Amsterdam, pp 61–90
Hendriks C (2009) Policy design without democracy? Making democratic sense of transition management. Policy Sci 42(4):341–368
Hendriks C (2011) The politics of public deliberation: citizen engagement and interest advocacy. Palgrave Macmillan, Basingstoke. http://SWIN.eblib.com.au/patron/FullRecord.aspx?p=851053
Inch A (2010) Culture change as identity regulation: the micro-politics of producing spatial planners in England. Plann Theory Pract 11(3):359–374
Inch A (2012) Cultural work', spatial planning and the politics of renewing public sector planning professionalism in England. Town Plann Rev 83(5):513–532
Innes JE (2010) Planning with complexity: an introduction to collaborative rationality for public policy. Routledge, New York
Jackson A (2013) An extra two million people across NSW by 2031. Sydney, NSW Government. http://www.planning.nsw.gov.au/DesktopModules/MediaCentre/getdocument.aspx?mid=1290
Lissandrello E, Grin J (2011) Reflexive planning as design and work: lessons from the port of Amsterdam. Plann Theory Pract 12(2):223–248
Manion J, Marketinfo (2013) Independent survey on the White Paper – a new planning system for NSW. https://docs.google.com/file/d/0B1phj5J6584wRE9rdzJycWppYU0/edit?usp=sharing. Accessed 7 Feb 2015
Markard J (2014) The energy transition: actors, strategies and coalitions. The Energy Revolution. TIK University of Oslo 26/9/2014. http://www.sv.uio.no/tik/om/aktuelt/aktuelle-saker/2014/jm-energy-transition.pdf
Markard J, Raven R, Truffer B (2012) Sustainability transitions: an emerging field of research and its prospects. Res Policy 41(6):955–967

Newton P (ed) (2008) Transitions: pathways towards sustainable urban development in Australia. Springer, Dordrecht
Newton P, Doherty P (2014) The challenges to urban sustainability and resilience: what cities need to prepare for. In: R P, Newton P, Pearson L (eds) Resilient sustainable cities: a future. Routledge, New York
NSW Government (2011) Multi-million dollar investment to bring integrity back to planning in NSW. Sydney, NSW Government. http://www.planning.nsw.gov.au/DesktopModules/MediaCentre/getdocument.aspx?mid=835
NSW Government (2013) A new planning system for NSW – white paper
NSW Parliament (2013) Planning bill 2013. https://www.parliament.nsw.gov.au/bills/DBAssets/bills/BillText/3227/b2012-088-d31-House.pdf. Accessed 4 May 2016
Piketty T, Goldhammer A (2014) Capital in the twenty-first century. Harvard University Press, Cambridge, MA
Powell W (2007) The new institutionalism. International encyclopedia of organization studies. Sage Publishers. http://web.stanford.edu/group/song/papers/NewInstitutionalism.pdf
Rotmans J (2005) Social innovation: Between dream and reality lies complexity. DRIFT Research Working Paper. Rotterdam, Erasmus University. https://ssrn.com/abstract=878564. Accessed 23 April 2015
Roy A (2010) Informality and the politics of planning. In: Hillier JAHP (ed) The Ashgate research companion to planning theory: conceptual challenges for spatial planning
Ruming K, Gurran N (2014) Australian planning system reform. Aust Plan 51(2):102–107
Searle G (2012) Is strategic planning in Sydney becoming 'Relational'?. Planning Institute of Australia 2012 National Congress | 29 April–2 May 2012 | Adelaide
Servillo LA, Van Den Broeck P (2012) The social construction of planning systems: a strategic-relational institutionalist approach. Plan Pract Res 27(1):41–61
Sharpe W (2013) Three horizons: a way of working with change sustainable learning – managing future change. http://sustainable-learning.org/2013/10/three-horizons-a-way-of-working-with-change/. Accessed 22 July 2015
Smith A, Stirling A (2008) Social-ecological resilience and socio-technical transitions: critical issues for sustainability governance. STEPS working paper 8. STEPS Centre, Brighton
Stein L (2012) A review of international best practice in planning law – for the NSW Department of Planning. http://www.planning.nsw.gov.au/Portals/0/PolicyAndLegislation/Review%20of%20International%20Best%20Practice%20in%20Planning%20Law.pdf
Treasury Australia (2010) Intergenerational report 2010: Australia to 2050: future challenges. Commonwealth of Australia, Canberra
Verma N (ed) (2007) Institutions and planning. Oxford, Elsevier
Walsh P (2016) Wicked city planning problems and planning system reform: a case study of Sydney. PhD thesis, Swinburne University of Technology – accessible from Swinburne Research Bank
Westacott J (2015) Australian unity great Australia day breakfast speech, Melbourne, 2015. Melbourne, Business Council of Australia. http://www.bca.com.au/newsroom/australian-unity-great-australia-day-breakfast-2015-speech-by-jennifer-westacott. Accessed 11 Feb 2015

Transition Management in Urban China: Co-creation of New Homes or a Monopoly Game?

Qianqing Lindsay Mai

Abstract This chapter places empirical applications and review of established Transition Management principles in the metropolitan areas in coastal China. Based on content analysis and semi-structured interviews, an overview of state envisioning processes of urban sustainability is first presented as a conflicting context for the disequilibria observed in the environmental governance system across the metropolitan areas. The interactions of observed disequilibria grounded locally potentially weaken the state-led envisioning processes. With the presence of challenges in state-led transitions, the chapter proceeds to case studies in Shenzhen, the city that is practicing the state-led sustainability visions, to analyse how local dynamics of resource and organisational landscape are integrated in the transition process.

Introduction

Recent research in sustainable and low carbon transitions have attributed that cities are transition seedbeds, and that changes can start in cities and get accelerated when expanding outwards (Bulkeley et al. 2010). When we consider changes to achieve urban sustainability, it involves not only reformed policies and development programmes; rather, it starts from changes in attitudes, perceptions, and the ways how we value urban living, and then it moves on to produce or improvise innovative ideas, and eventually it shapes the urban governance system that involves policy instruments, system innovations and sustainable practices. All these change elements should be put into account when it comes to steering and practicing management of transitions in urban experimental sites.

Historically in China, cities have been the tactical sites for experiments – to learn and to adapt the newly imported policy programmes, to try out the transplanted best practices, or even to test out some newly proposed forms of economies. The

Q.L. Mai (✉)
Department of Geography, Durham University, Durham, UK
e-mail: qianqing.mai@durham.ac.uk

N. Frantzeskaki et al. (eds.), *Co-creating Sustainable Urban Futures*, Future City 11, https://doi.org/10.1007/978-3-319-69273-9_12

economic reform in the late 1970s and 1980s have experimented the incorporation of liberal economy in a socialist political regime, which put the coastal conurbations in the centres of change to accommodate exogenous inputs of not only capitals and technologies, but also information and values. The market liberation reform substantially transformed the way of urban living in China. After over three decades of steady economic growth and improvement of the quality of urban life, these developed metropolitan areas located in a developing nation have reached a tipping point of economic growth that not only the local administrators and business investors, but also the national leadership are looking for structural cures for the locked-in negative environmental impact, the cures that evolve consistently with the transformed social dynamics to rebalance the paces of urban transitions.

Unlike in the Netherlands (as discussed in chapter "Transition Management: Guiding Principles and Applications", this volume), the particular notion and approaches to Transition Management were not integrated in China's national, nor local policy programmes. However, given the looming need for systemic changes towards sustainability, the management of transitions *per se* is embedded in the everyday business and learning among the local governments especially those of the developed urban areas along the China coast, and in the national government with its affiliate agencies which often seeks to 'coordinate' changes.

Meanwhile, the mission of Transition Management framework suits well with the logics of China's urban governance: with centralised state apparatus seeking to *influence* the transitions down in localities that demonstrate vast differences in geography, cultures and practices, it vacates spaces for the localised urban managers to search and experiment the effective practices gauging the local dynamics and opportunities.

An interdisciplinary application of Transition Management delivers a more explicit account of the *politics* involved in the processes of defining visions and boundaries of transitions, of empowering or exerting controls over actors, and of responding to crisis and enacting changes (Rotmans and Kemp 2008; Shove and Walker 2007). Such an empirical account of politics in transitions identifies possible trajectories of urban transitions in China by examining the obscured rationales encompassing political beliefs and social values alongside momentums of changes in the *state-society relations* that affect both the governing and the governed. The established tenets of Transition Management (unfolded in chapter "Transition Management: Guiding Principles and Applications", this volume) are applied in analysing both how the power structures in urban transitions are configured in coastal China, as well as how the implementation of sustainability is practiced by the mandated actors in cities that are experimented for socio-environmental and regulatory changes. The organic combination and application of these tenets also help locate the dilemmas in transitions and possibly provide structural solutions.

The empirical review of urban transitions in this chapter focuses on metropolitan areas in coastal China – administratively all of these metropolitans are classified either as 'deputy-provincial cities' (such as Shenzhen, Guangzhou, or Ningbo) or 'direct-controlled municipalities' (such as Shanghai, Beijing or Tianjini). Data was collected through documentary research, semi-structured interviews and field observation. A series of content analysis was conducted on public speeches of state leaders, newspaper articles (retrieved from selected keyword searches in media databases among articles published between 2005 and 2015), government publicity campaign materials, national and local policies and regulations, and teaching materials for civil service training. Eleven face-to-face semi-structured interviews were also carried out in Shenzhen between 2012 and 2015, respectively with local government officials and key informants from nongovernmental organisations concerning urban development, in addition to a fieldtrip observing the activities of one of the organisations.

This chapter starts from an overview of state envisioning processes of urban sustainability, alongside disequilibria in the environmental governance system across the metropolitan areas that potentially weaken the state-led envisioning processes. With the presence of challenges in state-led transitions, it proceeds to case studies in Shenzhen, the city that is practicing the state-led sustainability visions, to analyse how local dynamics of resource and organisational landscape are integrated in the transition process.

Envisioning the Urban Future: Enacting or Reacting to Changes?

Casting long-term thinking is a usual practice at the central level of the party-state governing apparatus, whereas at the local and subnational regional levels, reactive actions to visionary shifts and situational changes followed by preventives measures to emergencies and crisis have inevitably formed part of the realistic urban management strategies commonly shared among local governments given the complexities of normative, material and infrastructural landscapes in urban China. Disparities are apparent between the visions for future urban development constructed by the central leadership and the practices worked out by urban actors who exercise transitions in their routine businesses.

In this section, an overview of the urban transition visions and their rationales contextualised in the designated transition arenas are presented gauging their relative capacities of envisioning changes in urban milieus given the emerging disequilibria of urban stability. Tenets of Transition Management are partly reviewed in the face of the evolving sustainability images of urban China and the reconfiguring of powers in the change processes.

The governing priorities in the 'official ideology' evolved since the economic reform in the late 1970s have been to maintain socio-political orders and the stability of a party regime during times of socio-economic transitions (Feng 1997; Zhao 2008). Sharing this governing mentality, local governments (or urban managers) and state-sponsored intellectuals strive to maintain equilibriums in economic growth and socio-technical transformation, with or without explicit orientation towards sustainability. Actors are aligned and intermittently rearranged to render solutions for conflicts and persistent problems in urban China, wherein the often-contested areas for urban dwellers are ones that affect directly their quality of life, that is, the social and environmental issues arising in their living milieus.

State Visions for Future Cities

Without institutionalising a nationwide electoral democracy, maintaining administrative performance with output legitimacy is one feasible way to structure regime stability in China, whereas rendering sustainable images of post-reform urban development create a way of legitimating the governing powers to the urban population (Pow and Neo 2013). Advertisements commissioned by the local governments to advocate sustainable urban images, i.e. pedalling on the newly constructed greenways or practicing low-carbon living, are seen in the public transportation hubs or along the main roads in almost all metropolitan cities in China. These are nothing like the ideological propagandas on political posters or street billboards, to mobilise urban workforces with standardised graphic designs and communist slogans in the 1960s and 1970s, which were also commissioned by the city governments and managed by the then urban planning or environmental development agencies (Wang 1989). Each of the carefully constructed sustainable urban visions being publicly advertised nowadays is backed by a mandated transition plan, a set of theories or evidences in sustainable development and state building being the operative rationales that underlie the visions, and a team of authorised transition managers who figure – or are figured in – methods and practices of disseminating and critically assessing the visions (as shown in Table 1).

These three interrelated visions, as long-term thinking in managing the transitions for a state-endorsed urban future of sustainable development, cannot avoid conflicts within themselves. The visions for building a 'beautiful China' and 'low carbon cities' have met with competing rationales in their respectively implied political goals of consolidating nationalism and of rising localism in achieving sustainability.

Experimenting with low carbon visions in selected localities in itself promotes the logics of localism in sustainability transitions, which arguably are 'creating a more rational future with local governments leading the development of more sustainable communities, life and work styles' by incorporating their local knowledge and development needs, which thus will hand back 'political leverage to cities in reconstructing a new form of transformative local governance around the environmental agenda' (Marvin and Guy 1997).

This line of reasoning inevitably falls in conflicts with the nationalist arguments and the ecological modernisation theories – both being the rationales for envisioning a 'beautiful China' – with the former stressing on the central regime supremacy prioritised to maintain nationwide balance of orders and stabilities by offering 'a discourse of hope to ordinary citizens who are frustrated by the myriad

Table 1 State-endorsed urban visions

Labels	'Beautiful China' pathways	Civil(ised) cities	Low carbon cities
Transition team	National government components (Communist Party publicity, education, Communist Youth League);	A designated commission and secretariat for 'spiritual civilisation' led by a concurrent member of the Central Politburo;	National government, specifically Department of Climate Change, National Development and Reform Commission (NDRC);
	Xinhua News Agency;	Municipal governments that have applied and been awarded 'civil cities' (all of the sub-provincial cities);	Municipal agencies on building, energy, transport and economic development of the selected pilot cities;
	Communist Party Schools and Institutes of Public Administration at national, provincial, and municipal levels;		Local nongovernmental associations on energy efficiency, green buildings
Rationales	Nationalism: incorporated in the 'China Dream' state-building pledges by the current paramount leader (Xi 2013);	Moral authority for social orders and political stability, relives of urban grievance;	Socio-technical transitions with low-emission/ mitigation innovations;
	Anthropogenic metabolism in urban sustainability;	Social capitals for community attachment;	Strategic niche management to integrate innovations;
	Ecological modernisation: to build an 'ecological civilisation' with quality of life (Hu 2012);	'Liveable city' evaluation: GDP, employment rate, good governance, public safety, sustainable urban environment;	Climate-resilient economic growth or low-emission development;
			Localism in urban sustainability transitions
Transition plan (by a 'mandate')	It is a recurring agenda being co-expanded by party cadres and state media since an official report released at the end of the preceding national political leadership (Hu 2012)	A policy document was promulgated in 2003 by the designated central government commission that specifies criteria for 'Civil Cities' (Central Civil Commission 2003)	A low carbon plan was delivered by each selected pilot city for its own jurisdiction (with carbon audits, carbon emission reduction targets, action plans) in 2010 or 2012 (NDRC 2010, 2012)
Dissemination	Political education program designed for civil servants and students;	Multi-media publicity;	Public exhibitions and trade events of low carbon living;
	Multi-media campaign;		Multi-media advertising;
			School education toolkits

(continued)

Table 1 (continued)

Labels	'Beautiful China' pathways	Civil(ised) cities	Low carbon cities
Critical assessments	Commentaries published in state media and official party news outlet;	Triennial assessments of urban governance for 'Civil City Champions' since 2005;	Review workshops for low carbon managers from all pilot cities;
	Discussions in intra-party learning	Reviews in intra-party learning;	Self-assessment in municipal steering committee meetings

of societal and environmental crises' (Du 2014); whereas the latter underlines the importance of national regulations in transforming the development cycles to minimise their environmental impacts (Spaargaren and Mol 1992).

These competing rationales may confuse the transition managers in delivering the visions, particularly as to how far they can intervene and make flexible changes to the goal-setting process of the state-imposed long-term thinking, leading to conflicts between two of the Transition Management principles (chapter "Transi tion Management: Guiding Principles and Applications", this volume). The different power locus implied by the competing rationales may also lead to disparities of practices and reactions to environmental crises between national and local governments, which would be further discussed in the next section.

Disequilibria for Transitions: Unrests and Emergencies in Urban China

In light of the state-endorsed visions for urban sustainable futures, the governance of transitions in cities has taken on elaborate agendas of enacting environmental and ecological changes. At the same time, these urban transition agendas have encountered critical junctures of socio-environmental crises and system disequilibrium that are opening-up opportunities for arrangements of immediate and effective state or societal interventions. The effectiveness in sharing the state-endorsed visions is depending on how skilful and equipped the mandated transition teams are in response to these system uncertainties in the urban environment.

Disequilibrium 1: Environmental Emergencies

Industrial, or, manmade environmental disasters located close to urban agglomerations and with significant number of casualties, once happened, can be instantly broadcasted and communicated to a wide audience group with immense psychological impact. Crises as such can pose questions to the governing capacity of the local authorities involved, but also expand the possibilities for citizen participation, technological inputs and effective changes in the management of urban facilities. These environmental disasters are getting more visible in urban China given the widened channels of public scrutiny and media exposure. Recent instances include consecutive events of chemical explosions in two warehouses of Tianjin and Zibo – within the boundaries of two conurbations in the northeastern coast of China, which together accounted for nearly 200 deaths and injuries in the same month of August 2015 (Associated Press in Beijing 2015; Xiao and Chen 2015).

These tragedies have prompted discussion of not only the effectiveness of environmental resource management and inspection practices in identifying public safety risks of the existing production facilities and infrastructures in the urban milieus, but also a more liberated media regulatory system that enables higher transparency of government-held information and larger space for public debates (Fang 2015; Lee et al. 2007). Liberating the media controls has been a sensitive area of urban governance in China, as it is perceived amongst the centralised governing powers as putting the regime stability in volatile conditions. However, by media liberation, it may eventually empower individual citizens to participate in the scrutiny of local governments and industrial practices, whilst creating a sense of urgency for alternative urban development pathways to be purposefully rolled out. Provided that "transitions involve a range of possible development paths, whose direction, scale and speed government policy can influence, but never entirely control", the introduction of new urban sustainable practices to fit with the established visions can be expedited by unexpected or one-off events such as environmental emergencies and crises (Rotmans et al. 2001). However, such expedition process needs to be channeled through enhanced media transparency that bridges the information gap between top-down dissemination and bottom-up requests for disclosures.

At the moment, media campaign has been adopted as a common way of disseminating the state-endorsed sustainability visions for urban future (as in Table 1), not only in the state-dominated means of media, but also in the commercial means that are reserved for and commissioned by government agencies. The release of state controls on media thus put direct pressures on the effectiveness of state envisioning process. It requires the mandated transition teams to enhance the *diversity* in the components of the transition arenas, which in light of the Transition Management framework may ensure an equitable process and *rebalance* the power dynamics between the state and society given the impacts of system disequilibrium (chapter "Transition Management: Guiding Principles and Applications", this volume). Such diversity also encourages power pluralism in the urban governance around environmental issues that again goes back to the localism arguments and thus fragmentises the part of state-imposed visions that aim to promote nationalism.

Disequilibrium 2: Unlocking the Censorship

Despite stronger voices calling for media liberation triggered by the environmental emergencies, the national government in China is still gripping controls over various means of media by regularly renewing its censorship measures. Whereas the conventional means of media can be easily controlled with licensing on the release of newspaper and television broadcasting, new media such as digital news platforms and various forms of social media are not that straightforward in regulating, especially in urban milieus where 71.4% of Internet users in the nation are located and about 81% of them access the Internet through mobile phones (CINIC 2014). The censorship protocols are not always applicable to the extensive amount of free-flowing information available outside the state-built 'firewall' transmitted through the rapidly-developing information and communications technologies. The national regulatory body of the telecommunication industry, namely the Ministry of Industry and Information Technology, has to perform frequent upgrades to the techniques and devices it employs to maintain the restrictions on domestic web browsing, such as disabling the Virtual Private Network (VPN) – a major networking technology to hedge the Internet routing often provided by foreign institutions (multinational corporates or overseas universities) to their members – when the connections are established in the wireless network environment supported by the state-owned cellular providers.

Behind the state's attempts in moulding the digital infrastructures, it is the strong desires and rising awareness of urban residents to defy the state-imposed restrictions and to obtain a more diverse and fair share of information that drive the technology and power mixes in connecting cities to the world. A number of commercial consultancy reports have indicated a high rate of smartphone possession and usage across the urban areas in China. According to the traffic statistics generated by an Internet data provider company – Mobbo.com, for the smartphone users located in China, applications that provide or facilitate VPN services are one of the most downloaded and utilised ones. Specifically, five of the top twenty applications available on Android-Google Play Store and the top paid application in the category of utilities on Apple App Store recorded in China (excluding Hong Kong, Macau and Taiwan areas) were applications that connect users to the Internet with VPN services (data retrieved in September 2015).

This observed collective intention to break the virtual walls of information has disrupted the conventional ways of information sharing in metropolitan regions that are condensed with innovative ideas, built infrastructures and development opportunities. Such disequilibrium in the urban information system has expanded the breeding ground of citizens' informed rights for environmental wellbeing. In the same process, the sharing of urban sustainability visions by the state is intervened by the urban images captured by the citizens in a variety of alternative channels that are out of the state sanctions.

Disequilibrium 3: Environmental Protests

Environmental protests form a third major disequilibrium in the urban dwellers' value system of urban sustainability. As summarised in Table 2, environmental protests in China have been rapidly gaining scales in the recent decade, from a

Table 2 Major environmental protests in Chinese cities in the recent decade (2005–2015)

Dates	Location (city, province, region code[a])	Agenda	Scale (number of protesters)	Local government reactions (with interim solutions)
2015 Jun 22–27	Shanghai [B]	Anti-Para-Xylene production	10,000+	Imposing media censorship about the protest; deploying riot police; issuing reassuring public statements; *halting environmental approvals*
2015 Apr 12	Heyuan, Guangdong [C]	Anti-pollution (air) of a proposed power plant	5000–10,000	Deploying police; continuing communication and discussion
2015 Apr 6–8	Luoding, Guangdong [C]	Anti-incinerator	1000–10,000	Arresting protesters; *cancelling the approval of the incinerator construction plan*
2014 Mar 30–Apr 3	Maoming (–Guangzhou-Shenzhen), Guangdong [C]	Anti-Para-Xylene production	1000–10,000	Deploying riot police; detaining protesters; delivering public addresses and press conferences to condemn violent protest; *promising a public consultation*
2014 Sept 12, 14	Huizhou, Guangdong [C]	Anti-incinerator	1000–20,000	Deploying police and road blocks; detaining and dispersing protesters
2014 May 10	Hangzhou, Zhejiang [B]	Anti-incinerator	1000+	Deploying riot police; imposing media censorship; arresting protesters for violent behaviours; detaining opinion leaders; issuing public statements; *putting the plan on hold for public consultations and supports*
2013 Jul 15	Guangzhou, Guangdong [C]	Anti-incinerator	8000–20,000	Approving the planned protest; postponing the launch of construction; *relocated the siting in 2014*
2013 Jul 12	Jiangmen, Guangdong [C]	Anti-nuclear/–uranium plant	900–2000	Deploying police shield; *immediate announcement of project termination*

(continued)

Table 2 (continued)

Dates	Location (city, province, region code[a])	Agenda	Scale (number of protesters)	Local government reactions (with interim solutions)
2013 May 4, 16	Kunming, Yunnan [D]	Anti-Para-Xylene production	1000–2000	Delivering impromptus announcements; arresting protesters; detaining opinion leaders; imposing name registration of protesters; *releasing environmental assessment report*
2013 May 1, 11; Apr 21	Shanghai [B]	Anti-pollution (water) of a proposed battery plant	1000+	Having the private investor and project owner *cancelling the plan*; reclaiming the land reserved for the plant
2012 Oct 25–28	Ningbo, Zhejiang [B]	Anti-Para-Xylene production	1000+	Arresting protesters; deploying riot police; dispersing protesters with tear gas; issuing district government statement to review the proposed expansion of Para-Xylene production; *suspending the plan*
2012 Jul 1–3	Shifang, Sichuan [D]	Anti-pollution of a copper plant	1000–10,000	Dispersing protesters with tear gas; detaining and arresting protesters; *terminating the plant construction*
2011 Dec 20–23	Shantou, Guangdong [C]	Anti-pollution of a proposed coal power plant	500–10,000+	Deploying riot police with tear gas; closing down the nearby highway; detaining protesters; *withdrawing the construction plan*
2011 Aug 14	Dalian, Shandong [A]	Relocation of a breached Para-Xylene plant	10,000–70,000	Deploying riot police; *announcing immediate plant closure and relocation*
2011 Sept 15–18	Haining, Zhejiang [B]	Anti-pollution (water) of a working solar panel factory	500	Deploying riot police; *closing the factory; appointing an expert panel to examine and remedy the factory's accidental discharge*
2010 Jan 25	Foshan, Guangdong [C]	Anti-incinerator	400	Reassuring the public with a pending environmental assessment report
2010 Jan 19	Guangzhou, Guangdong [C]	Anti-incinerator	300	Continuing building the incinerator as planned

(continued)

Table 2 (continued)

Dates	Location (city, province, region code[a])	Agenda	Scale (number of protesters)	Local government reactions (with interim solutions)
2009 Nov 23	Guangzhou, Guangdong [C]	Anti-incinerator	800	Continued negotiations on the siting; *relocated the incinerator in 2014*
2007 Jun 1	Xiamen, Fujian [C]	Anti-Para-Xylene production	1000–5000	Imposing media censorship; suspending the original construction; conducting public hearing; *relocating the plant to a less populated city nearby*; emergency management of subsequent leak explosions in 2013 and 2015 after relocation

[a]Regions in China: [A] Northeast China Coast; [B] East China Coast; [C] South China Coast; [D] Southwest China Inland

group of several hundreds residents in one protest risen to around ten thousand demonstrators in each event. It has also become more frequent, and mobilised similar scale of protests in nearby cities. Protests and urban activism combined into one of the persistent problems in urban governance, resulted from the structural imbalances between changes in economy, society and environment. The different local governments involved with environmental protests have been learning from each other regarding how to devise various responsive actions to deal with the organised requests and environmental disputes. Facilitated with top-down policy mandates from the central and some provincial governments, this intra-governmental learning process has been organised in forms of civil service trainings, government-commissioned academic research projects, and an institutionalised support of 'emergency management expert group' (Guangdong Provincial Government 2016). Although such process is potentially changing the practices in the administrative system, it does not resolve the core problem in urban transition, i.e. structural imbalances of development, and cannot control the mounting urban activism on environmental issues.

This governing problem is further complicated by the strong rationality carried by the urban protesters, who have calculated the consequences and risks of their actions based on their environmental knowledge, the concerns on sustainable security of their urban homes, and experiences in dealing with the local governments. This 'rational resistance' in urban China (Zhang 2014) is weakening the state promises of urban sustainability. The presence of such rationality has further alerted the governing authorities, when a state-sponsored survey (Zhong 2013) with resident responses collected in 34 cities concluded that urban residents were very well ready for participating in organised petitions and protests against the local governments if there were polluting facilities being built within their living vicinity. The same survey also indicated that the new generation of urban population refused to settle these group disputes for economic compensation and 77.2% of them chose environmental protection over economic development (Zhong 2013), reflecting on the core issues of structural imbalances aforementioned.

Systematic learning from the experiences of these protests has been launched within the network of about 5000 entities (Yu 2014) of Communist Party Schools, Cadre Training Schools and Public Administration Institutes incorporated in different levels of government. For instance in Guangdong Province, where 10 of the 19 urban protests listed in Table 2 took place, the cadre education on emergency management was formally introduced in 2008 in a provincial policy document that ensures the fiscal coverage and progress of learning (Guangdong Provincial Government 2008). Part of the learning materials and video recordings of the national cadre training lectures have also been made publicly available online (Chinese Academy of Governance 2015), which cover governing issues incurred during emergencies and 'mass incidents' (the administrative synonym of protests) such as effective media communication and how to improve responsive capacities.

The Weakening of State-Imposed Visions of Urban Sustainability in China

The state constructed urban visions (Table 1) have been disseminated through administrative means and imposed on the urban population. However, with the presence of disequilibria of environmental emergencies, protests and the struggles for more open media, these state-led visions were fragmentised and weakened among its rising urban middle class – a majority in the urban population who are more willing to take their rational requests onto the streets, generally well-educated with knowledge of environmental hazards and with eagers to become better informed of sustainability issues with their wide use of various information and communication technologies. They are also the homeowners residing in cities, the major stakeholders who take up the ownership of constructing future cities. Their grievances and unrests with concerns on their sustainable living quality are jeopardising the state-led envisioning process.

The disequilibria display a crisis of class mentality and looming incompatibilities of values held between the state and society. Whereas the urban homeowners treasure their immediate wellbeing and the future of living quality in their cities, the local governments and state-owned enterprises on energy production seem to care more about the competitiveness of the local economy and profit performance. To manage transitions in Chinese cities thus needs to cope with the transformed mentality of the urban middle class.

The confluence of disequilibria forces the state to rethink its governance practices during the process of urban transitions and how to legitimise the operationalization of its long-term thinking over sustainable cities. In the next section, experiments of governing innovations with urban infrastructure niches are discussed to examine how to rebalance the power of the state given the rising urban environmental activism, by practicing the possible pathways of state-society co-visioning.

Experimenting with Urban Visions in Shenzhen: Adapting Changes

As one way of activating interventions at the operational level, experiments in urban management have been coordinated by the municipal governments that are seeking to steer the systematic socio-environmental changes from inside the urban milieus, to match with the state-led long-term thinking that are imposed on their agenda but fragmentarily disposed by the social dynamics. This section discusses how the implementation of the fragmentised long-term thinking affects the more flexible and localised short-term development goals of the urban sectors, and how such process reflects, neglects, or even contravenes the established tenets of Transition Management, in particular the niche creation and protection, the steering capacity in urban governance, in-group and out-group learning and reflexivity, as well as the extents of interaction, formality, and practices of stakeholder participation.

Experiment 1: Compromised 'Social Management'

Social management is a new genre of learning in the state-led academic study in China. State funding agencies at national and provincial levels govern the distribution of significant social science research grants, the major sources of revenues for higher education institutes in China. The grant application guidelines and priority research areas set out by these agencies often adhere to the concurrent ideological or governing principles designed by the party state leadership, which enable the political regime to steer the discussion inside the academic circle (Gong 2015). Meanwhile in the Party Schools instituted at different levels of government administration by the Communist Party of China, 'social management' has become an important topic of *learning* among incumbent political leaders, to devise and to experiment system innovations in state management.

The 'social management' discussion has emerged in this discourse as a state concerned governing issue, a way to maintain regime stability, which is defined among social science scholars in China, as 'the activity to regulate and coordinate social organisations, social affairs and social life' (Yu 2007). The governing and political tasks of 'social management' was frequently emphasised by the former state leader, Hu Jintao, who once addressed 'social management' as a governing approach that can ultimately strengthen the party leadership through measures such as improving both the state management and the self-development of 'social organisations', as well as enforcing comprehensive information network management on the 'virtual society' by institutionalising a 'steering mechanism' over the online public opinion (Xinhua News 2011). Since then, the tenets and experimental practices of 'social management' have been affecting urban living and the way how local governments approach urban management in China, which inevitably are shaping the sustainable urban transition in some Chinese cities.

Since the promulgation of Regulations on the Registration and Management of Social Organisations by the State Council in 1998, government-organised environmental nongovernmental organisations have become a dominant form of organising societal resources to promote state-sanctioned environmental agendas in urban China. They are often chaired by retired local governmental officials or university professors within civil-service establishments, or principally funded by government grants. These are also useful entities that the local governments may reach out for manpower support to complete some environmental tasks. Prescribed by the national regulation, they must be registered with and managed by two government departments. Such registry system has instituted administrative hurdles in establishing nongovernmental organisations that are independent from state influence. However, in the midst of 2012, the 1998 national regulation was revised for three cities – Guangzhou, Shenzhen and Shantou, by a provincial policy – a 'Plan to Foster the Development and Norm of Social Organisation Management in Guangdong', revoking the double registry requirement with government departments. As intervention at the tactical level of managing urban transitions, this policy has set out institutional basis to the experiment of citizen-based social innovations.

Thereby, regulatory niches have been created in the selected urban areas to protect the grassroots environmental activism in a still bounded but more open social space. The state power and the mandates of social management have been partly compromised to allow the possibilities of experimental governing practices and organisational innovations driven by urban residents, taking into account their high willingness of contributing to environmental nongovernmental organisations. According to the state-sponsored survey (Zhong 2013), despite an arguably low satisfactory rate on the urban government's environmental performance, the surveyed urban residents seemed quite willing to contribute to environmental nongovernmental organisations (ENGOs) – about 63.5% of them were either willing or extremely willing to make donation to ENGOs, and an even higher proportion – 77.1% – to volunteer for the ENGOs. In this context, independent ENGOs started to proliferate in the pilot cities of regulatory niches. For example, *Green Source*, a Shenzhen-based independent nongovernmental organisation operated by volunteers, was established in 2013 to participate in the city's sustainability transitions. Since its incorporation, this organisation has attracted hundreds of residents in Shenzhen to register and regularly participate in its network of voluntary activities.

Most of the voluntary activities are to enhance environmental sustainability at both community and city scales, while some of them also promote social capitals in neighbourhoods. For instance, workshops of hand making cotton baskets using recycled materials – such as colourful clothes from the old garments and unused disc containers – have been organised by the neighbourhood organisations in collaboration with the *Green Source* for its registered volunteers. The hand-made products from these workshops have been put on sale in Taobao – an online trading forum commonly used in China, and any profits gained from the sale are to be donated for charities or non-profit operational costs of the organisation. The state-sponsored vision of urban low carbon living is practiced through these 'zero-waste' workshops. Meanwhile through these community interactions, emotional attachments are gradually built among the volunteers based in the neighbourhood and the organisation. The bonding of social capitals in the neighbourhoods has then indirectly supported the 'civilised city' vision of the state.

The emotional bonding also enhances *Green Source*'s mobilisation power when it comes to activities that require large amount of manpower. One of these activities is the citizen-organised preservation of coastal mangroves in the outskirt of Shenzhen's urban agglomeration, some of which lay outside the municipal biodiversity protection areas. Due to the direction of river flows and ocean waves, there are all sorts of municipal waste floating along the mangrove coast. They are left on the wetland beach and damaging the growing mangroves. Given this situation, one of the leaders (i.e. the frontrunners as defined in chapter "Transition Management: Guiding Principles and Applications", this volume) of *Green Source* has been proactive in organising regular clean-up trips with volunteers to clear up the wastes tangled on the beach, and then to build up fences for the mangrove saplings that were planted also by the volunteers signed up with the *Green Source* organisation. During a fieldtrip to the Shenzhen wetland with this frontrunner of mangrove preservation, he demonstrated not only his knowledge in mangrove biodiversity and the ecological landscape of the area, but also how much he had been dedicated to his voluntary involvement and his leading role in initiating sustainability transitions with other volunteers. This frontrunner carries innovative ideas of coordinating and mobilising transitions. To mobilise the participants, barbeque outings were incorporated within those voluntary beach-cleaning trips. Social media platforms such as micro-blogging are also utilised to update the volunteers with the recent ecological damages of the area, and to advocate for changes in the practices of the local administration and the environmental agencies. The state-built vision of pathways to 'beautiful China' has also been softly landed on the ground through citizen-initiated ecological preservation practices.

Organisations like *Green Source* are forming interventions at the operation level of managing urban sustainability transitions. They are often led by frontrunners of sustainability practices, who are grounded in communities without official rankings or titles, and have gained trust from the networks of participants. These increasingly independent organisations have drawn in many voluntary participants, who are often well-educated rational actors that share the same missions with the organisations to practice urban environmental sustainability. The regulatory niches of 'social management' created by national and local administrators are experimental grounds for the state to learn how to better steer the urban transitions with a rapidly changing society. Having learned about the mobilisation capacity of these organisations, the local administrators have started to allow more space for these organisations and have opened-up the transition team to include these non-state participants.

Apart of commending some official awards to *Green Source*'s innovative practices, the local government has seldom intervened in the organisation's regular activities. Whereas for the national party-led government, the core discussion and learning from these experimental niches is on how to balance the state-society relationship during urban transitions, in particular whether the experimental zone of governing niches can be further expanded to cover the entire metropolitan regions without triggering instabilities and disorders, and how to remediate the fragmentised state-built visions for urban future with the convenient helps from these new-born social groups. The *Green Source* alone has substantiated and operationalized the three urban visions constructed by the state.

However, part of the urban state is still gripping controls on limiting the boundaries of transition arena, whilst some urban frontrunners are restricted to enter this space. For instance, the 'unorthodox thinkers' and frontrunners in environmental activism who are outspoken in the domestic social media platforms and have mobilised the urban environmental protests are most likely refused by the government-centred transition team. Their public speeches and statements, which have been identified by the state as having influenced and contributed to the urban protests, are often censored out in the state-controlled media and domestic Internet. Some of them had reportedly been physically detained by the authorities during or in the aftermath of the protests. An influential micro-blogger who ran an Internet consulting company was detained after the series of protests against Para-Xylene plant in Kunming in 2013. His published treads were removed from the social media platforms amid the protests (Boehler 2013).

Experiment 2: Co-opting for Urban Niches

Apart of the growing independent organisations, the existing government-organised nongovernmental organisations established for achieving environmental sustainability in cities are still important actors to be co-opted by the state in the government-centred urban transitions. They are often seen in the transition arena of urban land-planning niches that requires both professional and social inputs, for instance, the inducing of green building practices in planning and construction, and the building of inter-city urban 'green ways' for pedestrians and cyclists. In Shenzhen, these urban planning niches fall within a broader urban experimental framework of 'co-creating a national model of low carbon ecological city', an initiative that was handed down to the municipality directly by the national government in 2010.

During the initial planning process to set out the municipal agenda with short-term goals towards achieving the long-term vision in the national directive, *system analysis* was carried out by the lead change agent in the municipal government (i.e. Shenzhen Commission of Urban Planning and Land Resources), facilitated by its research agency, which looked at the city's energy and resource circulation system, infrastructural networks, land conditions of greenfield and brownfield developments, industrial landscape, existing spatial planning, etc. A municipal plan was then drafted and circulated across different departments in the municipal government that are responsible for different deliverables in the plan, for them to further negotiate in the goal setting process. A finalised *process plan* was then published in January 2011 on behalf of the entire municipal government (SZG 2011), which largely overlapped with the city's 'low carbon pilot' implementation plan subsequently promulgated by one of the other municipal departments (i.e. Shenzhen Development and Reform Commission) in May 2012 (SZ-DRC 2012) to fulfil another national government's requirement on the first-batch of pilot cities selected in 2010 (NDRC 2010; Table 1).

These plans shared similar goals for the landscape coverage of green buildings and the physical length of constructed 'green way'. For instance, by 2015, all the new government buildings (and about 40–50% of all new buildings) must fulfil the established standards of green buildings, meanwhile the construction of about 300 km of 'regional green ways', 500 km of 'urban green ways' and 1200 km of 'community green ways' must be completed (SZG 2011; SZ-DRC 2012). Apart of the fact-collecting and agenda-setting system analysis, formal *actor analysis* was also conducted. The 2011 plan has specified the implementation of each goal with a designated municipal department – the Housing and Construction Department to steer green building practices in the city, whereas the Commission for Human Habitat and Environment on green-way construction. The making and dissemination of these process plans are interventions at the tactical level of managing transitions, with negotiating, goal setting, collaborating and steering activities among government agencies.

However, *neither* of the two plans incorporated actor analysis of nongovernment actors to be potentially included in the transition arena. Instead, *informal actor analysis* has been conducted recurrently by the delegated municipal administrators, throughout the process of implementing the goals allocated to them from the process plans. For the Housing and Construction Department in reaching the target of green buildings, it identified the operational need of co-opting the existing government-organised associations to facilitate the processes of societal engagement and the gathering of professional inputs.

The Shenzhen Green Building Association was successfully launched in 2008 after overcoming the then administrative hurdles of social organisation registration. It was with the endorsement of the Housing and Construction Department that the nascent association was able to go through the queries posed by the registry department, regarding whether the proposed association can form a unique and useful representation of one industry, given the abundance of other construction-related associations already established in the city, and whether the development of green buildings adequately represented a niche industry (Interview, Executive Manager of Shenzhen Green Building Association).

Since then, the association has consolidated professional networks with urban planning institutes and national green building association, alongside its ongoing societal engagement activities to promote green buildings in the region. After the allocated target of green building expansion in 2011, the Housing and Construction Department has reached out to the nongovernmental Green Building Association for support in coordinating activities of research, social discussion, and site visits, in addition to the technical supports provided by its affiliated work unit – the construction science research institute (Interview, Building Energy Efficiency Project Officer in Shenzhen Housing and Construction Department).

On the 'green way' component, albeit the physical length is a hard target settled by negotiations among municipal departments with the planning and environmental science institutes regarding the feasibilities of urban land use and spatial arrangements, fostering its actual usage has become largely a 'social engineering project' (Interview, Coordination Officer on Green Way Construction in Shenzhen Human Habitat and Environment Commission). To activate the social movement of 'green way' living style when part of the infrastructures were still in construction, the coordinating officer initiated various contacts with the local media to broadcast the 'green way' construction, and disseminated internal referencing materials regarding the 'green way' progress to local enterprises. Given the leisure purposes of most urban and regional green ways, the coordinating agency also identified several registered hobby groups of cycling and photography as well as the retiree association, to collaborate on organising communal activities in venues (e.g. terraced recreation stations) provided by the green ways, so as to steer the mobilized usage of these innovative urban infrastructures.

Since urban niches – with green building practices and 'green way' usages in this case – focus on users and require technical supports with local knowledge, the co-opting of new and existing organised frontrunners in the professional fields and in the societal engagement activities is an intervention articulated with operational considerations adopted during the managing process of transitions. In this context, the transition is set towards the vision of low carbon ecological city localised in a particular municipal setting. Such operational intervention was not designed prior to the implementation. The steering capacity of governmental change agents was gradually built during their discovery of social resources, through which process the originally limited boundaries of government-dominated transition arena had been expanded. The state-led visions of low-carbon urbanism and 'ecological civilisation' (as in Table 1), despite being weakened in the dissemination process by the system disequilibria, can still be practiced on the urban ground with societal co-visioning enabled by broader stakeholder engagement.

Conclusion: Disequilibria, Diversity and Participation in Transitions

The envisioning and practices of urban transitions in China are configured in a concurrent societal transformation process with fundamental changes in values and mentality held by the urban middle class that emerges over three decades after the economic reform. The governments in China have been playing a proactive role in envisioning the social, political, environmental and economic futures of cities, with strong attempts of bringing in and steering the structural changes that can maintain the governing authority and legitimacy over the urban population. In managing changes and maintaining stability, the state is sensitive in the urban discontents and transformed dynamics in society.

Rampant environmental emergencies and urban protests, facilitated by a well-connected digital infrastructure that enables a virtual environment with liberated information and communication for urban residents, are evidently the major disequilibria in China's management system of urban transitions. These disequilibria remain the persistent problems for the state to keep urban governing crises at bay. Manifested in the complexities of urban dynamics, such disequilibria have also weakened the state-imposed urban visions, ones that carry political agendas of maintaining the state's moral authority, building shared images of nationalism, and maintaining steady local development (Table 1).

Nonetheless, these disequilibria have also expanded possibilities for a more *diverse* array of urban actors to participate and get engaged in the transition process. Whereas the management of urban transitions in China has been predominantly government centred, the disequilibria have brought the state into expanding the boundary of transition arenas and introducing governing innovations that enables co-envisioning and co-creation of future cities, in selected experimental urban sites. Rather than imposing administrative penalties on all the unorthodox thinkers and active protesters on top of some ad-hoc media censorship, the tenets and process prescribed by the Transition Management framework stands out as a more justifiable and legitimate way to engage changes and resolve the persistent urban problems. Some state-approved urban frontrunners have therefore entered the transition arenas to steer changes alongside the local government, given their local knowledge, community networks and innovative ideas in mobilising the urban population.

Urban sustainability transitions can also be expedited through proactive and innovative approaches of local governments and individual frontrunners at the micro level of socio-technical system in managing environmental crises and emergencies. For local governments to catalyse transitions with bottom-up pressures from the micro niche to macro regime levels, they can take a combination of different approaches: the 'defensive approach' by 'seeking to discredit other actors', the 'reactive approach of system improvement' and the 'innovative approach by contributing actively to a transition' (Rotmans et al. 2001). Currently, local governments in Chinese cities tend to rely more on the defensive and reactive approaches to yield short-term results. To accommodate the long-term thinking and co-visioning, a more proactive and innovative approach is inevitable in managing future cities in China. In terms of approaching emergencies (Disequilibrium 1), enhancing media transparency is a proactive approach by actively contributing to perception changes and empowering bottom-up requests for disclosures (Disequilibrium 2), to be distinguished from the usual reactive approaches (Table 2) or even defensive approaches (by discrediting activists or blocking the crisis-initiated transitions) to environmental activism (Disequilibrium 3). It might take more public administration resources to adopt the proactive and innovative approach but it is probably the more sensible way to govern urban environmental transitions in future China given the current rapidly evolving economic and technological context.

Competing agendas between building nationalism images and maintaining local development interests are also demonstrated in the state-imposed urban visions. The localised experiments to practice the possible pathways of managing transitions have provided protected learning space for the urban managers to incorporate national agenda with local dynamics. Whereas the state-impose visions may pose interventions at the strategic level, the goal-setting and steering activities in the local government form interventions at the tactic level of managing transitions, which then involve governance activities at the operational level that encompass both governmental and nongovernmental actors in the expanded transition arenas. Since the state visions are fragmentised and weakened at the strategic levels, the processing of transitions at the tactic and operational levels is seemingly more important to fill in practical implications of the distant visions.

On the operational level, discussion in section "Experimenting with urban visions in Shenzhen: adapting changes" has identified discrepancies between the tenets of Transition Management and the actual practices of managing transitions in the experimental site of Shenzhen. The process of goal setting has dominated the preparation of government-initiated process plan. As a result, the finalised process plan lacks a complete actor analysis of the transition arenas in the city, and does not include substantially any contingency protocols for adaptive and flexible changes. The transition process for the implementing administrators is merely to carry out the necessary transfers and communication of orders to meet with the rigid and narrow goal within a strict timeframe of task completion. For instance, the performance of 'green way' construction is merely measured by the physical length rather than comprehensively evaluated by its actual usage or by how it has improved the urban quality of life. The space for co-visioning at the local implementation level is rather limited, even for the coordination officers within the governmental arena.

Even though during the delivery of allocated policy goals, some transition managers may identify the needs of extending the engagement with frontrunners and organisational actors outside the government, the scale of *participation* as prescribed in the tenet of Transition Management is still inadequately supported by policy mandates as reflected in the process plan, and thus largely neglected in the current management style of urban transitions.

Given the analysis in this chapter, a set of toolkits for urban managers is compiled which might be useful to facilitate the adaptation to practicing transitions in Chinese cities that fits with their future development needs. Since the core-underlying problem identified is the structural imbalances among economy, society and environment changes, its remedy would be for the urban managers to focus their resources on adjusting the societal and environmental management practices. Listed in Box 1 below, these suggestions are categorised by the types of interventions in transition management process (see chapter "Transition Management: Guiding Principles and Applications", this volume).

Box 1: Practicing Urban Sustainability Transition Management in China

Orienting

- To carry out proactive and *in situ* system analysis by actively and regularly assessing the local dynamics;
- To assess and cope with the transformed mentality of urban population;
- To initiate complete actor analysis for the drafting of process plan and to diversify the transition team;

Agenda-setting

- To co-vision across levels of administration and to reconnect the state-imposed long-term thinking and locally-initiated short-term agendas;
- To formally engage with non-governmental frontrunners with their mobilisation powers and social capitals;
- To integrate contingency protocols for adaptive changes in the established process plan;

Activating

- To adopt proactive and innovative approaches in managing urban niches (e.g. transparent operation for engaging inputs and driving societal change);
- To maintain the diversity of actors and initiatives organised and protected in the transition arenas;
- To widen citizen participation while scaling up niches to urban practices;

Reflecting

- To conduct comprehensive program evaluation of niche implementation that captures the societal and environmental goals and outcomes of the experimentation.

References

Associated Press in Beijing (2015) Tianjin explosion: China sets final death toll at 173, ending search for survivors. The Guardian. Retrieved from http://www.theguardian.com/world/2015/sep/12/tianjin-explosion-china-sets-final-death-toll-at-173-ending-search-for-survivors

Boehler P (2013) Blogger foretold his own arrest after criticising rumour crackdown. South China Morning Post

Bulkeley H, Broto VC, Hodson M, Marvin S (2010) Cities and low carbon transitions. Routledge, London

Central Civil Commission (2003) Interim measures of the central commission for spiritual civilisation construction regarding the evaluation and commendation of national civil cities, civil villages and towns, and civil (work) units. (Civil Commission (2003) No. 9). Beijing: Retrieved from http://www.wenming.cn/ziliao/wenjian/jigou/wenmingban/201312/t20131231_1668724_2.shtml

Chinese Academy of Governance (2015) Public lectures. Retrieved 15 September 2015, from http://www.rmgpw.org.cn/portal/open_course!list.action

CINIC (2014) Statistical report on internet development in China. Beijing: Retrieved from http://www1.cnnic.cn/IDR/ReportDownloads/201404/U020140417607531610855.pdf

Du S (2014) Social media and the transformation of 'Chinese nationalism':'Igniting positive energy'in China since the 2012 London Olympics. Anthropol Today 30(1):5–8

Fang K (2015) How has Tianjin become a city 'without news'? Retrieved from http://cnpolitics.org/2015/08/tianjin-news/

Feng C (1997) Order and stability in social transition: neoconservative political thought in post-China. China Q 151:593–613

Gong R (2015) The logics behind the official university research topics in China, Commentary. Financial Time Chinese Edition. Retrieved from http://www.ftchinese.com/story/001063227?full=y

Guangdong Provincial Government (2008) Guangdong Provincial Government Office's scheme on further strengthening the promotion, education and training of emergency management. (GDPGO[2008]24). Guangzhou

Guangdong Provincial Government (2016) The work plan of emergency management expert group in Guangdong Province 2016. (GDPGO[2016]94). Guangzhou

Hu J (2012) Keynote report of the 18th national congress of the communist party of china. People's Daily, Beijing. Retrieved from http://paper.people.com.cn/rmrb/html/2012-11/18/nw.D110000renmrb_20121118_3-01.htm

Lee C-C, He Z, Huang Y (2007) Party-market corporatism, clientelism, and media in Shanghai. Harvard Int J Press Polit 12(3):21–42

Marvin S, Guy S (1997) Creating myths rather than sustainability: the transition fallacies of the new localism. Local Environ 2(3):311–318

NDRC (2010) Notice of work commencement on low-carbon pilot provinces and pilot cities. National Development and Reform Commission, Beijing

NDRC (2012) Notice of work commencement in the second batch of national low-carbon pilot provinces and pilot cities. National Development and Reform Commission, Beijing

Pow CP, Neo H (2013) Seeing red over green: contesting urban sustainabilities in China. Urban Stud 50(11):2256–2274

Rotmans J, Kemp R (2008) Detour ahead: a response to Shove and Walker about the perilous road of transition management. Environ Plan A Int J Urban Reg Res 40(4):1006–1012

Rotmans J, Kemp R, Van Asselt M (2001) More evolution than revolution: transition management in public policy. Foresight 3(1):15–31

Shove E, Walker G (2007) CAUTION! Transitions ahead: politics, practice, and sustainable transition management. Environ Plan A 39(4):763–770

Spaargaren G, Mol AP (1992) Sociology, environment, and modernity: ecological modernization as a theory of social change. Soc Nat Resour 5(4):323–344

SZ-DRC (2012) Medium and long term plan of Shenzhen low carbon development (2011–2020) (SZ-DRC[2012]512). Shenzhen Development and Reform Commission, Shenzhen

SZG (2011) Work programme of Co-creating a national model of low carbon ecological city between national housing and urban-rural development ministry and Shenzhen municipal government. Shenzhen Municipal Government, Shenzhen

Wang SZ (1989) Chinese modern design: a retrospective. Des Issues 6:49–78

Xi J (2013) Speech of the newly inaugerated Chairman of People's Republic of China at the closure of the First Session of the 12th Assembly of National People's Congress. Xinhua News Agency, Beijing. Retrieved from http://www.wenming.cn/specials/zxdj/xjp/xjpjh/201405/t20140504_1913938.shtml

Xiao H, Chen H (2015) Investigating safety risks can not tolerate any formalism. Xinhua News. Retrieved from http://news.xinhuanet.com/politics/2015-08/23/c_1116344000.htm

Xinhua News (2011) Hu Jintao: to firmly improve the scientific level of social management. Retrieved from http://news.xinhuanet.com/politics/2011-02/19/c_121100198.htm

Yu K (2007) The initial formation of civil society in China requires reforms in social management system. In: Hou Y (ed) Activating cells of the harmonious society – an institutional study of Yantian model. Central Compilation & Translation Press, Beijing

Yu K (2014) The Communist Party of China's cadre education and national governance. J Zhejiang Provincial Party Sch 3:5

Zhang A (2014) Rational resistance: homeowner contention against waste incineration in Guangzhou

Zhao Y (2008) Communication in China: political economy, power, and conflict. Rowman & Littlefield Publishers, Lanham

Zhong Y (2013) A survey on the environmental attitudes of urban residents in China. Public Opinions and Media Research Centre, Shanghai Jiangtong University, Shanghai

A Transition Management Approach for Shrinking Cities in the United States

Terry Schwarz

Abstract Transition Management is not a widely used approach for guiding urban development and change in the United States. However, the principles of Transition Management may prove valuable to planning and policymaking efforts in the US, particularly in the shrinking cities of the Rustbelt region. In these cities, population loss and deindustrialization pose daunting challenges, but also offer robust and multi-faceted opportunities to pursue more sustainable urban futures. Established planning practices in the United States are primarily geared toward managing growth. These practices are often ineffective in cities that are not growing. New, more flexible approaches to planning and community engagement are needed. The principles of Transition Management may prove especially relevant for shrinking cities in the US because the overarching idea that Transition Management is not about adapting, or correcting current conditions, but rather a focus on transformative change.

Introduction

The Transition Management approach is structured to influence the direction and pace of urban change toward more sustainable outcomes. Transition Management is a clearly defined framework, which nonetheless allows for flexibility in response to local conditions and priorities. In progressing through a series of planning phases, described in detail in this book (Wittmayer et al. chapter "Transition Management: Guiding Principles and Applications", this volume), Transition Management enables community members to understand the kinds of changes that will be needed for a sustainable future, devise initiatives that help lead to this future, and empower a diverse range of people to take action.

T. Schwarz (✉)
Cleveland Urban Design Collaborative, College of Architecture and Environmental Design, Kent State University, Cleveland, OH, USA
e-mail: tschwarz@kent.edu; http://www.cudc.kent.edu

N. Frantzeskaki et al. (eds.), *Co-creating Sustainable Urban Futures*, Future City 11, https://doi.org/10.1007/978-3-319-69273-9_13

While Transition Management is not a widely-understood planning term in the United States, the principles of transition management could potentially improve planning and community engagement efforts in the US, particularly in older shrinking cities that have experienced persistent population decline and widespread urban vacancy.

In the US, cities that are losing population tend to shun the label of "shrinking city" because of its negative connotations. Once a city has been categorized as shrinking, it can become more difficult to imagine opportunities for future development and growth. Because of this, the term "city in transition" instead of "shrinking city," as a way to denote the same concept in a more politically palatable way.

In 2012, the Planning Advisory Service of the American Planning Association published a report entitled, Cities in Transition: A Guide for Practicing Planners (Schilling and Mallach 2012). Although every city is always in transition to some degree, the report focuses specifically on cities that have lost 30% or more of their peak population. The changes these cities experience often include a loss of population density; an increase in vacant buildings and land; real estate disinvestment (either widespread or concentrated in certain neighborhoods); declining properties and tax revenues; under-used and under-maintained roads, sewers, energy and water infrastructure; concentrated poverty; and racial/social segregation.

This chapter explores some aspects of the Transition Management approach as it applies to United States' cities dealing with population loss and the cumulative changes associated with this demographic phenomenon. It will address the following questions to generate an understanding about the challenges of shrinking cities and how TM could support/inform planning responses to these challenges.

- What kinds of changes and challenges do shrinking cities experience?
- Are some currently used planning approaches in the US in some ways analogous to transition management?
- How can transition management approaches help shrinking/transitional cities better manage periods of transition?

These questions are explored in the context of Cleveland, Ohio—a shrinking/transitional city in the Great Lakes region of the US, commonly referred to as the Rustbelt.

Cleveland, Ohio: Past and Present

In the nineteenth and early twentieth centuries, Cleveland was a booming metropolis. The city's central location on the Great Lakes played an important role in its emergence as a commercial hub and transportation center. By 1940, Cleveland was the sixth largest city in the country. The city's population peaked in 1950 at 914,808 residents. But in the 1960s, industrial restructuring, economic decline, and growing racial tensions triggered the city's ensuing struggles with population loss, abandonment, and blight. Race riots in 1966 accelerated the patterns of out-migration and economic segregation that persist to this day. The city's economy floundered and by 1978, Cleveland became the first major US city since the Depression to default on its financial obligations.

A revival, of sorts, began in the late 1980s, fueled by the development of a downtown shopping mall and major league sports stadiums. In the city's strategic plan, downtown investments are complemented by a neighborhood-focused initiative to build 1500 housing units annually in an effort to restore the city's lost population, which had dropped below 500,000 by this time. Housing construction was incentivized by a 15-year, 100% property tax abatement on new residential development. Infill residential development occurred throughout the city the recent subprime mortgage foreclosure crisis in 2008 undermined the city's housing market. Over 46,000 foreclosures occurred within city limits between 2006 and 2012, affecting more than 25% of Cleveland's housing stock.

Foreclosures lead to abandonment. Vacant homes feed an aggressive black market economy in scrap metal. Once residents are evicted by banks and mortgage companies, scrappers rapidly descend and rip plumbing, siding, and anything else with resale value from the vacated structures. The scrapping process leaves houses in a condition where they cannot be rehabilitated in a cost-effective manner. Demolition becomes the inevitable next step. The city has over 29,000 vacant lots (more than 4200 acres in total) along with an estimated 8000–10,000 buildings awaiting demolition.

Despite the city's losses, Cleveland remains a vibrant and appealing place to live and work. Although total population continues to decline, demand for housing in the city's core downtown area increased 53% between 2000 and 2010. Many new housing developments are currently under construction. Downtown population is expected to double by 2020 to reach 25,000 residents. Moreover, Cleveland is home to world-class cultural institutions, major businesses and biomedical industries, and an efficient public transit system. Situated at the edge of Lake Erie, Cleveland has access to an ample supply of fresh water and an extensive network of public parks and green spaces.

Cleveland is in an improbable situation—thriving and declining at the same time. Transition management, if adopted in a systematic and consistent manner, could help city officials manage the current conditions of decline in ways that stabilize city neighborhoods and set up opportunities for future development that will ultimately lead to a healthier and more resilient city. To date, the city has taken a piecemeal approach to managing transitions–through scenario planning efforts and appreciative inquiry discussed later in this chapter. But first, it is important to review the changes and challenges Cleveland, and other cities in a similar situation, are facing.

Transitions in Shrinking Cities

Shrinking cities are a global phenomenon, impacting cities in Europe, Asia, South America, and the United States. For example, in the older industrial regions of Europe, urban growth has given way to economic stagnation and population loss in many cities. The process of shrinkage forces cities to contend with population loss, vacant buildings, economically distressed commercial districts, and underutilized infrastructure (Kabisch et al. 2006).

Although every shrinking city must contend with its own specific range of issues, Cleveland illustrates some common changes that most shrinking cities experience, in one form or another, and the challenges that result from these changes to real estate, infrastructure, and climate conditions.

Real Estate Changes As a city's population declines, buildings are abandoned and often demolished. In the case of Cleveland, the city has approximately 23,000 vacant parcels of land within city limits. This is equivalent to 14% of the city's overall land area (Cleveland City Planning Commission 2014). Excess real estate drains value and vitality from city neighbourhoods, as empty buildings and vacant land lead to further disinvestment and blight (Fig. 1).

Blighted conditions motivate existing residents to move to more stable areas, typically in suburban communities, if they have the resources to do so. The result is a hollowing out of the urban core, leaving behind the poorest and most vulnerable populations, including elderly residents and single-parent households.

Infrastructure Changes In cities that have lost substantial population, the capacity of existing infrastructure often exceeds demand. For example, in Cleveland the road and sewer network was designed for a city of over one million people. The city's current population is 390,000. Although there are fewer people and less tax revenue available to maintain this infrastructure network, decommissioning infrastructure is not a high priority in Cleveland for a variety of reasons. First, engineers and public works professionals are trained to maintain and expand infrastructure networks. The idea of removing infrastructure is contrary to the way they operate. Generally speaking, public officials prefer to incur the costs of maintaining the entire infrastructure network at some minimal level, even if it far exceeds current needs, rather than risk removing infrastructure that may have to be reinstated at some point in the future.

The issue is further complicated by the expense and physical difficulties associated with decommissioning infrastructure. There are significant capital costs involved in removing roads, sewers, and water lines. Public officials are hesitant to spend capital funds in the present to achieve a possible (but uncertain) maintenance cost savings in the future. Compounding their resistance is the fact that the future is unknown. The City of Cleveland, for example, has significant assets that point toward long-term growth and recovery. Surplus capacity–particularly in water, energy, and transportation infrastructure–is a competitive advantage that

Fig. 1 Vacant parcels in Cleveland, Ohio (Source: Cleveland Urban Design Collaborative, Kent State University 2015)

can be used to attract businesses and economic development to the city. Eliminating a surplus now could prove counter-productive over the long term.

There are physical constraints to downsizing infrastructure networks. Utilities, rail lines, and roadways operate on a fixed grid. It is difficult to remove components in depopulated areas without impacting the whole system. Water, sewers, roads, and power lines need to extend through depopulated areas in order to get to areas of a city or region where concentrations of people still live and work. Lastly, when cities are dealing with old infrastructure, redundancy is a benefit. When bridges fail, water and sewer lines break, and pumping stations need to come offline for maintenance, redundant aspects of an infrastructure network provide a back-up that enables a city to continue to meet the needs of residents and businesses in these emergency situations (Hoornbeek and Schwarz 2009).

Climate Change Although climate vulnerability in the US is most commonly associated with sea level rise in coastal cities, Cleveland and other inland cities near the Great Lakes experience climate change in the form of heavier and more frequent precipitation, greater fluctuation in seasonal temperatures, and an increased number of high heat days.

Cleveland reflects a common pattern in shrinking cities. Rapid development during the years of peak growth resulted in a dense concentration of buildings, roads, and other heat-collecting, impervious surfaces across the city, and a substantial depletion of the urban tree canopy over time. As the city's population subsequently declined, suburban outmigration and sprawling patterns of development resulted in an expanding urban footprint in which essentially the same number of residents who used to live in the city proper are now spread out across a large and expanding suburban region. This phenomenon of sprawl without growth has an adverse impact on local micro-climates and urban heat island effects. Additional challenges include increased flooding risks, storms of greater frequency and intensity, and measurable increases in heat-related mortality, particularly for households that access to lack air conditioning (Sheridan 2015).

Efforts to address climate change need to relate to the specific and unique conditions in shrinking/transitional cities. Outmigration from shrinking/transitional cities has typically led to concentrated poverty and racial segregation. As such, these cities have a responsibility to respond to social conditions and land use patterns that exacerbate the adverse effects of climate change, particularly in low-income neighbourhoods and communities of color, because recent history suggests that these populations are most vulnerable in the event of heat waves and flooding, as well as less extreme climate-related challenges such as escalating energy costs and disruptions in city services (Kresge Foundation 2015). New planning methodologies are needed to lessen overall demand for energy, anticipate and prepare for climate changes and shocks, and foster social cohesion in these cities.

How Can Transition Management Approach Help Shrinking Cities to Better Manage Periods of Transition?

The practice of urban planning in the US is founded on an assumption of growth, because most US cities have experienced continuous (if at times, irregular) growth throughout their histories. Shrinking cities, or, at least the open acknowledgement of population loss as an on-going urban phenomenon is relatively new and discussed largely within the context of a small cohort of cities in the deindustrializing regions of the country.

Because planning is most commonly used in the United States to manage growth, established planning approaches sometimes fall short when addressing the needs of cities that are not, in fact, growing. A traditional master plan organizes, supports, and manages future growth. Land use policies, infrastructure investments, and city services are then aligned with the master plan, which is updated as development occurs. With occasional exceptions, growth patterns follow the path of infrastructure investments and other public and private incentives as identified in the plan.

Shrinkage is a less predictable process. It is rare for a city to lose population all in one place. While a city may have concentrations of vacant and derelict buildings in certain neighborhoods, disinvestment may simultaneously occur in scattered areas and disorderly ways. Established practices, therefore, don't always work. Shrinking/transitional cities need new tools. The six principles of the transition management approach (Roorda et al., p. 10; Wittmayer et al. chapter "Transition Management: Guiding Principles and Applications", this volume) may be especially relevant for these cities because the overarching idea that TM is not about adapting or correcting current conditions, but rather a focus on transformative change. These principals include

1. **Get an insight into the system.** A traditional planning process would involve an analysis of demographic trends and real estate market potential, with a long-term vision designed to accommodate future growth. The situation in a shrinking/transitional city is often more complex and less linear. The transition management approach fosters an understanding of the dynamics and inter-linkages of multiple domains, actors, and scales, which is essential in cities where there are no clear or obvious solutions to current problems and great uncertainty about future conditions.
2. **Aim for system innovation in small but radical steps**. Daniel Burnham was one of the most influential US planners of the nineteenth and early twentieth centuries. He famously advised that one should "Make no little plans; they have no magic to stir men's blood" (Moore 1921, p. 147). In shrinking cities, the potential for big plans is diminished–or non-existent–due to a lack of resources and market demand. Smaller steps, as suggested by a transition management approach, and temporary interventions can enable community members to envision and experiment with a broader range of alternatives, potentially leading to new discoveries that might otherwise have been overlooked.
3. **Leave room for diversity and flexibility**. Early theoretical and practice-based models of planned shrinkage were framed as a top down response—re-concentrate remaining residents in smaller, mixed-use nodes, connected by transit (Dewar and Thomas 2013, pp. 1–14). Once people have been relocated, decommission the parts of the city, now devoid of people, and discontinue city services to these areas. Reuse vacated land for urban agriculture, reforestation, or other green uses. Variations of this approach have been explored, but not successfully fully implemented, in cities such as Youngstown, Ohio; New Orleans, Louisiana; and Detroit, Michigan (Beckman 2010, p. 387). Implementing an all-encompassing, cohesive master plan for a shrinking/transitional city is difficult because there is so much uncertainty about whether losses will continue, for how long, and to what extent. The market forces that shape urban development are disrupted and the supply of real estate exceeds development demand. Planning efforts under these conditions may be more effective if the approach is made up of a flexible menu of options, rather than a steady march toward a preordained future.
4. **Co-create**. Using transition management may prove a way to bring together a collection of diverse urban stakeholders in a shrinking city that could create a long-term vision for the strategic reuse of vacant land and blighted buildings,

and then forge new pathways toward achieving this vision. Some of these short-term actions thought of by participants may turn out to be dead ends, but the continuous feedback loops and learning that are part of the transition management process would help determine which actions can be effective on the short-term and should be expanded over time, along with those that should be discontinued. By thoughtfully coordinating hundreds (or thousands) of small–scale actions over years (or decades), a city could make measurable progress toward a collectively created future.

5. **Give room to change agents**. Inclusiveness is essential in a shrinking/transitional city because any kind of meaningful change must engage the people most affected, since they are often the ones who will need to implement these changes on the ground. For example, Cleveland is currently developing a Climate Resilience and Urban Opportunity Plan focused on climate resilience in four low-income neighborhoods. Cleveland is one of the poorest major cities in the US and it can be difficult to engage low-income residents in the abstract issue of climate change when they have the more pressing and urgent matter of basic survival to contend with. An actor analysis, as advocated by transition management, could reveal information about the underlying social dynamics in the target neighborhoods, and provide insights about residents' concerns and the key issues that would motivate them to participate in developing adaptation strategies, mitigation actions, and more effective approaches to fostering social cohesion.
6. **Facilitate social and institutional learning**. Long-time residents of cities that have experienced economic and population losses sometimes reminisce about a city's glory days, when growth was everywhere and seemingly endless. But this paints an overly optimistic picture of the past, when rapid growth may have occurred without much concern for social equity or the underlying ecological systems that were systematically paved over to make way for urban expansion. A shared learning effort to understand current problems and emerging opportunities will help community members, public officials, and local institutions make decisions based on knowledge rather than nostalgia. This will establish a more informed foundation for a future city that is environmentally sound and socially just.

Complementarity: Planning Approaches That Relate to Transition Management

Transition Management is not a widely used or understood urban planning approach in US cities. In fact, the term "transition management" is most often used in a corporate, rather than civic, context. For example, a company that is merging with another or expanding its mission might go through a Transition Management process in order to maintain productivity and customer relationships during the period of rapid change. In Cleveland, other approaches, including

Scenario Planning and Appreciative Inquiry have been deployed in an effort to manage transitions stemming from population loss and associated land use changes.

Scenario planning and Vibrant NEO 2040 Scenario Planning is a facilitated process through which community stakeholders consider possible views of their city or neighborhood described in narrative form (stories) that provide a context for decision making. Cleveland and its surrounding 12-county region have recently completed a scenario planning process to address population loss and changing regional dynamics. In the region, population growth is stagnant and sprawling suburban development has led to a gradual hollowing out of many inner-city neighborhoods and growing disinvestment in the city's inner-ring suburbs. Vibrant NEO 2040, a regional scenario planning effort, was launched 2010 in an attempt to address these challenges. It was funded by a $4.25 million federal grant through the Partnership for Sustainable Communities Initiative, a joint program of the federal Department of Housing and Urban Development, Department of Transportation, and Environmental Protection Agency.

The scenario planning process for Vibrant NEO 2040 engaged a wide range of stakeholders over multiple levels and to create shared visions and goals. Although the process didn't result in a truly bottom-up approach to governance, it did begin with the idea that a shared vision for the future, developed through an extensive community and stakeholder engagement process, could lead to a healthier, more economically and socially equitable future for the region.

Numerous public meetings were held across the region and *Imagine My NEO*, an interactive game, was created so that residents could consider several scenarios, including business as usual ("Trend" Scenario); regional growth following current patterns of suburbanization and de-densification ("Grow the Same" scenario); on-going population loss with new regional land use policies ("Do Things Differently" scenario"); and regional growth with new regional land use policies ("Grow Things Differently" scenario) (See Fig. 2).

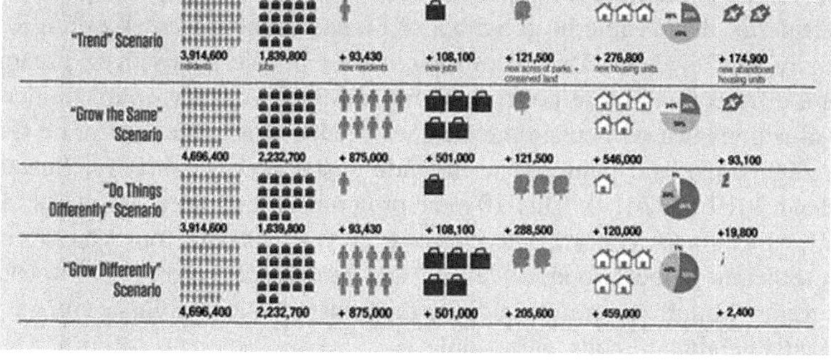

Fig. 2 Four future scenarios for Northeast Ohio (Source: Northeast Ohio Sustainable Communities Consortium 2010)

The strength of this scenario planning approach is that it presented clear, data-driven alternatives for the public to consider. But there are downsides to this planning process, including the cost. At $4.25 million, this is not a process that can be easily updated as conditions change in the region, or replicated in other metropolitan regions.

Another concern is that, while a comparison of scenarios clearly points to the advantages of "growing differently" there does not appear to be a widespread consensus among the residents of the region that this is the right approach. The highly polished presentations, reports, and videos prepared by the Vibrant NEO 2040 team have been met with skepticism and, sometimes anger, by conservative residents of outlying communities who are opposed to government intervention and change. Many of these residents see Cleveland and other central cities in the region as a drain on resources. And the economic and social forces that led to population dispersion and sprawling patterns of development in the region remain very much intact today. The Vibrant NEO 2040 process purported to be a community-driven process but, despite its good intentions, is perceived as a top-down approach that is at odds with the beliefs and values of many people in outlying, suburban areas of the region.

Lastly, the Vibrant NEO 2040 plan does not appear to have led to many actionable recommendations at the regional level. The *Vibrant NEO Tool Kit and Best Practices* guide contains an array of common sense strategies for regional land use, green space conservation, infrastructure investments, and sustainable development. But the plan has no regulatory force in a region where most communities have home rule authority and there is no funding available to incentivize the adoption of planning tools and best practices. As such, Vibrant NEO 2040 is a sensible approach to planning for the region's future, but it lacks pathways to implementation.

Scenario planning uses the power of community narratives to envision a more sustainable future, but lacks a specific mechanism for linking long-term visions to medium- and short-term actions, as advocated by transition management.

Appreciative Inquiry and Sustainable Cleveland 2019 Appreciative Inquiry (AI) is a positive, asset-based public participation strategy developed by Dr. David Cooperrider of the Weatherhead School of Management at Case Western Reserve University in Cleveland. The AI process invites people to discover, dream, and design a different and more positive future and then work backward to identify a series of actions that will create the changes needed to achieve this future (Fig. 3).

In 2009, Cleveland launched a citywide sustainability initiative, Sustainable Cleveland 2019 (SC2019). This 10-year program is organized around an annual summit, which focuses on a different topic each year. To date, topics have included energy efficiency, local food, advanced and renewable energy, zero waste, and clean water. Future summits topics include sustainable mobility, vibrant green space, vital neighbourhoods, and people.

SC2019 has a strong grassroots component that includes citizen-led working groups, more than 200 active volunteers, and hundreds of businesses and

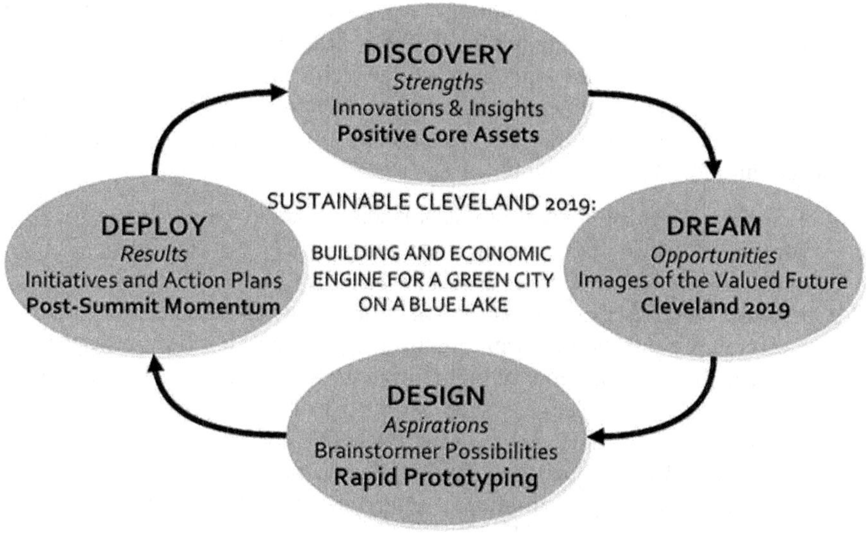

Fig. 3 Appreciative inquiry "4-D CYCLE (Source: Cleveland Office of Sustainability 2009)

organizations engaged in the work. At the annual summit, hundreds of people come together to "…co-create transformative projects that:

1. Focus our energy on the best of "what is" by designing projects that leverage our strengths and most unique assets;
2. Hone our strategies on specific goals, projects, and policies that generate momentum in our local and global communities; and
3. Create social, environmental, and economic value to build a sustainable economy" (Sustainable Cleveland 2009, 2019).

During the "dreaming" part of the summit, participants work in groups to complete the following exercise:

> *Tonight, after our session, you fall into a sound sleep. It's a good sleep. When you awake it is 2019, ten years from now. And while you were asleep Cleveland has created a green city on a blue lake. Our economy is thriving: clean manufacturing is saving and making money, natural resources are being used responsibly to great profit, and the dynamism is attracting up-and-coming businesses, new investors, and new citizens to Cleveland. You appreciate what you see. You are proud of Cleveland and the whole region—it has become what you most wanted. Cleveland has become the city that you believed it could be.*
>
> *Now please describe what you see—remember it's 2019. What do you see happening in business and industry? What do you see happening in terms of people and relationships? What do you see ecologically and in our relationships with nature and natural resources? How about our community, our schools, and public life? Our reputation nationally and globally? Taken as a whole, what's happening that is new? Better? What has changed? Can you describe some of the more compelling, exciting innovations?*
>
> *Finish this sentence: I will be most proud of Cleveland and our region when…..* (Sustainable Cleveland 2009, 2019)

Ideas generated from the dreaming exercise move forward into the design process. In this next step, summit participants brainstorm possible actions and programs to help achieve the outcomes they've envisioned. Some of these actions and programs are later implemented through collective citizen action, with the support of the Cleveland Office of Sustainability. Examples of ongoing efforts launched through SC2019 include:

- **Forest City Working Group** provides resources and expertise to assist in the reforesting of the City of Cleveland and the region. The working group has been actively involved in developing a detailed tree plan for the city.
- **Cleveland 2030 District** is a coalition of building owners, service professionals, and community stakeholders working together to reduce energy use, water use, and greenhouse gas emissions of buildings in the city, based on the targets of the Architecture 2030 Challenge, which has a goal of making all new buildings, developments, and major renovations in the US carbon-neutral by 2030.
- **Drink Local. Drink Tap.** is a non-profit organization that was created to encourage people to give up bottled water for tap water help to provide safe water access to people in need.

The SC2019 initiative provides a platform for community members to advance their ideas and establish consensus around priorities for action. The intent is to align hundreds of short-term actions with a long-term vision of "a green city on a blue lake." Although the array of individual actions in SC2019 may seem unfocused and unclear at times, the annual summit provides a shared point of reference and an opportunity to assess progress on an established set of goals.

A significant limitation of SC2019 is that this visioning process almost always leads to recommendations for policies and programs, but rarely for physical interventions that advance a vision a truly new kind of city. The simple, powerful language of "a green city on a blue lake" resonates with SC2019 participants and the broader public, but many people find it difficult to imagine the fine-grained details of what Cleveland would look like and how it would function if this vision were to be realized. Unfortunately, development decisions and infrastructure investments in the city frequently occur in ways that do very little to advance the intentions of SC2019.

Limitations of Transition Management in a United States Context

The primary limitation of Transition Management in a US context is that it is unfamiliar and untested in major US cities. One of the first US cities to explore a partial transition management approach is Flint, Michigan. Flint is a city ripe for reinvention. The city has lost nearly half of its peak population, declining from 197,000 residents in 1960 to just over 99,000 in 2014. Flint's growth was tied to the automotive industry. General Motors operated a manufacturing plant in the city that employed 80,000 people in 1978, dropping to 8000 by 2010. In the mid-2000s, Flint became known for its high crime rates and has been ranked among the "Most Dangerous Cities in the United States", with a per capita violent crime rate approximately seven times higher than the national average. The city was under a state of financial emergency from 2011 to 2015. It is currently in the midst of a public health emergency due to lead poisoning in the local water supply.

In 2014, Flint developed a Seven-Point Transition Management Plan, explicitly designed to help the city emerge from financial emergency and State oversight. The plan was a narrowly focused effort to stabilize the city's finances by eliminating debt, restructuring local government, and dealing with the legacy costs of public employee retirement and health benefits (City of Flint 2014). However, the development of Flint's Transition Management Plan coincided with an update of the city's comprehensive, long-range plan. Imagine Flint is the city's first comprehensive plan update in over 50 years. The process engaged over 5000 and was formally adopted in October of 2013 (City of Flint 2013).

Although this planning effort in Flint reflects some of the language of Transition Management, Imagine Flint is a fairly straightforward comprehensive plan, albeit one that deals with the difficult and complex issues of a shrinking/transitional city. In order for transition management to gain momentum in the US, one or more cities would need to adopt a true transition management approach in their planning efforts and demonstrate the outcomes that can be derived from this process. It might seem logical for a shrinking/transitional city to pursue an experimental transition management approach–since conventional planning approaches clearly are not working – and to become a leader in exploring new planning paradigms. But the inverse is more often true. Shrinking cities tend to be risk-adverse, often gravitating best practices from growing cities in an effort to align themselves with what is working elsewhere.

For example, Cleveland is currently exploring form-based codes, a land use and development practice that first became popular in the US in the 1990s as a way to shape development in growing cities and suburbs. Depopulating cities often align themselves with best practices from growing cities in an effort to emulate their success. Unfortunately, the planning strategies that work best in cities with strong real estate markets are sometimes ill-suited to the needs of cities that lack economic and population growth. As such, shrinking/transitional cities sometimes get caught in a cycle of plans that aspire to transformational outcomes but result in very little change on the ground. This creates cynicism among residents and a planning fatigue when neighborhood conditions repeatedly fail to improve.

For overcoming this limitation and for transition management approach to become a part of planning practice in the US, a city struggling with major transitions might have to take a risk and try something new. In this current era of epic uncertainty, where most cities are going to have to manage some kind of significant social, economic, or climatic change, the time seems right for a transition management approach.

More research in the field of transition management is needed, to compare the application of transition management practices in various cities and monitor outcomes over time. Conference presentations, training workshops, and the inclusion of TM in urban collegiate planning curricula could lead to the adoption of the transition management approach in US cities.

Conclusion

Transition Management is about self-discovery and learning-by-doing. This process is essential in shrinking/transitional cities. Many people find it difficult to imagine what the future will look like because it does not yet exist. Transition management, combined with place-based visualization efforts and temporary use strategies, could help community stakeholders to better understand their options, imagine a future that is very different from present-day conditions, and deploy multiple short- and mid-term actions that ultimately add up to a long term vision.

Transition and transformation do not occur easily in Cleveland. Residents tend to cling to what is known and familiar, rather than embrace things that are new and uncertain. Changes will sometimes occur in response to a natural disaster or other form of urban upheaval, but can also come about in reaction to fear or in anticipation of a reward or opportunity.

Transition Management has the potential to advance urban design practice and sustainability planning in US cities. In the older industrial cities of the Great Lakes region, transition management aligns especially well with the challenges these cities face relative to population decline, weak market demand, social isolation, and economic uncertainty.

While the term transition management is not commonly used among planners and urban designers in the US, the principles of transition management appear in other guises. In Cleveland, there is overlap in the form of scenario planning and appreciative inquiry, which help drive and support a citizen-based approach to urban regeneration.

References

Beckman B (2010) The wholesale decommissioning of vacant urban neighborhoods: smart decline, public-purpose takings, and the legality of shrinking cities. Cleve State Law Rev 58:387

City of Flint (2013) Imagine Flint. http://imagineflint.com/Documents.aspx. Accessed 1 Dec 2015

City of Flint (2014) 7-point transition management plan. Revised 19 Mar 2014. http://cityofflint.com/wp-content/uploads/7-Point-Transition-Management-Plan.pdf. Accessed 1 Dec 2015

Cleveland Office of Sustainability (2009) Sustainable Cleveland 2019 participant workbook

Cleveland City Planning Commission Land Use and Geographical Information System (GIS) data (2014)

Cleveland Urban Design Collaborative Mapping (2015) Kent State University
Dewar M, Thomas JM (2013) Introduction: the city after abandonment. The city after abandonment. University of Pennsylvania Press, Philadelphia
Hoornbeek J, Schwarz T (2009) Sustainable infrastructure in shrinking cities: options for the future. Northeast Ohio Research Consortium, Cleveland
Kabisch S, Haase A, Haase D (2006) Beyond growth–urban development in shrinking cities as a challenge for modeling approaches. Proceedings of the iEMSs Third Biennial Meeting: Summit on Environmental Modeling and Software. International Environmental Modeling and Software Society, Burlington
Kresge Foundation (2015) Bounce forward: urban resilience in the era of climate change. Island Press, Troy
Moore C (1921) Daniel H. Burnham, Architect, Planner of cities, vol 2. Chapter XXV "Closing in 1911–1912". Houghton Mifflin, Boston
Northeast Ohio Sustainable Communities Consortium (2010) Vibrant NEO 2040, A vision and framework for our future, Cleveland
Schilling J, Mallach A (2012) Cities in transition: a guide for practicing planners. Report number 568. American Planning Association Planning Advisory Service
Sheridan S (2015) An assessment of the impacts of extreme temperature on mortality in Cuyahoga county, Ohio. Cleveland climate resilience and urban opportunity plan. Cleveland Neighborhood Progress, Cleveland
Wittmayer JM, van Steenbergen F, Bach M, Frantzeskaki N, Loorbach D (2017) Transition management: guiding principles and applications, this volume

Context-Driven Transition Management as a Necessary Vehicle for Sustainable Urban Futures in Suriname

Daniël A. Lachman, Martinus R.J. Panday, and Deryck J.H. Ferrier

Abstract The capital of Suriname has experienced enormous growth over the past decennia. De decline of industries in other urban settlement have resulted in migration to the capital along with population growth and ongoing country immigration. With the absence of sound urban and spatial planning, the result is urban sprawl where increasing population and traffic density create ongoing specialization of services in the capital with continuously decreasing action radius; there is a tendency where people opt to live further from the city core but which makes the utilization of services in the city's core cost-prohibitive. The answer to this is the creation of new urban settlements which build on new industries with increasing returns. However, this implies a break from existing paradigms with respect to strategizing, policy, planning, tourism, logistics, urbanism, etc. This requires a transition for which management has been broken down in several steps in literature. However, transitions tend to differ along geographical parameters; for instance, regimes tend to be relatively stronger in developing countries. Thus, transition management needs to fit the context which requires assessment tools such as Quinn's culture model and the STEEPL and OPTIMISM frameworks to anticipate pitfalls and opportunities during the transition.

D.A. Lachman (✉)
Applied Intellect, Paramaribo, Suriname

Institute for Graduate Studies and Research, Paramaribo, Suriname

FHR Lim A Po Institute, Paramaribo, Suriname
e-mail: danny_lachman@yahoo.com

M.R.J. Panday
University of Applied Sciences, Paramaribo, Suriname

D.J.H. Ferrier
Centre for Economic and Social Scientific Studies, Paramaribo, Suriname

N. Frantzeskaki et al. (eds.), *Co-creating Sustainable Urban Futures*,
Future City 11, https://doi.org/10.1007/978-3-319-69273-9_14

Introduction

Suriname is situated in the Northeast of South-America with around 500,000 inhabitants (Fig. 1) that relies on export of oil, gold, bauxite and wood, and (relative small-scale) tourism. Last decennia Suriname has experienced an ongoing concentration of economic activity in and around its capital, Paramaribo (Fig. 2); this results in migration to the capital, traffic jams and exuberate travel times, economic activity impacting the livelihood of neighborhoods, skyrocketing property prices, lack of employment and the emergence of slums, etc.

Paramaribo is factually the sole urban conglomerate in Suriname, while other – much smaller – settlements, such as Albina, Moengo, Apoera and Nickerie are declining in size due to migration to the nation's capital. Furthermore, many villages in the interior (Fig. 3) do not have any controlled urban development despite the gold rush attracting (short-term) economic activity. This all leads to an accelerating migration rate to the capital.

Simultaneously, Suriname will most likely be hit hard by climate change (Magrin et al. 2007; Sims et al. 2007); its close proximity to the ocean, make it prone to rising sea levels while extreme weather events can easily disrupt infrastructures. Thus, ongoing demographic and economic concentration in and around Paramaribo results in several tensions and inefficiencies, while increasing impact in the wake of climate change.

This unquestionably calls urgently for new urban settlements, which implies that a radical new direction needs to be taken, since Paramaribo as the center of economic and administrative activity forms the ruling paradigm and urban planning has been barely existent for years. This requires irrevocably a transition effort which will impact different dimensions such as behavior, urban planning, trade and industry, and government policy.

Developments and Trends in Suriname's Urban Environment

Last few decades economic activities in regions other than Paramaribo have been in decline. Meanwhile, Government services (e.g. academic education, hospitals, etc.) have primarily been exclusive to Paramaribo. The result is that over the last four decades people are increasingly flocking – from especially the larger settlements, see Fig. 4 – to Paramaribo in search of work, education, health care, career opportunities, etc.

There were plans to revert this, but those failed to be realized. Currently, no other visions, strategies, policies, or plans exist within Government, political parties or NGOs to fill the void. Migration to Extended-Paramaribo and its population growth occur while basic infrastructure schemes in Paramaribo have virtually been unchanged for decades (even centuries). Furthermore, service centers (e.g. lawyers,

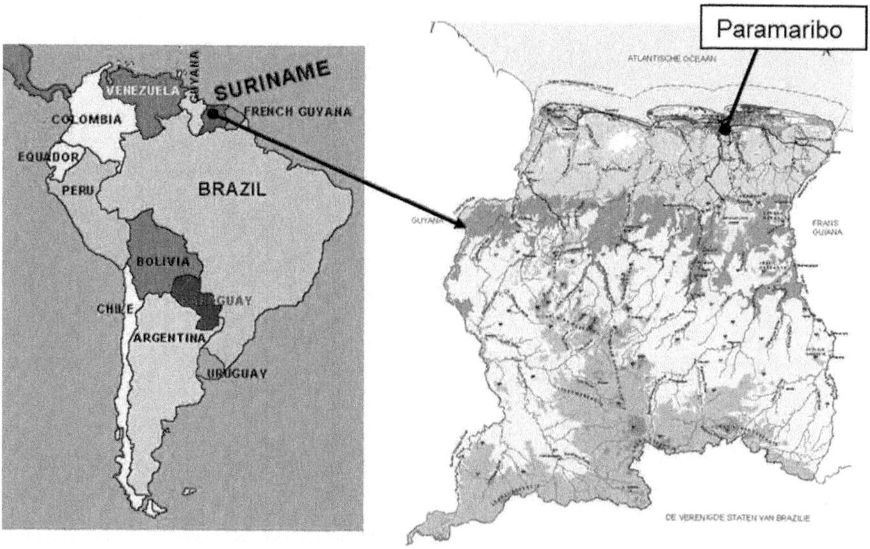

Fig. 1 The country of Suriname

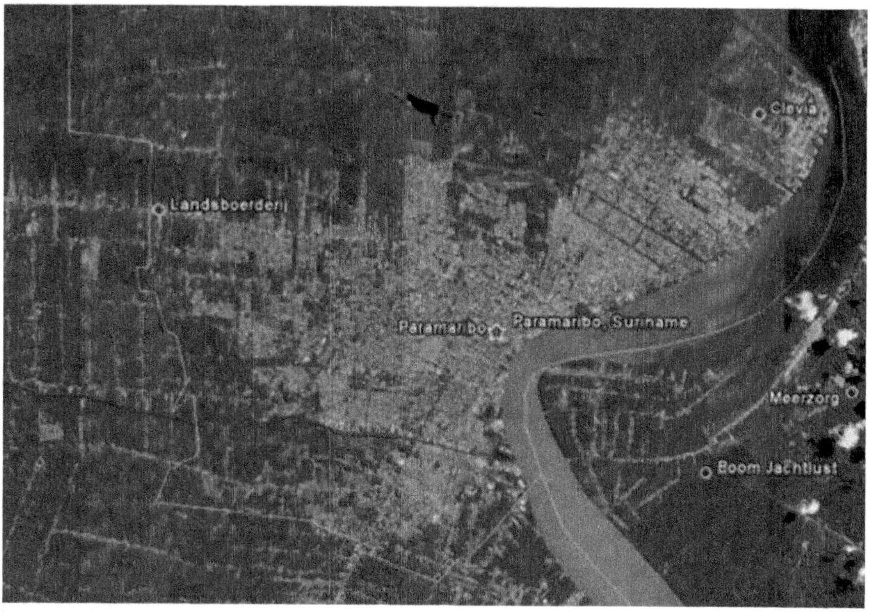

Fig. 2 Concentration in Paramaribo

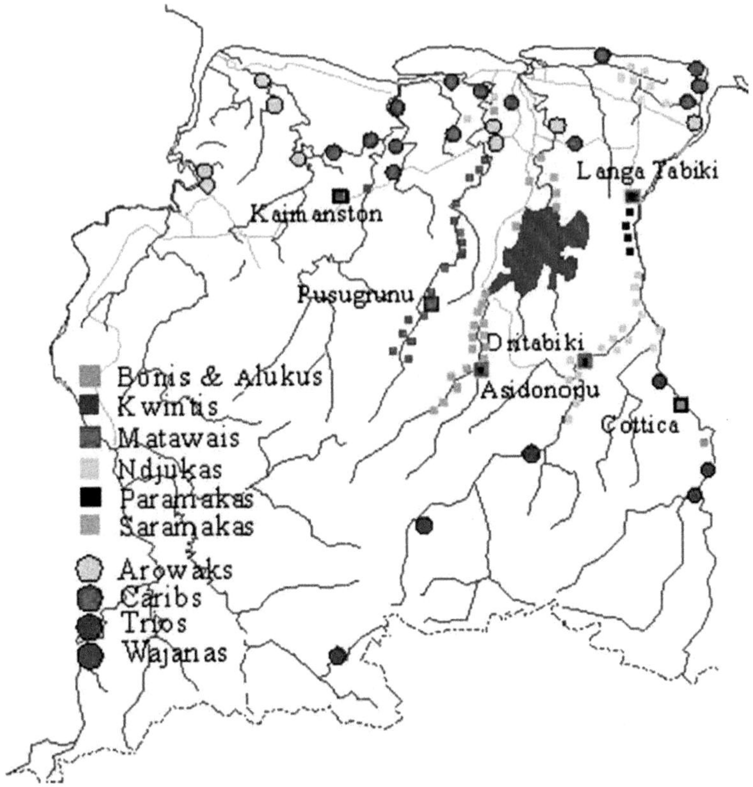

Fig. 3 Various Surinamese indigenous and maroon villages

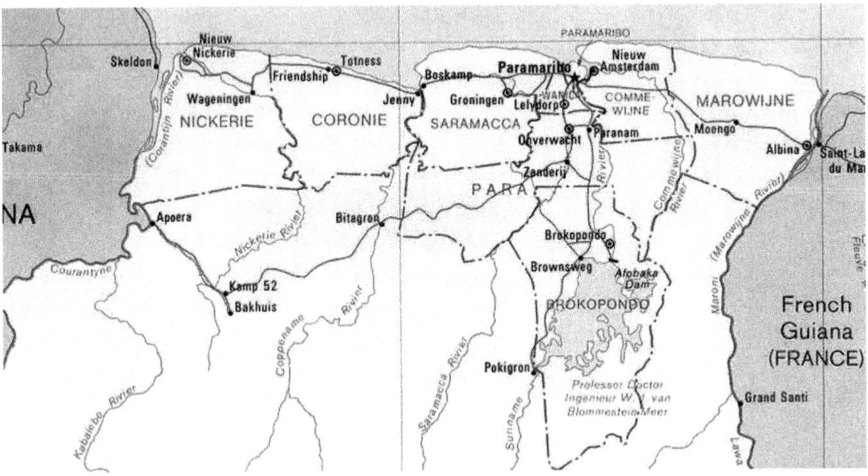

Fig. 4 The larger settlements in Suriname

government offices, shops) have mostly remained in Paramaribo's center without spreading out of Paramaribo's centers. This has led to a situation where:

- People flock to Paramaribo;
- More than 70% of the population lives within the Greater-Paramaribo area;
- Property values skyrocket;
- People travel greater distances to reach particular services;
- Travel consumes a lot of time;
- Conflicting interests and demands clash (e.g. a shopping mall situated in a neighborhood).

These trends and developments are embedded in the economic, geographic and urban profile of Suriname and Paramaribo, and any change away towards an alternative implies a drastic restructures of these profiles. Such a profound and large-scale transition requires peculiar attention and sound management, hence the governance of such a transition requires deep scrutiny and needs to be fully fleshed out.

Main Conditions for the Development of New Urban Settlements

The most important condition for the inception of new urban settlements is the availability of sufficient socially acceptable employment. This should be preferably situated in industries with increasing returns, usually high value-adding sectors with a strong, difficult-to-copy knowledge (or other significant asset) component in the value chain.

Another important condition for new urban settlements is anonymity (Riesman 2001): settlements need to have enough inhabitants to enable them to choose between several service suppliers. Furthermore, anonymity guarantees sufficient individual freedom making it possible for people to choose their own lifestyles and engage in work deemed low in class (but nevertheless important for proper functioning of urban settlements), like garbage collection.

Finally, the structural lay-out of new urban settlements and designation of urban locations for specific functions need to be planned in detail well in advance with the future (a time span of a couple decades) in mind. This means that in an early stage of development sufficient space for internal centers with bundled concentrations of service supply (e.g. shopping centers, industries, neighborhoods), and for infrastructure (e.g. walking and bike lanes, parking space, highways with enough lanes) with future expansion possibilities need to be reserved.

Agglomeration of Extended-Paramaribo has been progressing for decades without above mentioned conditions being met. Therefore, further migration to Extended-Paramaribo will ultimately result, in social disruption, risk concentration, inefficiencies, etc., in the future, which cannot be countered nor prevented without the creation of new urban settlements to reverse the migration phenomenon.

An Evidence-Substantiated Vision of Possible New Urban Settlements in Suriname

Recently, a path towards new urban settlements in Suriname has been conceived. Figure 5 provides the main constituents of this scheme drawn on the map of Suriname.

Tapajai Hydropower Project

A large water reservoir exists south of Paramaribo, the Brokopondo reservoir (Boksteen 2008a, b), which was built to generate hydroelectricity and spans

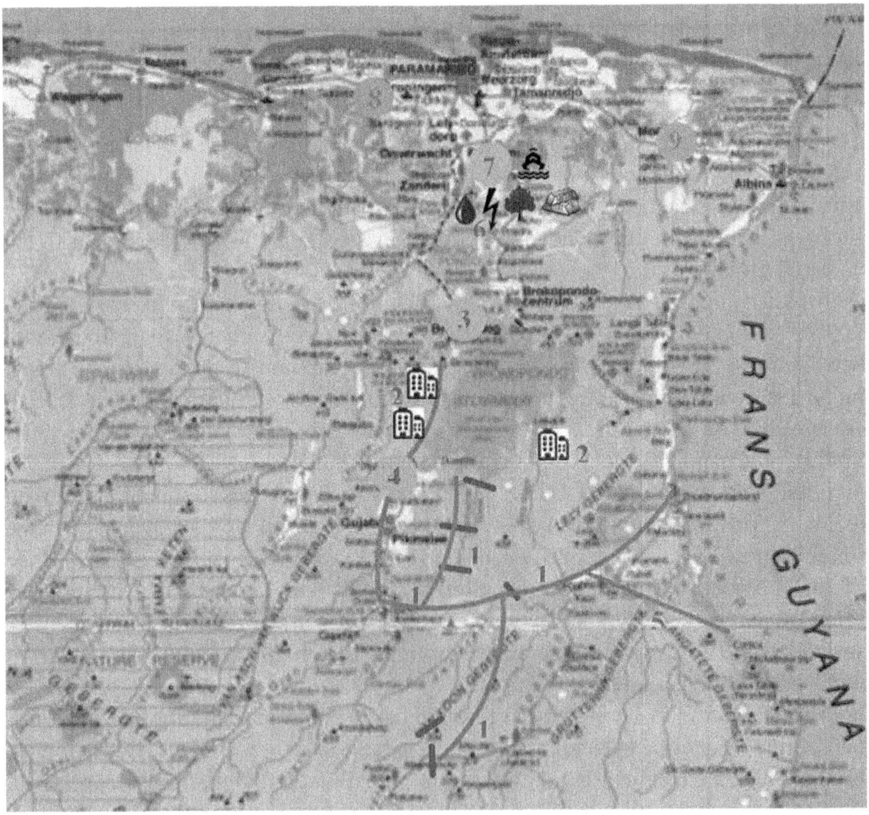

Fig. 5 Lay-out of a scheme for new urban settlements in Suriname

1560 km^2 and the associated power station has an installed generating capacity of 189 MW, with an annual average amount of firm power delivered around 93 MW.

The Tapajai Hydropower Project (THP) is an option that diverts the Tapanahony River with a number of dams via the Jai Creek into the Brokopondo reservoir, thus increasing the average level in the reservoir and consequently increasing the average hydroelectricity generating potential. This would increase the footprint of the reservoir with 5%, but also the firm power output with 182 MW. This firm output would be realized with an expansion of the existing hydro power facility and five new power stations along the Tapanahony River – Jai Creek trajectory. To achieve this, around 129 km of new roads and 170 km of project roads need to be constructed (1 in Fig. 5, where the red lines depict the roads, and the blue stripes represent the dams), which access a great deal around the Brokopondo reservoir. The water, energy and especially the roads associated with the Tapajai Hydropower Project will serve as the feeding ground for the next step towards the inception of new urban centers, the Brokopondo Tourism Project.

Brokopondo Tourism Project

Scheduled airlines (common air carriages) are public air transport systems with high service standards, fixed published schedules (regardless of passenger occupancy) and published fares by a major or regional air carrier (Cooper et al. 2008). Charter airlines (low cost airlines) have lower service standards, operate predominantly from regional airports (which have low-cost services compared to international airports) and often seek discounts on airport charges by contracting airport services in bulk and offer point to point destination services (Cooper et al. 2008). Some charter airlines are even vertically integrated with holiday companies or tour operators (Harrison and Enz 2009). They operate on destinations that either have an established popularity, or, the potential of becoming a popular holiday destination. Accommodation facilities as well as tourism supply services (e.g. restaurants, bars, car and bike rentals) should be near the core attraction in the destination. (Vertically) integrated tour operators generally buy both accommodation (hotel rooms) and flight seats in bulk to compose competitive holiday packages with both flight and hotel included at the lowest possible cost.

Suriname has a number of world-class tourism resources, such as vast tracts of wilderness, numerous large pristine river systems, extremely productive coastal ecosystems and unique cultural and historic attractions. Nevertheless, the Surinamese tourism industry faces serious challenges in growth due to expensive travel tariffs form both Europe and North America and the lack of sufficient provision of tourism products and services in its core product, namely the untouched wilderness (Suriname Tourism Foundation 2013): in 2013, of the 250,000 visitors, just 40,000 were leisure travelers ("real" tourists) to the country according to data from the Suriname Tourism Foundation.

Hotels and other tourism products and services are concentrated in Paramaribo while Suriname's core product is located in the interior. Local tour operators offer trips to the interior but the costs of transportation and accommodation are relatively high compared to other jungle destinations in the region and abroad. Some lodging is offered in the interior, but the majority lack sufficient quality in both their products and services, which (in conjunction with the high transportation costs) makes Suriname's core product uncompetitive.

The Brokopondo Lake's accessibility by road and the generated hydroelectricity can bring a solution in both the accommodation capacity in the core product as well as the provision of additional tourism products and services. In this concept, the aim is to centralize and concentrate accommodation (aiming at 1200 to 1500 hotel rooms from the start) and supply services around the Brokopondo lake (2 in Fig. 5; these are arbitrary picked locations) in order to persuade international tour operators to include Suriname in their packages (charter flight plus accommodation) (Sheoratan et al. 2013). The Government should allocate land around the lake for the development of tourism products and services, while existing tourism entrepreneurs in Paramaribo will be given the opportunity to franchise in the area. This is where the electricity, water and infrastructure of the Tapajai Hydropower Project play a significant role as the foundation upon which these investments are built. The tourism entrepreneurs need to project a forecast of the tourism products and services they plan to offer and subsequently liaise with international (vertically integrated) tour operators to get an indication of potential tourism supply. After adjustment of the projection and contracts with (vertically integrated) tour operators, investment in the franchises in the Brokopondo lake area can commence. First studies indicate that this concept should be able to annually generate 500,000 tourists to Suriname (Sheoratan et al. 2013).

Aerodrome Atjoni

Atjoni is a village at the southwestern part of the Brokopondo reservoir and functions as a logistics hub between the coastal area and the villages in the hinterland by means of boat transport. The Aerodrome Atjoni would further signify Atjoni as a logistics hub by constructing an international airport (fit for a Boeing 727/737 category aircraft) in that area (Fig. 6), which would carry out national and regional flights. This would align with the Brokopondo Tourism Project, since tourists will now also be able to depart from the reservoir and visit other destinations in Suriname or destinations abroad.

Urbanization of Brownsweg and Atjoni

With the Tapajai Hydropower Project, the Brokopondo Tourism Project and the Aerodrome Atjoni, a wealth of new services and products are delivered which inadvertently implies creation of a significant amount of employment in (primarily) industries with increasing returns that offers a wide range of social-economic

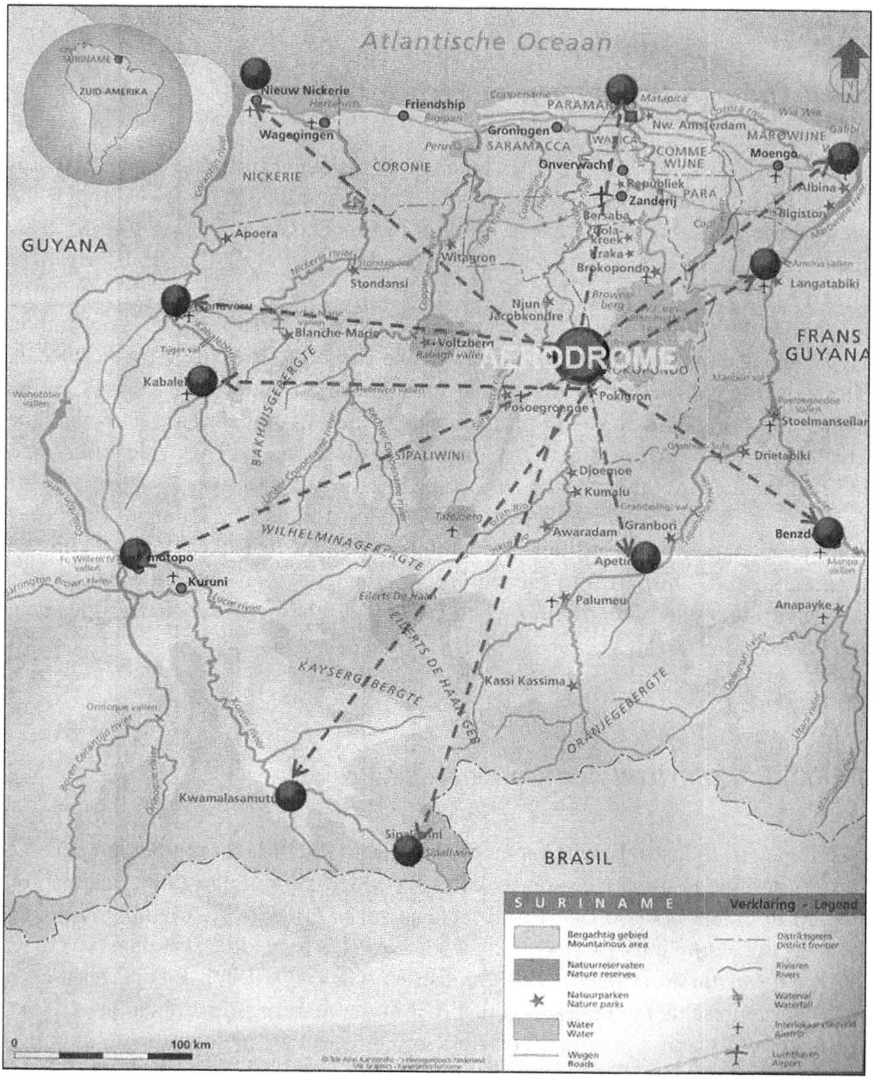

Fig. 6 Aerodrome Atjoni (CESWO 2011)

opportunities in terms of job scope (from relative low-key positions to higher-management). Given that Atjoni already functions as a logistics hub and the Browsweg area currently is the fastest growing conglomerate in Suriname, they are the most likely candidates for urbanization (3 and 4 for respectively Browsweg and Atjoni in Fig. 5) through planned conception, communication, involvement and disciplined execution of the conditions necessary for urban settlements (of which the employment condition has been satisfied in the previous steps).

Road Through Drietabbetje to French-Guyana

A road, fit for commercial vehicles, must be constructed from the Tapajai Hydro-power Project road trajectory underneath the Brokopondo reservoir, through the area where Drietabbetje village is located, to French-Guyana (5 on Fig. 6). This road will lead via French-Guyana to the northern region of Brazil. This part of Brazil, situated between the Trombetas and the Jari River, is increasingly industrializing, and its economy is expected to grow substantially in the near future. However, it exports are hampered due to the distance from the Atlantic Ocean. The most logical solution is the port at Paranam in Suriname (6 in Fig. 5); of the entire Guyana shield it is the port closest to the Atlantic Ocean, and with the greatest draft. With the port open to third parties, this will provide direct and indirect business to other Surinamese companies.

The French have already expressed their interest to use the Paranam Port; they want to have their own facilities close to the port so they can easy load and unload their cargo and transport it to French-Guyana, and are even willing to construct a road, starting at the border with French-Guyana all the way to their facilities near the Paranam Port.

Paranam Industrial and Commercial Project

The Paranam Industrial and Commercial Project (PICP) aims to relocate existing lumber and wood mills from (especially) Extended-Paramaribo to Paranam (6 on Fig. 5) to create economies of scale and increase efficiencies by moving wood/lumber processing closer to locations where wood is logged. It also plans to create further downstream (value adding) industries (CESWO 2009). Figure 7 shows a map that represents the realization of PICP in the rearranged Paranam area. Also, the opportunity exists to attract investors willing to invest in downstream alumina/aluminum industries which will be fed with the alumina refinery output. The wood/lumber and alumina/aluminum industries will be fed by water and energy provided by the electricity power house at Paranam and can use the port facilities there to export their products.

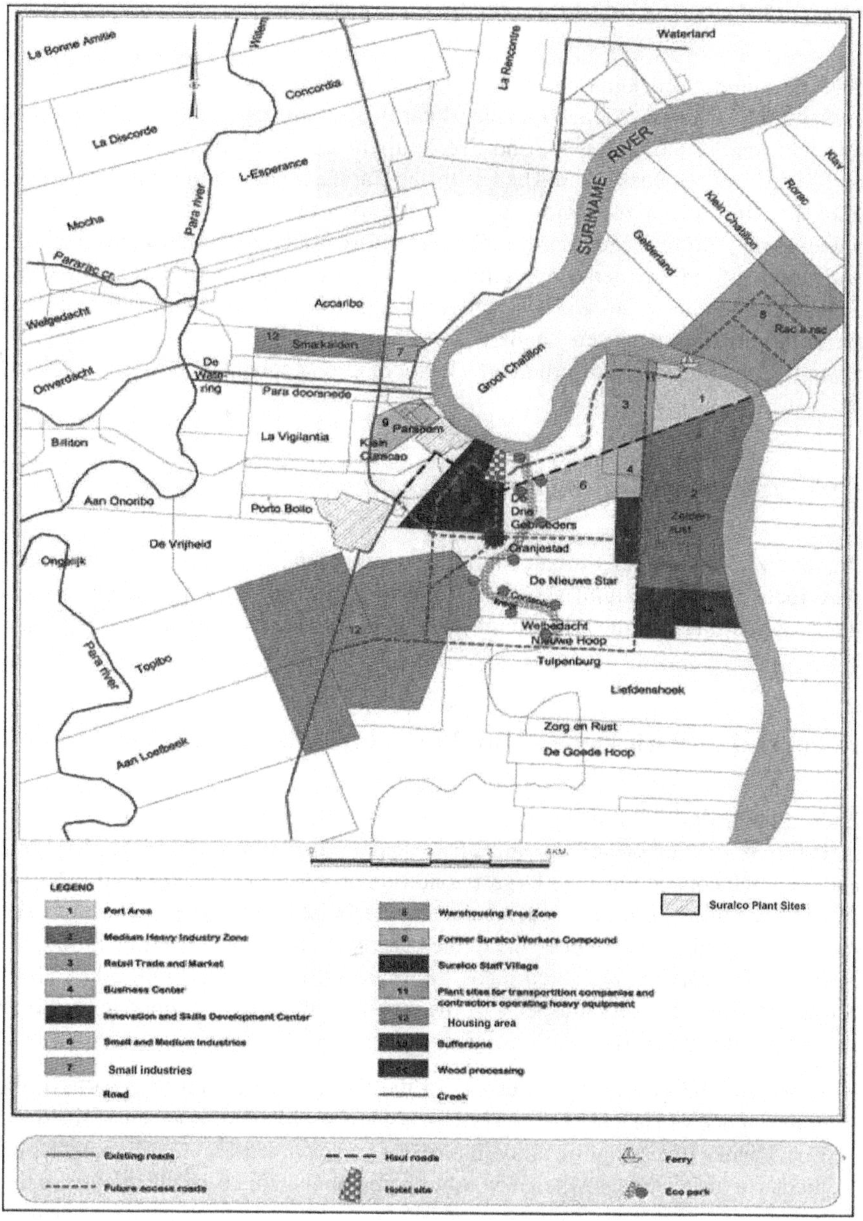

Fig. 7 Rearranged Paranam area for the Paranam commercial and industrial plan (CESWO 2009)

Urbanization of Paranam-Overwacht-Zanderij

Bauxite mining and alumina refining require relatively low-skilled labor which explains why Paranam (depending on primarily this industry) has never been able to thrive beyond its existing shape and size. With the advent of downstream industries and high-level services established with the Paranam Industrial and Commercial Plan, handling of cargo coming from neighboring countries, and expansion into value-adding alumina/aluminum industries, this can be changed by introducing a wide range of employment opportunities which can stimulate migration out of the Extended-Paramaribo area. These provides can transform the Paranam area (and the adjacent Onverdacht en Zanderij area) into an urban settlement (7 in Fig. 5), again taking conditions for urban development into account.

Urbanization of Groningen and Moengo

To sanitize Paramaribo's urbanization, Groningen and Moengo can contribute if they focus on their natural assets for industrial/commercial exploitation, respectively agriculture and the wood/lumber industry (respectively 8 and 9 in Fig. 5).

Recognizing Path Dependent Trajectories and Paradigms

Even though the different pieces described fit well into each other to thrive the inception of new urban settlements, the endeavor itself is nevertheless challenging and resource consuming. The largest condition to have this idea realized is buy-in from stakeholders, such as locals, environmentalists, investors, etc. The difficulty lies in overcoming vested interests in the current status, which actually are impediments to the earlier mentioned ideas, projects and concepts (Unruh 2000, 2002; Sovacool 2009). Important aspects of the current status that will impede to realize the proposed urban settlements realized are:

- Lumber/wood processing industries in the Extended-Paramaribo area have the convenience of easy access to a wide range of suppliers of goods and services;
- Local banks invest significantly in property in Extended-Paramaribo which puts them at a disadvantage when new urban settlements will be established since this will lower property prices in Extended-Paramaribo and thus imply a loss on their investments;
- Parties dealing with and having vested interests in the port in Extended-Paramaribo (Nieuwe Haven) will fear losing the focal point for cargo loading and loading to the Paranam Port. The same analogy can be drawn with respect to the Aerodrome Atjoni versus the international airport Zanderij and the local airport at Zorg and Hoop;

- The main tourist attraction in Suriname is the jungle; if therefore the Brokopondo reservoir is outfitted with tourist accommodations and affiliated goods and services, businesses in Extended-Paramaribo (especially those that have not franchised to the Brokopondo Lake) will fear losing customers;
- The ruling existing business model in the Surinamese tourism sector entails providing tourists with a predetermined experience; whereas international trends indicate that tourists want to make their own choices, while being in the main attraction;
- The provision of services (e.g. governmental services, banking, insurance, etc.) is primarily confined to Extended-Paramaribo; if new urban settlements need to be realized in the targeted areas for, a degree of decentralization of these services is required. Though some banks and insurance companies have started doing this, there are still great lengths to go. This is a necessary condition for people to opt to move out of the Extended-Paramaribo area without sacrificing their conveniences. This can be achieved through the introduction of agents, mobile services, internet services or simply physical establishments in new urban environments.

This list is far from exhaustive and conclusive, but provides a glimpse of the breadth and depth of ruling paradigms, beliefs, conventions, vested interests, etc. that need to be managed if new urban settlements as proposed here are to be established. The next section digs into a proposal how such management should be undertaken.

Context-Driven Transition Management as a Tool to Engage in the Creation of New Urban Settlements

The proposed realization of new urban settlements in Suriname implies a fundamental change in structure (e.g. organizations, institutions), culture (e.g. norms, behavior) and practices (e.g. routines, skills); this easily takes one or two generations to achieve. The time span might seem like something beyond the horizon, however the centralization of functions in Extended-Paramaribo renders the entire country extremely vulnerable to climate change, and creates all sorts of inefficiencies with respect to resources (time, energy, water, raw materials, etc.); thus, new urban settlements are needed sooner rather than later. Such a dramatic change impacting structure, culture and practices is nothing short of a full-fledged "transition".

Given the scope and impact of the transition, it is important that the main reasons behind the transition and the intended outcome of the transition are agreed on a political level; respectively the strategic agenda and the strategic plan are conceived. A seminar/discussion (with inputs from NGOs, multilateral institutions, unions, universities, etc.) can provide these deliverables. Once the agreement is reached, the intention needs to be elevated to a law which needs to pass parliament and thus guarantees that there is a wide range of long-term support for the transition.

Next, a formal team must be assembled tasked with leading and managing the transition: the transition arena, a small network of various frontrunners with respect to the spatial decentralization of urban development but also with those in favor of the transition but that have vested interests in the situation that needs to be transitioned from (e.g. the banking sector, traditional tourist tour operators, Paramaribo Port). The transition arena is tasked with developing a vision, goals, roadmaps, milestones and are also competitively rewarded based on the progress they achieve. They have access to resources and can direct the Government's executive branch to a particular degree, and – considering the importance and urgency of the task at hand and their influence over the executive branch – are directly responsible to the head of Government. The transition arena can have the supervision over other groups which are each respectively responsible for a part of the plan to transition towards the future (e.g. a team that focuses on the realization of the Tapajai Hydropower Project, another that is responsible to have the road to French-Guyana realized).

The transition itself is unprecedented in the Surinamese context, neither is there a predetermined path that needs to be walked; managing the transition can therefore not be set in stone with respect to exact outcomes, timeframes, relationships between different developments, and neither can it account for unforeseen and unexpected events which are surely bound to occur with such an endeavor. Therefore, with respect to the transition itself, the transition arena must exercise a form of transition management which is a form of reflexive and participative governance that attempts to manage transformative change (i.e. influence the speed and direction of change) towards the normative goal (new urban settlements) by combining long-term thinking with short term action (thus complementing conventional policy) through a process of searching, experimenting and learning; it does not rely on blueprints, but on iterative decision making (Lachman 2013). Key aspects of transition management are (Loorbach and Rotmans 2006):

- Experimenting and learning to guide variation and selection (learning-by-doing and doing-by-learning) while not chasing "silver bullets" (thus keeping all options and the playing field open), but still having the desired end result in mind;
- Obtaining stakeholder (from multiple domains and levels) input through inclusion and involvement;
- Complementing conventional policy (which has a short-term focus) with long-term thinking with the aim of the establishment of new urban settlements which abide by the conditions necessary to thrive urbanism;
- Continuous reflection (monitoring, evaluating, improving) on all levels;
- Bringing system innovation alongside system improvement.

Transition management is executed on three levels (Loorbach and Rotmans 2006; Loorbach 2010, also Wittmayer et al., Chap. 4, this volume):

1. Strategic level: development of a vision, goals, roadmaps, milestones, etc.;
2. Tactical level: implementing a transition agenda towards the desired goal with the consent of regimes, by aligning them with the long-term goal;
3. Operational level: experiments are used, rather than chasing "silver bullets", to stimulate learning and thereby guide variation and selection.

The three levels described above follow a cyclical path consisting of problem structuring and envisioning (strategic level), agenda building and networking (tactical level), experimenting and diffusing (operational level), and executes continuous monitoring, evaluating and adjusting on all levels (Loorbach et al. 2008). A conceptual tool that is necessary for transition management is the "transition scenario" which is a narrative of plausible developments which may result in a plausible end-state (Nekkers 2006). Transition scenarios help to anticipate sudden changes and deviations from trends, keep options open, contribute to learning (Sondeijker et al. 2006; Wiek et al. 2006), and – particularly important here – align and engage multiple stakeholders (Lachman 2011).

The following actions, shaped by the methodology used by Lachman (2014) to assess energy system transition management in developing countries, act as a guideline with respect to the road to be taken by the transition arena (and/or its subgroups) to address the requirements on a strategic, tactical and operational level on the way towards new urban settlements:

1. Formulate a vision and mission based on the seminar that was held with the aim of gaining political (and other social groups) support;
2. Create a set of scenarios that relate how events external to Suriname will influence urban life in Suriname (e.g. climate change, global financial stability, etc.). These scenarios are fully developed stories of the future with a beginning, middle, and end, which tell challenging, diverse, and plausible stories that are relevant to the issue of the need and urgency to establish new urban settlements.
3. Create a normative scenario that plausibly fits in the external scenarios created in the previous step. The scenarios thus far generated have the purpose as a whole to communicate to all stakeholders the need and urgency for the decentralization of urban living and the proposed solution and path forward to achieve this. The set of scenarios also helps to maintain focus for all stakeholders involved;
4. Create subgroups, each tasked with a specific portion of the transition plan, and lay out a multi-year plan with objectives and targets (compensation of members are based on the achievement of these objectives and targets);
   - Each subgroup identifies regimes and niches, concepts used by Rip and Kemp (1998), in order to form a picture of the Multi-Level Perspective (Geels 2002) – although time-consuming, use can be made in this regard of the methodology used by Lachman (2014) – within the field they are responsible for (e.g. establishing the Brokopondo Tourism Project, urbanization of Paranam);
   - Each subgroup fleshes their part of the transition out in more detail and manages the execution of that part;

- Each subgroup enhances selection pressure on the regime through economic instruments and regulation (Geels 2002) and stimulates the emergence and diffusion of niches, for instance in the case of the Brokopondo Tourism Project, by providing tax incentives to investors in that area. This implies that the subgroups should have direct links with the executive branches of Government;

5. Monitoring the subgroups;
6. Reiterate the scenario building exercise every 3–5 years (van der Heijden 2005; Nekkers 2006) and re-evaluate functioning of the transition arena (based on deliverables).

The steps above are depicted in Fig. 8.

These elements of transition management and its execution can be typically found in literature. However, every tool, idea, approach, concept, etc. must fit within its context to succeed. An important condition for transition management to succeed is therefore the extent to which efforts to transition fit within the context in which they are executed. There are a vast amount of examples of expensive consultants which transfer ideas to developing countries but which fail miserably simply because the context has not been taken into account (Chaffey and White 2012). Thus, transition management should be context-driven; specific characteristics from the context (which can be on a macro or micro level, depending on the scale and impact of the transition) need to be heeded in such a way that the functions of each aspect of transition management can come to full fruition on all execution levels. The specific

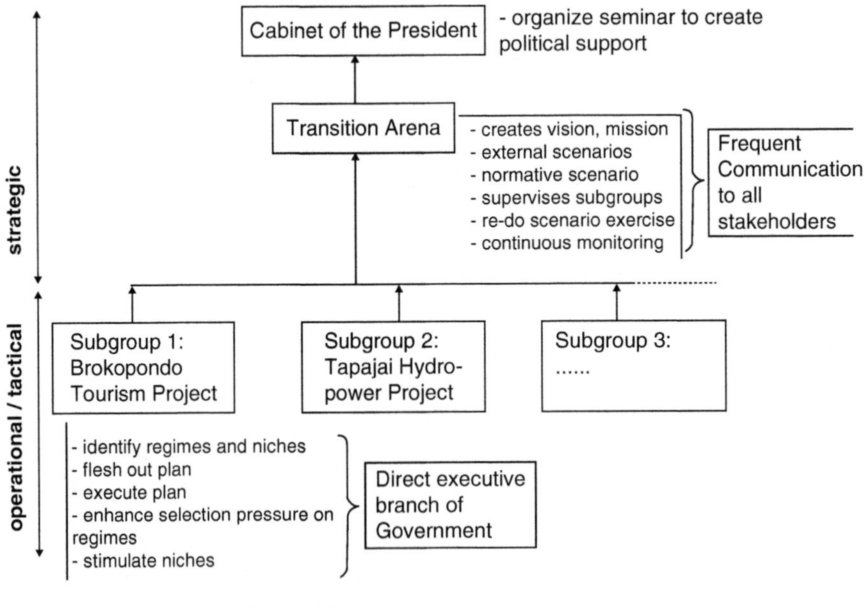

Fig. 8 Structure to engage in transition management

characteristics from the context can be found in the environmental, social, legal, political, economic, and technological spheres (van der Heijden 2005).

Suriname ranks as a middle-income country, but is skewed towards the typical characteristics of developing countries. It is therefore mandatory to identify and anticipate possible traps in the execution of transition management lying within these characteristics. Developing countries tend to have cultures that have relatively high power distances in relations (de Waal 2007; Cameron and Green 2014), which implies that aspects such as status and hierarchy play an important role. This makes it more difficult for niches to emerge to contest with existing regimes, so there is little innovative thought that goes against the viewpoint of those with higher status ranks. For a significant part this explains why developing countries are risk averse, and this might impede proper functioning of the transition arena; those within the arena with higher status will overwhelm others and (either consciously, or subconsciously) prevent them from speaking their mind. The transition arena will therefore have a bias from the start. Another aspect of the culture of developing countries is that it is low on individualism, i.e. people value loyalty to their group above their own interests), which coincides with a high context and relationship-oriented culture, in which people deem everything as part of a greater whole and therefore relationships are important (de Waal 2007, Cameron and Green 2014).

These characteristics make it very likely that during transitions there will be a strong bias in favor of regimes, which impedes the transition itself. This implies that because of the characteristics of developing countries, regimes are more rigid than their Western counterparts, a conclusion also reached by Lachman (2014) through observations in several developing countries, see Fig. 9. In effect, that means that crucial elements of transition management, such as the cohesion of the transition arena, obtaining stakeholder input, continuous reflection, enhancing selection pressures on regimes, and stimulating the emergence and diffusion of niches are relatively more difficult.

This likelihood towards bias is further evidenced by noting that the earlier mentioned characteristics of developing countries firmly place these countries on the right side of Quinn's organizational culture model (Cameron and Quinn 2011), while transition management's outward focus and flexible and governance approach push it towards the "Adhocracy" quadrant (Fig. 10). This again reinforces the notion that in cultures of developing countries people are less likely to against the so-called regimes, and therefore effectively don't have a level playing field for niches. Within transition management efforts (and within the transition arena specifically) interests of regimes are less likely to be challenged.

In order to deal with this challenge, open, honest, clear and transparent communication on and between all levels within the transition management movement is an essential condition. Thereto, a clear set or rules and codes need to be established – and frequently reiterated – which not only guides this communication but enables its open, honest, clear and transparent nature. Next, it is of the utmost importance when embarking on transitions that the characteristics of the context and their resulting pitfalls for the transition efforts are made explicit upfront to all stakeholders involved (as done earlier in the case of developing countries). Use

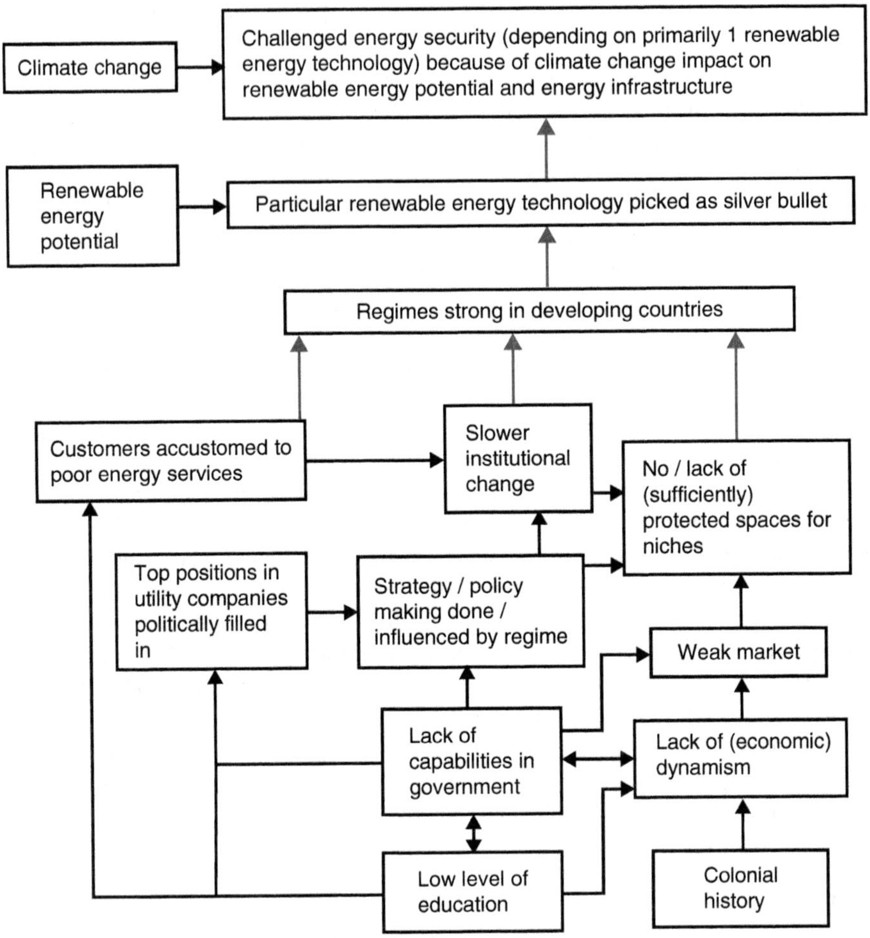

Fig. 9 Factors leading to rigid regimes in developing countries (Lachman 2014)

Fig. 10 Quinn's model for organizational culture (Cameron and Quinn 2011)

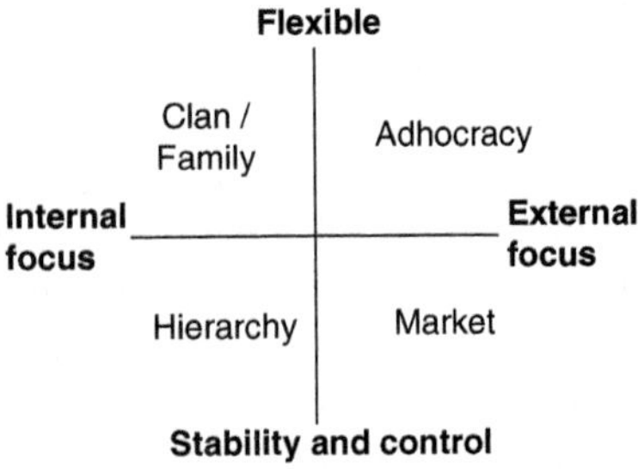

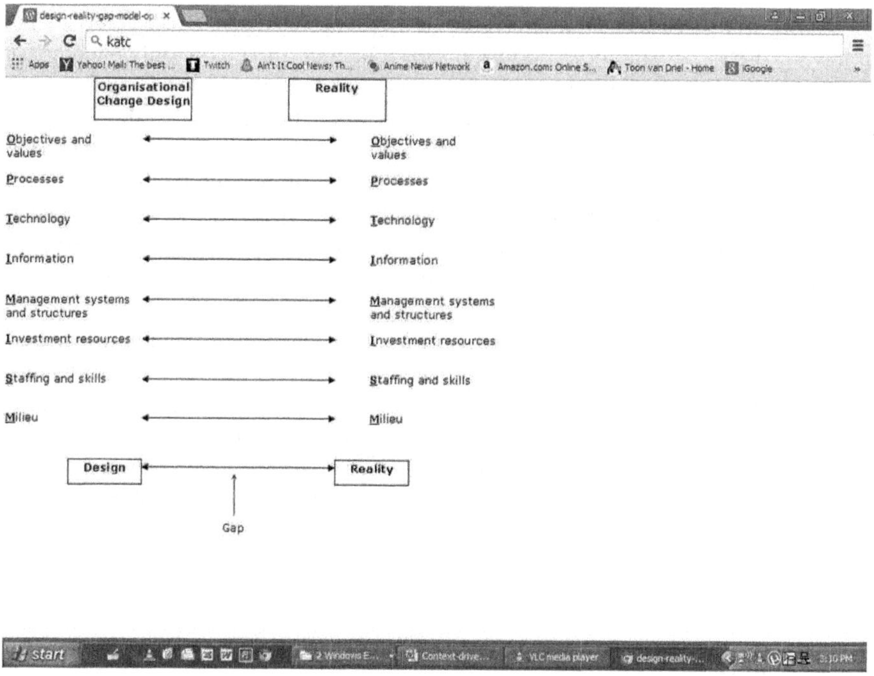

Fig. 11 Gap analysis using the OPTIMISM framework (Chaffey and White 2012)

can be made of the so-called STEEPL (abbreviation for social, technological, economic, environmental and legal) dimensions and/or the OPTIMISM (abbreviation for objectives and values, processes, technology, information, management systems and structures, investment resources, staffing and skills, and milieu) dimensions, or other similar frameworks, to highlight avenues where problems can be expect. Each dimension within the framework will be scrutinized in the case of on the one hand the assumptions, beliefs, values, viewpoints, etc. underlying the transition management methodology, and on the other hand the context in which the transition needs to take place. An example is provided in Fig. 11 (Chaffey and White 2012).

With the gap analysis done, the way the transition is managed can be tailored to suit the context, for instance by including foreign expertise within the transition arena to guarantee a so-called objective stance or by tinkering with the way communication is done. Another practice which can add to the progress of the transition in contexts strongly biased towards regimes, is setting normative goals at the beginning of the transition endeavor, like for instance setting a fixed target for the amount of migration out of Paramaribo achieved in 10 years or the establishment of sufficient employment for at least two urban settlements in 10 years. These should be high-level target which everyone involved in transition management must be committed to; they need to have SMART objectives (supporting of these high-level targets) for which they are rewarded upon achievement.

As mentioned earlier, continuous monitoring and feedback on all levels of the transition are of paramount importance. This reflection should not only encompass the progress of the transition, evaluation of the involved members' objectives and comparisons with the progress of other similar transitions, but also the extent to which the countermeasures to deal with the outcomes of the gap analysis are still in place and working properly, i.e. the extent to which the playing field is kept level within the transition arena, the transition effort in general, and thus between regimes and niches. To this end, external watchdogs can be used as an extra set of eyes from outside the context to create or ordeal with respect to the playing field being level.

To summarize, it is necessary, even before the execution of transition management efforts, that the characteristics of developing countries (in this case Suriname) which might impede transitions are explicitly stated upfront with countermeasures deployed to deal with them. However, it must be noted here that this notion not only holds for developing countries, but also for the Western world where there is a great amount of diversity within different spheres between these nations. More research is required on this subject matter.

Conclusions

It is clearly evident that developments and trends with respect to the ongoing migration to and concentration of services in Extended-Paramaribo will induce inefficiencies, higher costs, social unrest, and high risk and associated impact in the face of climate change. Thereto, new urban settlements are required which are based on essential conditions, of which social-acceptable employment is one of the more difficult ones to realize. Fortunately, Suriname has the conditions to do this, but this will require breaking down existing structures, paradigms, habits and processes, and disciplined execution for at least one or two generations; this is nothing short of a transition.

Transition management has been quite fleshed out over the years, both in academia and in practice. However, managing the transition might be impeded by characteristics specific to Suriname/developing countries. These characteristics tend to favor existing regimes at the cost of innovative activity, despite using transition management recipes as the transition arena, continuous feedback, etc. Therefore, these characteristics need to be identified upfront in an open and honest setting in order to device countermeasures to deal with them. These countermeasures will complement efforts to transition and effectively make transition management fit in the context in which the transition takes place. Still, more research is required with respect to the impact of contexts (on different scales) on the dynamics within transition management. This is essential since every effort to transition is defeated even before it is started if the transition management is not rooted in the context; context-driven transition management bound in the Surinamese context is therefore a necessary condition to achieve success on the road towards new urban futures, and is a necessary condition for any transition undertaken in any place.

References

Boksteen L (2008a) Tapajai hydro plan. Een mogelijkheid voor de vermindering van onze afhankelijkheid van thermische energie. S.n., Paramaribo

Boksteen L (2008b) Tapajai hydro plan. Informatie over het Tapanahoni-Jaikreek omleidingsproject. S.n., Paramaribo

Cameron A, Green M (2014) Making sense of change management, 3rd edn. KoganPage, London

Cameron KS, Quinn RE (2011) Diagnosing and changing organizational culture: based on the competing values framework, 3rd edn. Jossey-Bass, San Francisco

CESWO (2009) Paranam industrial and commercial project. Design for a new urban, industrial port and service center in Suriname, Centre for Economic and Social scientific Studies, Paramaribo

CESWO (2011) Traffic and transportation study. Centre for Economic and Social scientific Studies, Paramaribo

Chaffey D, White G (2012) Business information management, 2nd edn. Prentice Hall, Harlow

Cooper C, Fletcher J, Fyall A, Gilbert D, Wanhill S (2008) Tourism principles and practice. Harlow Publishers, London

de Waal A (2007) Strategic performance management, A managerial and behavioral approach. Palgrave Macmillan, Hampshire

Geels FW (2002) Technological transitions as evolutionary reconfiguration processes: a multi-level perspective and a case study. Res Policy 31(8–9):1257–1274

Harrison JS, Enz CA (2009) Hospitality strategic management: concepts and cases. Wiley-Blackwell, Somerset

Lachman DA (2011) Leapfrog to the future: energy scenarios and strategies for Suriname to 2050. Energ Policy 39(9):5035–5044

Lachman DA (2013) A survey and review of approaches to study transitions. Energ Policy 85:269–276

Lachman D (2014) Metamorphosen. Energy system transition management in developing countries. PhD-thesis, University of Suriname, Paramaribo

Loorbach D (2010) Transition management for sustainable development: a prescriptive, complexity-based framework. Gov Int J Policy Admin Inst 23(1):161–183

Loorbach D, Rotmans J (2006) In: Olsthoorn X, Wieczorek AJ (eds) Managing transitions for sustainable development. Understanding industrial transformation. Views from different disciplines. Springer, Dordrecht, pp 187–206

Loorbach D, van den Brugge R, Taanman M (2008) Governance in the energy transition: practice of transition management in the Netherlands. Int J Environ Technol Manag 9(2/3):294–315

Magrin G, Gay García C, Cruz Choque D (2007) In: Parry ML, Canziani OF, Palutikof JP, van der Linder PJ, Hanson CE (eds) 2007: Latin America. Climate change 2007: Impacts, adaptation and vulnerability, Contribution of Working Group II to the Fourth Assessment Report of the Intergovernmental Panel on Climate Change. Cambridge University Press, Cambridge

Nekkers J (2006) Wijzer in de toekomst. Werken met toekomstscenario's. Uitgeverij Business Contact, Amsterdam

Riesman D (2001) The lonely crowd: a study of the changing American character. Yale University Press, New Haven

Rip A, Kemp R (1998) In: Rayner S, Malone EL (eds) Technological change. Human choice and climate change, vol 2. Battelle Press, Columbus, pp 327–399

Sheoratan R, de Kroon W, Sheoratan A, Sheoratan S (2013) Een prefeasibility studie voor de centralisatie van toerisme rond het Brokopondo stuwmeer. Suriname Business forum, Paramaribo

Sims REH, Schock RN, Adegbululgbe A (2007) In: Metz B, Davidson OR, Bosch PR, Dave R, Meyer LA (eds) 2007: Energy supply. Climate change 2007: Mitigation, Contribution of Working Group III to the Fourth Assessment Report of the Intergovernmental Panel on Climate Change. Cambridge University Press, Cambridge

Sondeijker S, Geurts J, Rotmans J, Tukker A (2006) Imagining sustainability: the added value of transition scenarios in transition management. Foresight 8(5):15–30

Sovacool BK (2009) Rejecting renewables: the socio-technical impediments to renewable electricity in the United States. Energ Policy 37(11):4500–4513

Suriname Tourism Foundation (2013) Marketing strategies: jumpstarting the tourism industry of Suriname. Suriname Tourism Foundation, Paramaribo

Unruh GC (2000) Understanding carbon lock-in. Energ Policy 28(12):817–830

Unruh GC (2002) Escaping carbon lock-in. Energ Policy 30(4):317–325

van der Heijden K (2005) Scenarios. The art of strategic conversation, 2nd edn. Wiley, Chichester

Wiek A, Binder C, Scholz RW (2006) Functions of scenarios in transition processes. Futures 38 (7):740–766

Urban Transition Management as a Democratic Practice? The Case of Rotterdam's Waterfront Regeneration

Shivant Jhagroe

Abstract This chapter examines the democratic aspects and challenges associated with urban transition management. It does so at a conceptual level and by critically reflecting on the application of transition management in the context Rotterdam's sustainable waterfront regeneration. As such, it aims to shed a more reflexive light on the increasing academic and public interest in transition management.

Introduction

Managing transitions is highly complex and involves conflicts between different problem definitions, visions and possible solutions. Therefore, transitions always generate winners and losers. This chapter focusses on the practice of Urban Transition Management (UTM) with a particular focus its democratic aspects and challenges.

In the past decade, Transition Management (TM) has been criticised for overlooking issues of legitimacy, democracy and politics (Shove and Walker 2007; Smith and Stirling 2010). Typical democratic challenges associated with transition management include tensions between shot-term electoral cycles and long-term sustainability engagements, conflicts between institutional and local cultural legitimacy, and attracting actors beyond the 'usual suspects'. In recent years, the relationship between transition management and democracy has been taken up and explored both conceptually and empirically (e.g. Hendriks 2009; Jhagroe and Loorbach 2015). Even though transition management discourses and practices have been analysed in the context of nation-state arrangement and policy networks, democratic Transition Management has rarely been addressed at the urban level. This is surprising, since contemporary cities express unsustainable systems and practices which are increasingly subject to social struggle and

S. Jhagroe (✉)
Industrial Engineering and Innovation Sciences, Technical University Eindhoven, Eindhoven, The Netherlands
e-mail: S.S.Jhagroe@tue.nl

N. Frantzeskaki et al. (eds.), *Co-creating Sustainable Urban Futures*, Future City 11, https://doi.org/10.1007/978-3-319-69273-9_15

reconfigurations e.g. regarding urban smog, biodiversity, and housing (Coenen et al. 2012). Urban regions seem to be particularly relevant spaces to account for the decentred nature of transition management practices. How, then, can the democratic challenge of transition management be understood in urban contexts? To what extent can projects that are informed by urban transition management be considered as democratic practices? And how can urban transition management be applied in a way that advances sustainable urban futures in a democratic fashion?

The objective of this chapter is to examine on the democratic aspects and challenges associated with urban transition management. It does so at a conceptual level and by critically reflecting on the application of transition management in the context Rotterdam's sustainable waterfront regeneration. As such, it aims to shed a more reflexive light on the increasing academic and public interest in transition management. The text is structured as follows. section "Transition Management and its democratic aspects: towards a conceptual framework" discusses how transition management can be understood in relation to democratic politics at a conceptual level. Drawing on insights from critical urban research, I explore a conceptual framework of democratic transition management. Contrary to many critiques, I argue that transition management is not necessarily undemocratic, but that democratic politics can enrich transition management discourse. For a more focused analytical approach, I translate these insights at four transition management levels in urban environments (Loorbach 2007, 2010). Section "The Case: Making Rotterdam's Waterfront Sustainable" introduces the empirical case. It reconstructs when, why and how a transition management discourse entered and shaped Rotterdam's long-term sustainability-oriented waterfront plans and its short-term focus of experimental projects. It zooms in on a specific strategy called the 'Floating Communities Strategy'. Section "Case reflection: Democratic waterfront regeneration?" reflects on the empirical case using the democratic transition management framework. It illustrates that Rotterdam's waterfront is an interrelated set of physical and symbolic spaces imbued with social, legal and economic struggles that advance technological and neoliberal urbanism, often downplaying efforts to democratise Rotterdam's waterfront area. The final section concludes with some reflections on the democratic issues associated with the application of transition management.

Methodologically, this chapter follows four steps. First, a democratic transition management framework is developed, combining insights from transition management and critical urban research. I use four analytical levels from transition management theory (cf. Loorbach 2007) to empirically investigate the democratic aspects of transition management: strategic, tactical, operational and reflexive. This framework offers a set of analytical focus point to reflect on empirical materials. It focusses on the transition management process, the actors involved, as well as on substantive results and broader socio-political implications. Analysing democratic aspects of transition management assumes significantly different or conflicting, problem definitions, viewpoints and claims. To sensitise different interpretations, my starting point is an interpretive epistemology. This epistemology informs qualitative research methods to collect and analyse empirical materials. Second, a single case study method is used to empirically explore the democratic aspects of transition management. Following Hesse-Biber and Dunleveay, case study research is particularly relevant to explore the subject matter and "provide the researcher with a holistic understanding of a problem, issue, or phenomenon. . .because the case is investigated from many different angles and pays attention to many different dimensions of the issue" (2011: 256). Third, the case reconstruction is based on the collection of dozens of policy documents, website texts and interviews with key actors and stakeholders (policy actors, architects, contractors, resident organisations). Fourth, these empirical materials are then reflected upon with the democratic transition management framework. As a researcher, I have not been directly involved in the transition management process. However, some of my (ex-) colleagues at the Dutch Research Institute For Transitions (DRIFT) have played a role in interactive stakeholder meetings in the 2000s.

Transition Management and Its Democratic Aspects: Towards a Conceptual Framework

In previous years, transition scholars have critically reflected on sustainability transitions and their management (cf. Jhagroe and Loorbach 2015). Shove and Walker argue that "there is a politics to transition management, a playing out of power of when and how to decide and when and how to intervene, which cannot be hidden beneath the temporary illusion of 'post-political' common interest claims of sustainability (Swyngedouw 2006)" (2007: 5). In a similar vein, Smith and Stirling state that it "is unclear how transition management processes sit in relation to prevailing policy institutions and political activities" (Smith and Stirling 2010: 9). The *democratic politics* of transition management seem to be a particularly critical challenge. Researchers that focus on the democratic politics associated with transition management call for thinking *beyond* existing paradigms of democratic institutions and deliberative democracy (Hendriks and Grin 2007; Hendriks 2009; Voß et al. 2009). For example, conflicts and tensions that emerge between in (traditional) democratic institutions and more open societal governance to pursue sustainability goals. Hendriks proposes more 'polycentric democratic landscapes', referring to the fractured and dynamic character and sites of democratic deliberations and political decision-making. Some 'democratic solutions' proposed by Hendriks are to: (1) explicate linkages between official democratic representations and transition networks; (2) include a wide variety of elites, institutions, actors and materialities in the political process; and (3) foster the public character of transitions and long-term politics. This perspective on democratising transition management is also evidenced by Voß et al. who argue that that alternative democratic practices should be facilitated by convening "different transition arenas for dissenting voices, rather than to get everyone around the same table" (Voß et al. 2009: 293).

These scholars suggest that productively conceptualising democratic transition management requires a bold move away from traditional conceptions of democratic governance. Clearly, a main challenge in the context of transition management, are inherent tensions underlying existing democratic institutions and procedures. For example, opposing democracy against technocracy assumes modernist and liberal schemes of a public *demos* that is unfree as soon as it is ruled by technical experience and expert knowledge (Jhagroe and Loorbach 2015). Following some suggestions in the field of transition research (Shove and Walker 2007; Hendriks 2009), I argue that more critical approaches on democratic thought and practice can enriched our understanding of democratic transition management (Laclau and Mouffe 1985; Rancière 2004). Such critical perspectives on democracy move away from traditional notions of democracy (e.g. representative democracy, pluralist democracy, deliberative democracy) and a priori 'democratic foundations' (parliaments, ministers, political parties, stakeholder networks, etc.). Alternatively, it challenges the very fundaments of alleged democratic institutions and deliberative processes. Notions like 'consensus', 'agreement', 'contract', 'support' or a 'win-win solution' often build on exclusions and produce new antagonisms. This understanding of democratic institutions radically decentres and rescales democratic practice and political agency. But how can this understanding of more radical democratic transition management be rendered tangible?

Cities as Sites to Grasp Democratic Transition Management

I argue that cities and urban regions offer fruitful entry points to overcome traditional conceptions of democracy and reframe democratic transition management. Much of the criticism on transition management relates institutionalised democratic arrangements, or arenas and networks organised along sectoral lines. However, since the late twentieth century, reflexive eco-modernisation has materialised in specific urban settings (e.g. resisting modernist planning, increasing bio-diversity, reducing carbon emissions). Since the early 1990s, such concerns have been captured under labels such as 'sustainable development' and 'sustainability'. Cities have been targeted as spatial nodes where many modernisation concerns could be addressed and tackled. A wide range of initiatives, policy programmes, commercial businesses strategies and citizen engagements aimed at intervening in cities to solve issues of (un)sustainability. Predictably, cities and urban environments are increasingly addressed in research on socio-technical transitions (Bulkeley et al. 2010, 2014) and complex sustainability-led transitions (Dixon et al. 2014; Murphy 2015).

In reviewing the role of space and geography in transition literatures, Hansen and Coenen conclude that "generally, a distinction can be made between studies that take a perspective in line with more traditional approaches in *economic* geography, stressing the positive influence of geographical proximity in stimulating network formation (e.g. Coenen et al. 2010), and studies that draw heavily on the relational approach, highlighting that space is *socially* defined (e.g.)" [italics, by author] (2013: 13). Coenen, Benneworth and Truffer argue that "scales are actively constructed through socio-spatial struggles by actors seeking to achieve their ends" (2012: 976). The 'geographical turn' in transition research suggests that the democratic challenge of transition management can be examined in terms of uneven socio-economic spatial developments. The move away from nation-state arrangements, sectoral system delineations and institutionalised democracy highlights how transition wins and losses are distributed geographically. Instead of taking national or local electoral politics and legal-administrative bodies as starting points, multiple spaces and power-laden nodal points can be taken as a point of departure. This inter-spatiality of transitions and the inherently uneven dynamics and multi-territoriality of transition management practices is often overlooked. In order to conceptualise democratic transition management, it is critical to understand urban spaces and sites as power regimes that are always subject to struggle and politicisation.

Insights from Critical Urban Research

The field of critical urban research has developed interesting insights in this context. Lefebvre, a well-known neo-Marxist urban theorist, maintains that space is produces socially and structured in a specific manner (e.g. in the name of public health or housing), always at the expense of other ways to organise the urban. Instead of seeing space as a given, we should look at the ways in which spaces are produced, reproduced and contested. For Lefebvre, the production of space is indeed political and imbued with contradictions, conflicts and the struggle for hegemony (Lefebvre 1996, 2003). Space is a strategic tool of control and shape reality in specific (and therefore ideological) ways. Resonating with the critical work of Lefebvre, some have explicitly used the work of radical political thought. For example, Hankins and Martin employ the radical political theory of Rancière to understand urban democratic politics. They state that "Rancière's 'police' includes, for urban politics, the ordering and naming of territories such as neighbourhoods as specific, bounded, often very local sites of participation in urban governance" (Hankins and Martin 2012: 28). Such urban sites of government, then, become interrelated spaces of "ordering, stabilizing and partitioning the sensible" (ibid). Following Rancière's approach, this means that urban policing depoliticises urban challenges and urban life. Democratic political dynamics "enable dissensus" and "enable the interruption of ordering" (e.g. destabilising categories of territory, residents, etc.). Adopting a post-Marxist frame, Hankins and Martin exemplify how the shift from modern industrial urbanisation (with its class struggles) to post-industrialization (with its depoliticised identity politics) did not coincide with the disappearance of socio-urban antagonism (Davidson 2014: 191). Rather, urban antagonisms have been depoliticised and rendered invisible.

What these critical urban scholars do is pay attention to the 'dark side' of urban governing regimes. They argue that urban government relies on a form of policing the city that backgrounds contestation and conflict. They do not argue against urban management per se, but draw attention to the importance of being able to call into question the technical ways in which urban spaces and populations are governed. To understand the city as a given, or a site for management, is not democratic, but rather 'demographic'. In that, urban management relies on policing of given groups and populations, not on democratic politics. So, from a critical urban perspective cities and urbanisation processes are highly contingent and always prone to radical critique. As Brenner states: "(. . .) critical urban theory emphasizes the politically and ideologically mediated, socially contested and therefore malleable character of urban space – that is, its continual (re)construction as a site, medium and outcome of historically specific relations of social power" (ibid). This also means that stable conceptions of town/country or urban/rural rely on hegemonic frames of 'the city'. We know these labels as territorial city boundaries, nation-states and other historically produce 'givens'. For Brenner, critical urban scholarship focuses on the "wide-reaching engagements – theoretical, concrete *and* practical – with the planetary dimensions of contemporary urbanization across diverse places, territories and scales" (Brenner et al. 2011: 227).

This particular conceptual sensibility enables us to enrich (urban) transition management research regarding the challenge of democracy, especially because urban sustainability transitions are virtually impossible to 'manage', because they denote various struggles "across the multiplicity of networked relations that (re) make that city" (Davidson and Martin 2013: 6).

A Democratic Transition Management Framework

In order to more systematically understand democratic transition management, I inject insights from critical urban research at the four levels of transition management (Loorbach 2007; Roorda et al. 2014). If we translate these insights into the transition management framework, it does not change basic structure, but rather extends it. These extensions and reframings can be summarised as follows (Table 1).

A democratic transition management approach does not replace a transition management approach, as developed by for example Kemp et al. (2007) and Loorbach (2007, 2010). It adds insights from critical urban research, drawing attention to the ways in which hierarchical power relations, uneven spatial developments and forms of resistance are part and parcel of urban transformations. This also implies that all aspects of transition management practices should be subject to critical scrutiny and public contestation. One of the key insights here is that

Table 1 Transition management and its democratic aspects

Level	Key aspects of urban transition management	Aspects of democratic urban transition management
Strategic	Vision development, strategic discussions and long-term goal formulation about the sustainable city	Challenging the urban status quo and creating radically alternative imaginaries type of urbanisation (fostering equality, liberty, etc.)
Tactical	Achieving goals in a specific urban context, steering activities, negotiation, collaboration, etc.	Struggling against incumbent networks and alliances and mobilising the non-institutionalised
Operational	Short-term, everyday decisions, specific projects and experiments to materialise a sustainable city	Countering dominant urbanisation projects and materialising radically alternative imaginaries in local settings
Reflexive	Monitoring, assessing and evaluation activities of other levels (adaptation and adjustments)	Disputing hegemonic modes of monitoring and valuation, employing non-dominant measurement techniques and schemes

management and democracy are opposite forces in urban transitions. Formulating long-term goals of what entails 'a sustainable city' is contrary to disagreeing why a sustainable future is worthwhile, what its contours look like, or what actors should (not be) included in the transition management process. Furthermore, only safeguarding democratic contestation at the strategic and tactical level is undemocratic, unless contestation and disagreement is *also* addressed at the operational and reflexive level. This might sound counter-intuitive as concrete actions are often considered technical issues of implication and monitoring. However, the power of a radical democratic perspective on urban transitions is that contestation and foregrounding marginalised voices should not be abandoned, especially when technical assessments and practices are employed.

This slightly redefines, or rather extends a more orthodox understanding of urban sustainability transitions and transition management, in the sense that it highlights the uneven developments and spatial struggles associated with the transition towards sustainable cities and urban livelihoods. An urban sustainability transition is a radical shift in how urban spaces are organised, moving towards sustainable forms of socio-economic and socio-environmental regimes and practice. Transition management is defined as (the more or less) guided practices associated with re-defining, re-scaling and re-evaluating these urban regimes and practices. Democratic transition management, then, highlights all dissent and antagonistic voices and forces related to transition management practices (mainly outside institutionalised democracy), seeking to challenge and dismantle dominant urban (unsustainable) regimes and practices. Despite this focus, democratic transition management is based on the linkages between democratic transition practices and management transition practices. Having conceptualised democratic transition management, we can now move to the empirical case.

The Case: Making Rotterdam's Waterfront Sustainable

It is instructive to briefly introduce the UTM context of Rotterdam's waterfront regeneration. Rotterdam is a major port city in the Netherlands. In the last decades, Rotterdam has been faced with rapid developments and a number of serious concerns (cf. Van Eijndhoven et al. 2013; Frantzeskaki et al. 2014). Skyrocketing container output/throughput ratios, new maritime activities and port urbanisation reshape many port cities around the world. This global port city network is shaped by technological innovations, inter-port competition, global market dynamics, and local city-port interfaces. In this context, industrialised port cities in the late twentieth century have increasingly been confronted with many challenges revolving around local water management and climate change, urbanisation, de-industrialisation and environmental concerns. And, in a world where "90% of trade volumes occur by sea, ports and maritime transport remain more important than ever (...)", it is safe to say that the global port cities network push port authorities, political actors and urban planners to be strategic and creative (Ducruet 2011: 1). One of the responses of port cities is called 'urban renewal' or, specifically focussed on port-city spaces, 'urban waterfront regeneration'. Waterfront regeneration can be understood in various ways, e.g. as the 'successive stage of waterfront development' (after the primitive city-port, expanding city-port, industrial city-port, retreat from the waterfront, and redevelopment) (Hoyle et al. 1988) or perhaps as "part of a broader growth-oriented strategy agreed by local elites to re-image their cities in an increasingly competitive urban system" (Bassett et al. 2016: 1757–1758), or even as the 'postmodernisation' of urban waterfronts (Norcliff et al. 1996; Klamer and Kombring 2004). In any case, urban waterfront regeneration is a historically situated and particular strategic governance response to contemporary port city challenges. Port history is not a singular history, but a dual that combines both maritime history and urban history. Urban waterfront governance projects are celebrated as a 'worldwide urban success story', but also seen as recent expressions of elitist capital accumulation with facilities and spatial development for the 'happy few' (Harvey 1990; Daamen 2010). In this sense, urban waterfront governance does not only refer to a set of economic interventions, but might also create unequal distribution effects triggering numerous social and political questions.

Since the late twentieth century, the restructuring of Rotterdam's port became increasingly pertinent. A number of plans and concerns came together around the early 2000s, opening up institutional space to re-imagine the old city ports. A newly gained political authority of the local government, a 2000 *Vision and Dare* covenant, but also international 'policy tourism' in Hamburg and London made Rotterdam rethink its waterfront: "1) the corporatization of the Rotterdam port authority; 2) the financing of the port's Maasvlakte 2 expansion plan, and 3) the founding of the Rotterdam CityPorts Development Company (OMSR)" (Daamen 2010: 76). This last decision laid the foundations for an alternative framing of Rotterdam's city-port nexus. The case of Rotterdam's waterfront regeneration is particularly interesting because it represents a complex transformation of urban spaces, partly advocated by transition management discourse and practices. As Frantzeskaki et al. (2014: 407), Rotterdam's urban port transition "presents a number of successes in realising an innovative vision for a sustainable urban transition. In the urban context, a sustainability transition is conceptualised as a long-term process of fundamental change of systems of provision coupled with radical shifts in ways of organising (structures) and operating (practices) (in) a city as well as altering assumptions and perceptions (cultures) towards more sustainable trajectories (Frantzeskaki and de Haan 2009; Coenen and Truffer 2012; Nevens et al. 2013)" (Fig. 1)

In the context of my focus on the democratic features of transition management, it is important to note that Rotterdam's waterfront regeneration vision was a priority on the policy and political agenda, but also expresses "differing and conflictual perspectives" (ibid). Interestingly, the renewed interest in old port areas "created an

Fig. 1 Geographical positioning of Rotterdam's Stadshavens project
(Sources images: https://commons.wikimedia.org/wiki/File:Netherlands_in_Europe.svg and http://mirt2016.mirtoverzicht.nl/mirtgebieden/project_en_programmabladen/481.aspx)

opportunity for rethinking urban sustainability and for creating or restoring new sense of place (Tidball and Stedman 2013: 297)" (ibid). Often, such considerations focus on economic restructuring, but also on and socio-environmental innovation, offering a few (not much) entry points for renewed social and political engagements vis-à-vis waterfront regeneration. Even though Rotterdam's waterfront regeneration has been used as a case of (urban) transition management by various scholars (e.g. Van Eijndhoven et al. 2013; Frantzeskaki et al. 2014). However, rarely has the democratic aspects and challenges associated with this transition management case been addressed.

Imagining a Sustainable Waterfront

In the early 2000s, a number of innovative projects and new initiatives emerged under the label "City Ports" (in Dutch: *Stadshavens*) in Rotterdam. These projects have been clustered with the aim to re-structure the Rotterdam city-harbor area. In 2007, the city of Rotterdam Council and the Port of Rotterdam Authority established a dedicated temporary office to develop an overall strategy for the entire area. This office (Bureau Stadshavens, or BS) worked relatively independent but adopted a transition management approach, also informed by DRIFT researchers, linked to the Erasmus University Rotterdam. The action research oriented approach of transition management enabled various stakeholders to reframe conflicts and the stalemate in more positive and affirmative terms, opening-up new perspectives for the city-port nexus. BS started to organise a broad transition arena process involving over 150 participants that were personally invited. The selection was based on achieving a broad representation of different backgrounds (multi-actor) and personal competences and skills. These actors were often non-experts with unconventional visions and expectations about the long-term goals of the city ports of Rotterdam (cf. Van Eijndhoven et al. 2013). Over the course of 6 months, this network discussed in various forms and around different themes which kind of sustainability challenges should be placed centrally in the urban water development strategy and which type of development and planning process was needed. The discussions were facilitated by BS based on the idea that the area could not be developed in a traditional way through blueprint planning but should be developed in a more experimental and organic way through public-private collaborations and extra-institutional projects (Frantzeskaki et al. 2014). The main ambitions of this strategy were to develop the area into a vital working and living environment centred around sustainable innovations in the areas of water, energy and mobility (Project Bureau Stadshavens Rotterdam 2008a: 2). Even though the plans seemed to be rather ambitious and long-term oriented, the contingency of these plans was also acknowledged. As one of the institutional strategic actors said: "Making a port vision and city port vision is nice, but as soon as the ink dries, there are developments in the world around us one has to account for"[1] (Interview E).

In 2008, the city of Rotterdam Council and the Port of Rotterdam Authority proposed their common effort to restructure the harbour area with the document *Creating on the Edge,* formulating a strategic vision. This vision was accompanied with a more specific implementation programme, covering a wide range of actors, initiatives and efforts, targeting an area of about 1600 ha around the Maas River at the intersection of the Rotterdam harbour and the city. The vision for regenerating the Rotterdam city-port nexus is presented as both a matter of urgency and opportunity (Programme Bureau Stadshavens Rotterdam 2011: 18). To this end, some legal requirements and regulatory procedures were bypassed in order to create space for experimentation. This exception was made possible by the so-called Crisis and Recovery Act.[2] This national legal framework enabled infrastructural projects and economic growth to unfold in times of crisis. It accommodated for instance 'floating projects' as it rendered flexible and viable spatial and developments plans (in Dutch: *flexibele bestemmingsplannen*). Additionally, knowledge institutes were involved in the design and creation of specific buildings and constructs (e.g. RDM Campus, Hogeschool Rotterdam, Alberda College).[3] The Stadshavens Rotterdam website states that it:

> ...wants to develop into a quality port and an excellent location, not only for port and transport related industry, but also for innovative businesses and knowledge institutes. Rotterdam is also creating an image of itself as a trendsetter in the fields of sustainable energy and climate adaption, with the aim of attracting professionals and pioneers keen to try out these new trends. Stadshavens can provide them with everything they need for setting up their businesses, along with exceptional residential developments, cultural amenities and good educational facilities.[4]

Importantly, each of these areas has its own history and characteristics, also in terms of different nuisance categories (in Dutch: '*geur, stank en geluidscontouren*'). A number of ports and docks were considered as potential places for development. As many stakeholders told me, the simplest and most traditional solution was to reverse the nineteenth century excavation, and to simply fill the docks for urban planning. However, this was too expensive and the urban river would also lose its unique character. Alternatively, the docks and river remained intact. The programme finally targeted at four geographical districts, together covering the 1600 ha. The four areas are: (1) Rijn-Maashaven; (2) Merwe-Vierhavens; (3) Waal-Eemhaven; and (4) RDM-Heijplaat. These areas

[1] In Dutch: "Havenvisie of stadshavensvisie maken is leuk, maar de inkt is nog niet droog of er zijn ontwikkelingen in de wereld om ons heen waar je ook rekening mee moet houden".

[2] In 2007, the minister of Infrastructure and Environment commissioned a working group to investigate why infrastructure projects take 'so long' from design to implementation and how these processes could be realised faster.

[3] For an extensive reconstruction of how the Stadshavens programme emerged at an institutional level see Daamen (2010).

[4] See: http://stadshavensrotterdam.nl/eng/vision

were considered as too heterogeneous for one single approach, so the areas are related strategically. The programme articulates *five strategies* that cut across the four areas (Project Stadshavens Rotterdam 2008a, b):

Reinventing delta technology: providing space for experimentation and innovation for energy transition and water management (e.g. for an energy neutral city-port nexus);
Volume and value: optimising existing space through innovative logistics for industrial growth and more space for environment and landscape;
Crossing borders: (re-)connecting city and port both physically and socially (e.g. by providing cultural and sport facilities near the water);
Floating communities: creating floating housing, workplaces, recreation facilities to land scarcity and withstand climate change, also restructuring the urban economy and creating new social networks;
Sustainable mobility: creating a multi-modal system of mobility that also connects public transportation on land and water.

Virtually all strategies utilised flexible networks and coalitions to develop these areas. Instead of using publicly funded programmes, small incremental steps were taken together with heterogeneous actor networks. As someone put it: "Forming coalitions is a powerful political means to intervene"[5] (interview A). These coalition partners, however, were often not local citizens and NGO's but private businesses. Non-state engagements in most of the City Ports strategies were considered in terms of "funds for innovation and investments" in order to "get the city moving again" (interview B). As mentioned before, in order to illustrate specific TM activities and operations, focus on one of the five city ports strategies: the Floating Communities Strategy.

The Floating Communities Strategy: A Blue Revolution

The vision of the Floating Communities Strategy is framed in relation to the fact that on third of the port area consists of water. Water "gives the city life its own unique character" (Project Bureau Stadshavens 2008a: 14). The potential of unused spaces of water is highlighted: "Basins where port industry has moved away are ideal locations for floating housing and workplaces. This will further enhance the quality of City Ports Rotterdam and benefit all the residents of Rotterdam" (ibid). This vision is supported by arguments about what these floating buildings will actually do. "Floating constructions, like work spaces and pavilions, serve as boosters for development. Floating restaurants and other public facilities will be the centre of attraction in the Rijnhaven". (...) "There's a superb view from the many footpaths and cycle routes along the waterside, which pass by cultural

[5]In Dutch: "Coalitievorming is krachtige interventiepolitiek".

attractions, floating cafes and restaurants and water sports facilities, including an open-air swimming pool" (ibid). On the City Ports' website the main idea of the Floating Communities Strategies is presented as follows:

> Space will become available in Stadshavens for 5,000 homes on or beside the water. This extraordinary residential environment should attract professionals and pioneers[6] who want to stand out from the crowd. These people are needed to inject new life into Rotterdam's economy. The SS Rotterdam, a floating complex of conference, leisure and catering facilities, will be the first step in the process.[7]

Next to these explicit references to the strategic vision, a variety of other discourses are intertwined with this future image. A number of organisations produced a brochure about the floating cities (Urgenda 2008: 9). This brochure nicely mentions how different urban design concepts relate to different images of a future floating city[8] (Fig. 2).

This imaginative picture is accompanied by the following text:

> Different town planning typologies are conceivable varying from a compact core to more dispersed configuration. Investigating the two extremes reveals how density affects infrastructure and spatiality. A compact city needs fewer infrastructures but the sense of space and contact with

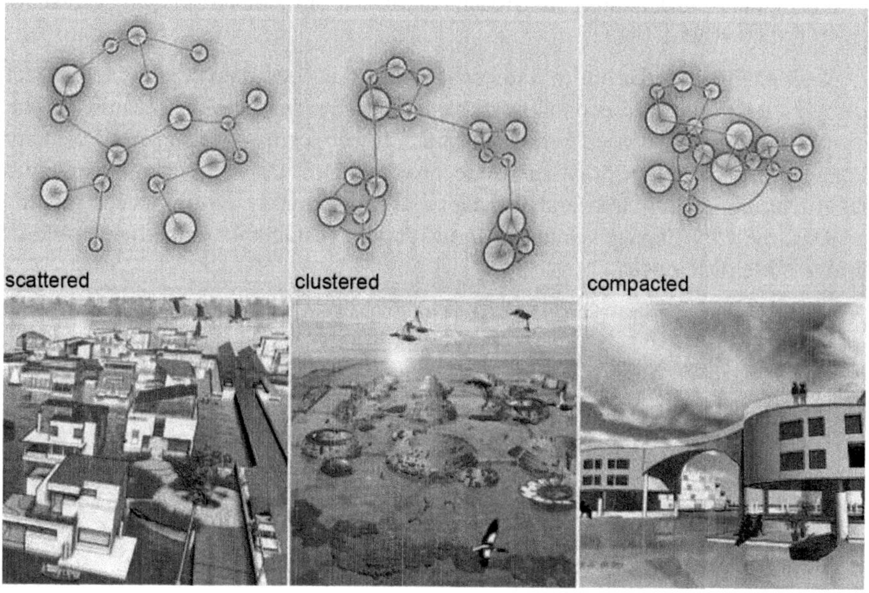

Fig. 2 Urban design concepts of floating cities

[6]In Dutch the word *hoogopgeleide groepen* is used, which literally means 'higher educated groups'.

[7]See: http://stadshavensrotterdam.nl/eng/floating-communities.

[8]Brochure of Urgenda about 'the Floating City', 2008 (images designed by DeltaSync).

> the water are more limited. This is the opposite for a dispersed model where spaciousness comes at the cost of more infrastructures. The dynamic nature of the Floating City makes it possible to begin with a dispersed plan and to develop this further into one or more centres. The best possible use of the available water surface is thus continually achieved (ibid).

This text describes how houses on water are imagined and made into liveable spaces that connect Rotterdam city and the harbour. The floating community strategy imagines future residents as follows:

> Rotterdam could make a name for itself with floating residences on tidal waterways in a number of Stadshavens outside the dikes. Recent research has shown that there is significant market potential for this sort of housing. Its special appeal could persuade highly educated people to stay in Rotterdam, as well as attracting new people from elsewhere. These groups are important for restoring Rotterdam's economy (Project Bureau Stadshavens Rotterdam 2008b: 17).

It is clear that this refers to the 'target audience' of the floating houses and the socio-economic status the future 'floating community'. The type of people that are attracted here are referred to as 'pioneers'.

> Pioneers will want to come and distinguish themselves from the masses, people will come to eat, play sport or sleep on a floating facility and businesses will set themselves up on a floating work island. Floating construction can therefore provide a boost for the development of the area (2008b: 17).

So, not only does the future water spaces near the Rotterdam city harbour attract pioneers for living, and economic and leisure activities shape the meaning of the new harbour. As mentioned before, sustainability concerns are also part of the City Ports strategy, implying that there are possible conflicts between the economic re-structuring of this area and the everyday high quality living spaces. In this context, the City Ports programme argues how a sustainable transition is able to bridge these differences.

> The development of the districts will create a sustainable transition for Stadshavens. The City Council, the Port of Rotterdam Authority and the State are putting together a strategy for sustainability, which should help to resolve any conflicts arising where living and working environments converge (Project Bureau Stadshavens 2008a: 4).

The implementation of the Floating Communities Strategy is also supported by imagining the exact location of) the floating constructions. Since the 2010s, more and more reports were published to imagine ad propose living on water as a feasible project. Consider, for instance, the image of floating communities in the City Ports implementation programme (Project Bureau Stadshavens 2008b: 17).

Water was often framed in terms of a 'unique experience' of nature. The Nassauhaven, not part of City Ports area was one of the first areas that was considered to experiment with a small living community (12–14 houses). These houses were also envisioned as 'sustainable', in that they should be powered by solar panels and façades of glass. Many technical studies were carried out to analyse the expected impact of the City Ports plans. These environmental impact assesments were executed by engineer bureau of the city along the lines of different domains and expertise (e.g. sound, smell, sound, culture, etc.) and finished in 2010

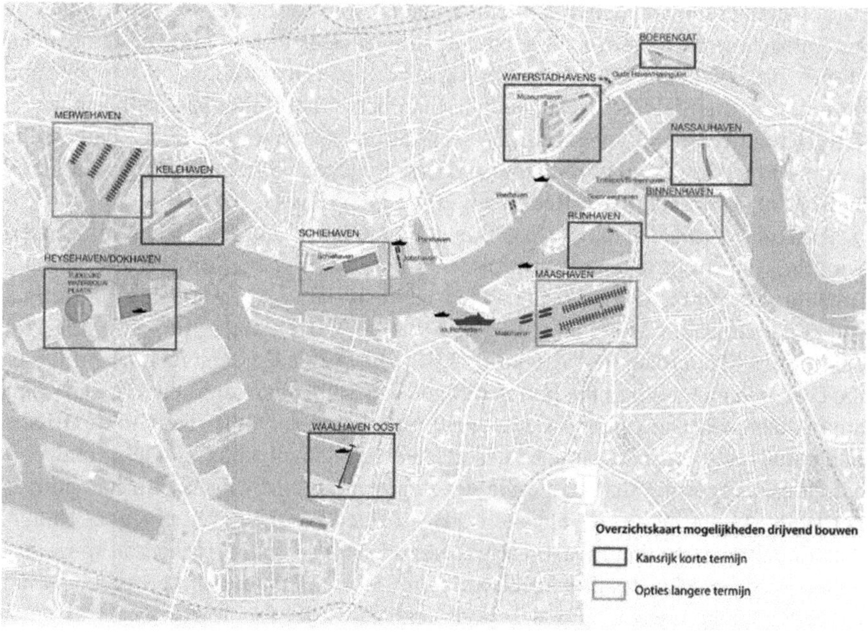

Fig. 3 Spatial mapping of possible locations for floating objects

(Municipality of Rotterdam 2010). These studies visualised, mapped and calculated the horizon of 'responsible' project activities (Fig. 3).

It is important to note that the Floating Community Strategy and the efforts make Rotterdam's city port sustainable, fit into a broader set of urban sustainability project initiated by city authorities. As many policy makers suggest, the City Ports programme and the associated transition management practices fit in a wider strategy to make Rotterdam a cleaner, greener, attractive, climate adaptive and more liveable city. This is also clearly evidenced by an official report called "Investing in sustainable growth" (in Dutch: *Investeren in duurzame groei*) (Municipality of Rotterdam 2011). Numerous other examples seek to establish similar results, employ different means in different areas in the city, for example sustainabvle houses (e.g. making thousands of houses sustainable through isolation, solar panels), sustainable mobility (e.g. stimulating bike use and electric cars), sustainable energy (e.g. using surplus warmth of industrial steam pipe), sustainble food (e.g. stimulating urban farming and community garden), water proof and climate change initiatives (e.g. green roofs, and water proof squares) (Municipality of Rotterdam 2011; interview C, D, E, F). In many instances, these activities are set up by a state actors, semi-public organisation, business or residents. Additionally, some of these 'sustainability initiatives' are connected to the Floating Community Strategy (e.g. sustainable energy, sustainable mobility). Obviously, this is not only a concern for the city of Rotterdam. Many delta and port cities are confronted with similar concerns and have taken up similar initaitives. This is why Rotterdam is also part of an international network of cities, and is often considered as a frontrunner by journalists and cities across the globe regarding issues of becoming sustainble and climate proof (interview C). Crucially, this also means that notions of economic and port 'development' and 'investment' are omnipersent in Rotterdam's urban plans as well. Urban (re)development is not an implicit objective, but rather part of the design of the City Ports programme and the city plans more generally in the 2000s and 2010s (Municipality of Rotterdam 2003; interview G, B, A, N).

As a more cynical anekdote that circualtes among certain professionls associated with the waterfront regenretaon projects is that the City Port programme is just a small project for the city to 'make up for the Maasvlakte 2'. Even though this might sound pessimistic, there is an telling image in the 2010 Annual Report of the Port Authorities (Rotterdam Port Authority 2010: 15). An image depicting the entirity of the port area, categorises the port activities, e.g. bulk, chemicals, etc. The city port area is depicted and categorised as well, but simply as "other activities". This is intersting, as the urban port area is related to a small slice of the overal port activities, suggesting that economic development and growth-based economics is pertinent. As put in the city plans for 2014–2018: "For the development of Rotterdam, it is important that the typically Rotterdam-minded enterpreneurial spirit gets ample space" (Municipality of Rotterdam 2014: 2). Some of the participating consortia could use City Ports lessons and networks for new projects in other international contexts (interviews G, H).

An interesting critique in this context comes from local residents and business owners or the lack of including them in City Ports projects more generally (interviews I, J, K). Another challenge related to the Floating Communities Strategy is that a number of docks (150) had to move to a different location due to the phasing out of port activities (Schuttevaer 2011). Inland navigation organisations had to relocate further down the estuary, at the expense of the 'high quality location' of the docks and the convenient locations for the families spending time on these docks. Since ships have become (and still are becoming) much bigger than before, the number of docks will also decrease in the City Ports area (interviews L, M).

Case Reflection: Democratic Waterfront Regeneration?

What does the city ports regeneration case tells us about transition management as practice? To what extent is this transition management process also democratic? To be clear at the outset, this transition management context contains only some democratic traces and elements (Jhagroe and Loorbach 2015). It is difficult not to note that these transformative practices mainly render possible a new socio-economic ordering of the old port areas. The waterfront emerged as a field of vision to combine various concerns and experiment with new economic activities, connecting local schools to businesses and advancing new technologies. The city ports were imagined as clean delta areas that were attractive for creative classes and new start-ups. This ongoing 'boulevardisation' of port cities (e.g. building floating homes and offices) seems to be a trend to combine concerns around climate change, carbon-neutral energy, increasing environmental quality, economic restructuring and creative classism. Land and water, then, become part of a new waterfront governing culture that moves away from classical urban planning and foregrounds contingency, surprise and out-of-the-box thinking. This seems to create an entrepreneurial waterfront, creative-class lifestyles and an environmental-friendly future. So, at a general level, the Stadshavens projects do not seem to put emphasis on fundamentally challenging the status quo and transform the waterfront area through democratic forces. Let us reflect on the four transition management dimensions as presented earlier to tease out the specific local dynamics.

Urban Transition Strategy and Democratic Aspects

At the level of the urban transition strategy, it is clear that a long-term vision on a 'sustainable waterfront' has been created that integrates opposing ideals and waterfront interests. This imagination creates a flexible umbrella that combines a general focus (on economic restructuring and socio-environmental innovation) and enables cooperation between business, architects, political agendas, policy actors and residents (also cf. Van Eijndhoven et al. 2013; Frantzeskaki et al. 2014: 414). But, to what extent is this strategy democratic? Are radically alternative imaginaries created and are counter-hegemonic discourses cultivated to transform the waterfront? This is the case only to some extent, due to the fact that the waterfront was projected as a renewed service-oriented port economy (not industrial) as well as an urban environment meant for work, living and leisure. Growth is the underlying rationality, while sustainability is seen as the paradigm that can stimulate innovative and robust future developments. The coupling of sustainability and innovation made the agenda for regeneration attractive and inspiring for community groups (such as NGOs, social entrepreneurs and scientists) and for corporate actors (businesses, construction actors, SMEs). This appears as a critical success requirement for urban regeneration (Coaffee and Deas 2008: 174; Shipley 2000; Shipley and Newkirk 1999). At the same time, the development of the vision in a participatory way revealed opportunities for public, private and community actors.

Urban Transition Tactics and Democratic Aspects

At the level of urban transition tactics, a number of alliances were established. The Stadshavens Project Office and the RCI, for example, included a number of actors and organisations and coordinate the strategic long-term goals in terms of the five spatial strategies and associated local projects. In what way does this tactical level count as a democratic practice? Are non-institutionalised networks and alliances mobilised? The role of the local government is crucial in this context. With its participation in these partnerships, the local government (Rotterdam Municipality) actively took a meta-governance approach in "coordinating different forms of governance" (Jessop 1997: 442). The local government also created a productive governance space for "coordinated governance" (Mueleman 2010: 55). The Floating Communities Strategy embodies transitions tactical alignments, incorporating both traditional and innovative actors. However, few radically innovative and counter-hegemonic actors and discourses were facilitated, as there was a focus on networks and connections between governmental agencies, technical experts, architects and contractors. Tactically, this transition management process seemed to accommodate institutional actors more than residents and non-institutional groups.

Urban Transition Operations and Democratic Aspects

The Floating Communities Strategy and projects exemplify urban transition operations. The Floating Pavilion project involves different stakeholders (architects, investors, policy actors, etc.) that seek to experiment new building concepts and flexible planning ideas in vacant port areas. Despite technical, legal and financial concerns, the flexible network and semi-open process of 'trial-and-error' can be considered as a 'success' (Frantzeskaki et al. 2014: 414). But in what way did these specific initiatives materialise alternative imaginaries? Again, this is the case in only some extent. Despite the open-ended aspirations, many material waterfront regeneration issues were controlled and managed by institutional measures. As plans for the Floating Community Strategy increased, local and national policy procedures were utilised in tandem with water-based architectural techniques were used to build a floating test case including solar panels and self-supportive water systems (floating pavilion). Next to the activities of architecture bureaus, contractors, building companies, more formal and legal procedures were employed (for some this was 'too much bureaucracy'). A range of unprecedented architecture plans and urban planning techniques were introduced such as 'slow urban planning' and 'floating urbanism'. One of the tools that was dealt with professionally was promoting City Ports programme and floating objects through city marketing.

Urban Transition Reflections and Democratic Aspects

Local actors and scholars that studied this case can agree on the fact that little reflection has been taking place (Frantzeskaki et al. 2014). Arguably, due to the flexible and open-oriented character of the transition process so far, few monitoring and evaluation documents have emerged. This is not only a general transition management concern, but also a democratic challenge. As Frantzeskaki et al. argue: "Questions like "who is going to live in the future floating neighbourhood?" may be crucial questions to consider in the near future in order to ensure that the values depicted in the sustainability vision are safeguarded during the scaling-up of the successfully developed projects at present" (2014: 414). This also means that hegemonic modes of evaluation cannot be challenged and that radically alternative forms of evaluation and monitoring are lacking. In this sense, the reflexive level is undemocratic in two interrelated ways: it lacks a systemic evaluation of the transition practices, and it (therefore also) lacks critical examination of dominant economic and technocratic evaluation criteria.

Conclusions

This chapter focused on the democratic challenges associated with transition management context. I proposed a transition management frame that accounts for democratic aspects and challenges. Instead of relying on traditional and institutional democratic heuristics, I used insights from critical urban research. The case of Rotterdam's waterfront illustrates that and how transition management materialises in an urban context.

A number of general insights and democratic challenges can be distilled. First, there is an inherent tension between management practices and democratic practices in a transition context. This is no surprise, given that transition management navigates between urban regimes and institutions on the one hand and alternative urbanisation visions and practices on the other hand. This general democratic challenge re-enters at the four transition management levels. At the strategic level, the long-term waterfront objectives might bring opposing visions and interests together, it thereby obfuscates underlying conflicts about the meaning of waterfront areas and urban life. At the tactical level, many usual actors and knowledge schemes seem to dominate the waterfront regeneration projects (architects, business interest, policy actors), marginalising residential voices. The specific practices of the Floating Communities Strategy (the operational level) depended on circumventing technical institutional rules (e.g. on housing, safety, financing) in order to experiment with floating houses and new spatial plans, mainly for the creative class.

Perhaps most challenging, the lack of any systemic evaluation at the reflexive level seems render impossible a critical scrutiny and assessment of the proposed objectives and alternative evaluation criteria. Future applications of transition management with a similar historical and socio-spatial context (of urban regeneration) should consider the democratic challenge of transition management, from the first meeting onwards, especially regarding monitoring and evaluation. This does not deliver a more 'effective process' or 'better outcomes' per se, in fact, it might even complicate things and make cooperation and urban governing more complex. What it does do, however, is to question underlying socio-economic and socio-environmental assumptions of 'a sustainable city' by challenging the taken-for-granted stakeholders and frontrunners, abstract long-term sustainability visions, and the gains of local transition projects. Importantly, each and every transition management context is different, which means that such democratic issues and concerns differ per case.

These democratic transition management challenges also have a broader spatial dimension that need to be addressed in future transition management practices. As critical urban scholarship suggests, urban landscapes are inherently decentred and hierarchical, with centres and peripheries, creating unequal access to green, healthy and liveable environments. In Rotterdam, there has been a historical division marked by the schism between the wealthier Northern parts (connected to the Holland region) versus the poorer Southern parts (often populated by port workers and new immigrant families). In the Stadshavens case this is relatively evident, as the on-going boulevardisation of Rotterdam's waterfront seems to create an expansion of a vivid city centre for a new creative class and residents from privileged areas. Even though few actual floating houses, offices and services are realised, this future is not illusionary, but even part of the actual plans to attract high-income groups that can live and work on or near Rotterdam's waterways. Arguably, for many residents from Southern parts of Rotterdam, the regeneration of the waterfront (including its floating community aspects) is part of the city centre and its expansion. Connected to the city ports regenerations, as we have seen, is the *Maasvlakte 2* expansion and the environmental damage that is officially 'compensated'. These types of eco-gentrifications, however, are not a natural process, but actively pursued in numerous ways (especially the focus on so-called 'economic restructuring'). The historical disparity between Rotterdam's northern and southern areas remain and perhaps even intensify, despite the well-intended waterfront regeneration projects.

It is crucial to note that the rise of Rotterdam's sustainable waterfront is work-in-progress. It is a contingent process full of socio-economic and spatial struggles. Despite the relative 'openness' of the new waterfront, various entrepreneurs and social initiatives struggle to gain a legitimate position in this new waterfront economy. The 'open source' and 'bottom-up' character of some regeneration projects make the 'here and now' indeterminate. Transition dynamics of the waterfront and the democratic management therefore relies on the ways in which the waterfront is imagined and enacted in the future. This is not only a topical issue for the practice of transition management, but also for conceptual work in the field of sustainability transitions research. The democratic question remains a challenge for transition management and sustainability transitions. Uneven power relations and spatial developments can simply not be addressed and solved with a guideline or a checklist. They should reminder transition researchers and practitioners that democratic challenges are here to stay.

Interviews

	Position and organization	Date of interview
Interview A	Policy actor municipality of Rotterdam	24 February 2014
Interview B	Policy actor municipality of Rotterdam	4 December 2013
Interview C	Policy actor municipality of Rotterdam	29 November 2013
Interview D	Policy actor municipality of Rotterdam	20 November 2013
Interview E	Policy actor municipality of Rotterdam	28 November 2013
Interview F	Policy actor municipality of Rotterdam	21 February 2014
Interview G	Consultant at architect organisation	10 January 2014
Interview H	Policy actor municipality of Rotterdam	13 January 2014
Interview I	Works for resident association	4 March 2014
Interview J	Works for residential interests	27 February 2014
Interview K	Researcher	15 January 2014
Interview L	Works for inland navigation interests	27 June 2013
Interview M	Actors working for Port Authority and municipality of Rotterdam	14 November 2014
Interview N	Works for port industry and businesses	4 November 2013

References

Bassett K, Griffiths R, Smith I (2016) Testing governance: partnerships, planning and conflict in waterfront regeneration. Urban Stud 39(10):1757–1775

Brenner N, Madden DJ, Wachsmuth D (2011) Assemblage urbanism and the challenges of critical urban theory. City 15(2):225–240

Bulkeley H, Broto VC, Hodson M, Marvin S (eds) (2010) Cities and low carbon transitions. Routledge, New York

Bulkeley HA, Broto VC, Edwards GA (2014) An urban politics of climate change: experimentation and the governing of socio-technical transitions. Routledge, New York

Coaffee J, Deas I (2008) The search for policy innovation in urban governance lessons from community-led regeneration partnerships. Publ Pol Adm 23(2):167–187

Coenen L, Truffer B (2012) Places and spaces of sustainability transitions: geographical contributions to an emerging research and policy field. Eur Plan Stud 20(3):367–374

Coenen L, Benneworth P, Truffer B (2012) Toward a spatial perspective on sustainability transitions. Res Pol 41(6):968–979

Daamen T (2010) Strategy as force: towards effective strategies for urban development projects: the case of rotterdam city ports. IOS Press, Amsterdam
Davidson M, Martin D (eds) (2013) Urban politics: critical approaches. Sage, Vancouver
Davidson M (2014) Is class relevant to urban politics? In: Davidson M, Martin D (eds) Urban politics: critical approaches. Sage, London, pp 189–205
Dixon T, Eames M, Hunt M, Lannon S (eds) (2014) Urban retrofitting for sustainability: mapping the transition to 2050. Routledge, New York
Ducruet C (2011) The Port City in multidisciplinary analysis. In: Alemany J, Bruttomesso R (eds) The Port City in the XXIst century: new challenges in the relationship between port and city. RETE, pp 32–48
Frantzeskaki N, De Haan H (2009) Transitions: two steps from theory to policy. Futures 41 (9):593–606
Frantzeskaki N, Wittmayer J, Loorbach D (2014) The role of partnerships in 'Realising' urban sustainability in Rotterdam's city ports area, The Netherlands. J Clean Prod 65:406–417
Hankins K, Martin D (2012) The neighbourhood as a place for (urban) politics. In: Davidson M, Martin D (eds) Urban politics: critical approaches. Sage, London, pp 23–41
Harvey D (1990) The condition of postmodernity: an enquiry into the conditions of cultural change. Blackwell, Oxford
Hendriks CM (2009) Policy design without democracy? Making democratic sense of transition management. Policy Sci 42(4):341–368
Hendriks CM, Grin J (2007) Contextualizing reflexive governance: the politics of Dutch transitions to sustainability. J Environ Policy Plan 9(3–4):333–350
Hoyle BS, Pinder DA, Husain MS (1988) Revitalizing the waterfront: international dimensions of dockland redevelopment. Wiley, Hoboken
Jessop B (1997) Capitalism and its future: remarks on regulation, government and governance. Rev Int Polit Econ 4(3):561–581
Jhagroe S, Loorbach D (2015) See no evil, hear no evil: the democratic potential of transition management. Environ Innov Soc Trans 15:65–83
Klamer A, Kombrink D (2004) Het Verhaal van de Rotterdamse Haven. Een Narratieve Analyse. Stichting Economie & Cultuur, Erasmus University of Rotterdam, Rotterdam
Kemp R, Loorbach D, Rotmans J (2007) Transition management as a model for managing processes of co-evolution towards sustainable development. Int J Sust Dev World Ecol 14 (1):78–91
Laclau E, Mouffe C (1985) Hegemony and socialist strategy: towards a radical democratic politics. Verso, London
Lefebvre H (1996) Writings on cities. Blackwell, Oxford
Lefebvre H (2003) The urban revolution. University of Minnesota Press, Minneapolis
Loorbach D (2007) Transition management: new mode of governance for sustainable development. Doctoral thesis, Dutch Research Institute for Transitions (DRIFT), Erasmus University Rotterdam
Meuleman L (2010) The cultural dimension of metagovernance: why governance doctrines may fail. Publ Organ Rev 10(1):49–70
Municipality of Rotterdam (2003) Rotterdam Zet Door. Op Weg naar een Stad in Balans. Municipality of Rotterdam, Rotterdam
Municipality of Rotterdam (2010) Werken aan Talent en Ondernemen
Municipality of Rotterdam (2011) Investeren in Duurzame Groei. Programma duurzaam. Municipality of Rotterdam, Rotterdam
Municipality of Rotterdam, district North (2014) Noord is Noord. Een Boekje Vol Inspiratie. Municipality of Rotterdam, district North, Rotterdam
Murphy JT (2015) Human geography and socio-technical transition studies: promising intersections. Environ Innov Soc Trans 17:73–91
Nevens F, Frantzeskaki N, Gorissen L, Loorbach D (2013) Urban transition labs: co-creating transformative action for sustainable cities. J Clean Prod 50:111–122

Norcliffe G, Bassett K, Hoare T (1996) The emergence of postmodernism on the urban waterfront. J Trans Geogr 4(2):123–134
Project Bureau Stadshavens Rotterdam (2008a) 1600 ha. Creating on the edge. Vijf Strategieën Voor Duurzame Gebiedsontwikkeling. Project Bureau Stadshavens Rotterdam, Rotterdam
Project Bureau Stadshavens Rotterdam (2008b) 1600 ha. Implementation Programme. 2007–2015. Project Bureau Stadshavens Rotterdam, Rotterdam
Rancière J (2004) Disagreement: politics and philosophy. University of Minnesota Press, Minneapolis
Roorda C, Wittmayer J, Henneman P, Steenbergen, Van F, Frantzeskaki N, Loorbach D (2014) Transition management in the urban context: guidance manual. DRIFT, Erasmus University Rotterdam, Rotterdam
Rotterdam Port Authority (2010) Jaarverslag 2010. 'Wereldklasse doen!'. Havenbedrijf Rotterdam N.V, Rotterdam
Schuttevaer (2011) Binnenvaart Wil Maashaven Niet Opgeven. https://www.schuttevaer.nl/nieuws/havens-en-vaarwegen/nid15370-binnenvaart-wil-maashaven-niet-opgeven. Visited in 2012
Shipley R (2000) The origin and development of vision and visioning in planning. Int Plan Stud 5 (2):225–236
Shipley R, Newkirk R (1999) Vision and visioning in planning: what do these terms really mean? Environ Plan B 26:573–592
Shove E, Walker G (2007) CAUTION! Transitions ahead: politics, practice, and sustainable transition management. Environ Plan A 39(4):763–770
Smith A, Stirling A (2010) The politics of social-ecological resilience and sustainable socio-technical transitions. Ecol Soc 15(1):11
Stadshavens Rotterdam (n.d.) [http://stadshavensrotterdam.nl/eng/vision]. Visited in 2012–2015
Tidball K, Stedman R (2013) Positive dependency and virtuous cycles: from resource dependence to resilience in urban social-ecological systems. Ecol Econ 86:292–299
Urgenda (2008) The floating city. Into an ocean of opportunities. Stichting Urgenda, Rotterdam
Van Eijndhoven J, Frantzeskaki N, Loorbach D (2013) Connecting long and short-term via envisioning in transition arenas. In: Edelenbos J, Bressers N, Scholten P (eds) Connective capacity in water governance. Ashgate Publications, London, pp 172–190
Voß JP, Smith A, Grin J (2009) designing long-term policy: rethinking transition management. Policy Sci 42(4):275–302

Part IV
Reflecting on the Impacts of Transition Management in Cities

Chapter 16: Holscher – So what? Transition management as a transformative approach to support governance capacities of cities. This chapter synthesises the insights from the various book contributions to critically reflect on what transition management offers when applied in and for cities. In this book, transition management has been suggested as a transformative governance approach in and for cities. Transition management seeks to provide impulses for systemic change by creating spaces for developing innovative ideas and practices and for empowering actors to develop transition experiments. It also holds the potential to fundamentally challenge existing planning and governance systems by aligning a diversity of actors, networks and initiatives under a shared and long-term sustainability vision and questioning existing roles and responsibilities of actors. We explore how transition management thus contributes to building *transformative capacity for systemic change* and *orchestrating capacity for the coordination of self-organisation.* On the basis of the book contributions we discuss the potentials, opportunities and challenges of transition management as a prescriptive and transformative governance approach. We conclude by identifying implications for transition management to strengthen governance capacities for navigating sustainability transitions in cities.

So What? Transition Management as a Transformative Approach to Support Governance Capacities in Cities

Katharina Hölscher

Abstract This chapter synthesises the insights from the various book contributions to critically reflect on what transition management offers when applied in and for cities. In this book, transition management has been suggested as a transformative governance approach in and for cities. Transition management seeks to provide impulses for systemic change by creating spaces for developing innovative ideas and practices and for empowering actors to develop transition experiments. It also holds the potential to fundamentally challenge existing planning and governance systems by aligning a diversity of actors, networks and initiatives under a shared and long-term sustainability vision and questioning existing roles and responsibilities of actors. We explore how transition management thus contributes to building *transformative capacity for systemic change* and *orchestrating capacity for the coordination of self-organisation*. On the basis of the book contributions we discuss the potentials, opportunities and challenges of transition management as a prescriptive and transformative governance approach. We conclude by identifying implications for transition management in light of the objective to strengthen governance capacities for navigating sustainability transitions in cities.

Introduction

Cities have become a key loci for searching for alternative solutions and approaches to contribute to sustainable and resilient societies. On the one hand, social, environmental and economic crises and challenges from both urbanisation and globalisation converge in cities and contribute to the pressure for rerouting their current development pathways towards sustainability, resilience and liveability (Leichenko 2011; Rosenzweig et al. 2011; Rink et al. 2015; Frantzeskaki et al., chapter "Understanding the Urban Context and Its Challenges", this volume). On the

K. Hölscher (✉)
Dutch Research Institute For Transitions (DRIFT), Faculty of Social Sciences, Erasmus University Rotterdam, Rotterdam, The Netherlands
e-mail: holscher@drift.eur.nl

N. Frantzeskaki et al. (eds.), *Co-creating Sustainable Urban Futures*, Future City 11, https://doi.org/10.1007/978-3-319-69273-9_16

other hand, cities have emerged as cradles of innovation aimed at providing alternative solutions to tackling issues such as climate change, air and noise pollution, unsustainable energy and mobility (Grimm et al. 2008; Nevens et al. 2013; McCormick et al. 2013; Castán Broto and Bulkeley 2013). This demonstrates the diversity of opportunities and conditions in cities that manifest in "not only the quantitative weight of cities in terms of urban population, metabolism and form", but also in "their qualitative momentum as networked places that concentrate the conditions, resources and opportunities for change" (Wolfram and Frantzeskaki 2016: 1).

Scientists and practitioners search for new governance and planning approaches that boost cities' governance capacities to strategically tap into and mobilise their innovative potentials (Nevens et al. 2013; McCormick et al. 2013; Wolfram and Frantzeskaki 2016). Traditional approaches in cities continue to rely on planning approaches designed and implemented from highly specialised silos that operate based on narrowly defined, short-term goals (Friend et al. 2014; Adil and Ko 2016; Malekpour et al. 2015; Shaw et al. 2014). This results in planning programs that respond only partially to relatively isolated sustainability problems. Against this context and considering the system-wide implications and impacts of unsustainability problems and climate change in cities, new approaches are required to complement existing practices. Amongst others, these should consider system-level interventions and long-term planning horizons. For those approaches to be incorporated in urban planning, city officers need to invest in capacity-building and policy learning through engagement and co-creation with urban stakeholders. Via processes of knowledge co-production and enrichment, cities will be able to strengthen their capacities for innovating and for transforming towards sustainable urban futures.

This book has presented and reflected on transition management in and for cities as a practice-oriented governance approach that includes a portfolio of tools to influence governing activities in such a way that they boost and coordinate the navigation of sustainability transitions (Wittmayer et al., chapter "Transition Man agement: Guiding Principles and Applications", this volume, Frantzeskaki et al., chapter "Introducing Sustainability Transitions' Thinking in Urban Contexts", this volume). Transition management starts from the recognition of complex interactions of political, cultural, technological, institutional and economic elements of a system that create lock-ins and persistent problems. It seeks to progressively build up long-term and radical system change (Rotmans et al. 2001; Rotmans and Loorbach 2010). Transitions refer to fundamental nonlinear changes in cultures (e.g. values, attitudes), structures (e.g. institutions, physical infrastructures) and practices (e.g. behaviours, routines) (Frantzeskaki and De Haan 2009). Emerging transitions can be influenced, guided and accelerated to promote sustainability (Loorbach 2010; Loorbach et al. 2015; Frantzeskaki et al. 2012). While transitions cannot be 'managed' in the traditional meaning of the word, transition management provides a portfolio of tools to guide actors in long-term thinking, contextualise transitions' thinking, co-develop narratives of transformative action, develop and experiment with alternative ideas and solutions and connect with other actors (Wittmayer and Loorbach 2016).

The preceding chapters have provided an introduction to the theoretical foundations and practical process guidelines of transition management to promote urban sustainability transitions, contributed critical lenses on how transition management relates to urban planning, on actors and roles in transition management and urban sustainability transitions as well as the democratic legitimacy of transition management in cities. The heart of this book presents in-depth reflections on experiences with applying transition management in both developed and developing countries' cities. These illustrate how transition management as a governance experimentation approach integrates proposed process steps, content, context and potential solutions while maintaining a dialectic interface to advance urban sustainability transitions. The exploratory discussions of challenges and promises of the transition management methodology add further insights about how the process could relate to and support transition governance in diverse urban contexts.

The book contributions have established how transition management could act as a transformative governance approach in and for cities. Transition management seeks to provide impulses for systemic change by creating spaces for developing innovative ideas and practices and for empowering actors to develop transition experiments. It also holds the potential to fundamentally challenge existing planning and governance systems by aligning a diversity of actors, networks and initiatives under a shared and long-term sustainability vision and questioning existing roles and responsibilities of actors.

In this final chapter we synthesise the insights from the various contributions to critically reflect on what transition management offers when applied in and for cities. We explore to what extent and how transition management contributes to building new types of governance capacities to transform existing urban strategies, structures and processes as well as governance approaches. Transition scholars identify several governance tenets to navigate sustainability transitions (Loorbach et al. 2015; Frantzeskaki et al. 2012; Kemp et al. 2007; Wittmayer 2016). These essentially come down to enabling and strengthening a diversity of innovative impulses for change as well as to coordinating multiple self-organising actor networks across scales and sectors to promote synergies for achieving the large-scale aspired systemic changes. While showing how transition management supports urban governance for sustainabilty and indicating changes that are required in existing governance configurations, the analysis of governance capacities also serves to highlight potential limitations of transition management.

Transition management contributes to building and enhancing governance capacities by motivating and supporting the processes through which actors define, implement and coordinate innovative and common strategies, projects and approaches. Building on the theoretical underpinnings of transition management and the case studies presented in this book we can identity two different types of capacities created from transition management processes in cities. Firstly, transition management seeks to support *capacities for disruptive innovation* (transformative capacity) by creating space for innovation and experimentation and anchoring innovation. Secondly, it seeks to enhance *capacities for the coordination of self-organisation* (orchestrating capacity) to establish synergies between a variety of actors and networks in line with a common orientation towards a sustainability transition (see also Hölscher et al. 2017a).

In the remainder of this chapter, we first provide a brief overview of the case studies presented in this book in terms of their contexts, addressed transition challenges, main actors involved and types of governance capacities created and enhanced. The subsequent two sections reflect on how the two identified kinds of governance capacities are supported and enhanced by transition management in light of both its theoretical and methodological ambitions and the insights from the empirical case studies. The vast amount and diversity of cases that are presented in this book offer a good basis for reflecting on the contribution of transition management to governance capacities in and for cities. On the basis of these insights we discuss the potentials, opportunities and challenges of transition management as a prescriptive and transformative governance approach. We conclude by identifying implications for transition management to strengthen governance capacities for navigating sustainability transitions in cities.

Transition Management in Different Contexts: The Cases Presented in This Book

The contributions of this book describe (potential) transition management applications to a variety of urban contexts and to diverse transition challenges (Table 1). This offers a comprehensive basis from which the applicability of transition management in and for cities can be explored. In particular, the case examples illustrate how transition management integrates process steps, content and potential solutions in relation to particular contexts to boost local capacities.

The cases presented in this book represent small- and medium-sized cities such as Ghent, Aberdeen, Ludwigsburg, Rotterdam and Cleveland, large cities like Sydney and Chinese metropolitan areas, as well as an example from an urban neighbourhood in Rotterdam. Two cases discuss (the potentials of) transition management in developing country contexts, namely La Botija, Honduras, and urban settlements in Suriname. The transition challenges differ widely across case studies, including challenges of urbanisation and urban planning, urban water management, low carbon and energy transitions, shrinking cities and urban regeneration as well as sustainable neighbourhood development. This shows that in theory, transition management is expected to be applicable to all kinds of contexts and problems.

While the empirical case studies in Part II of the book offer insights into the contributions of transition management interventions in concrete governance settings, the exploratory case studies in Part III provide indications of how transition management could support existing planning processes by building new types of governance capacities.

Table 1 Overview of the case studies in this book

	Aberdeen (UK) (Maas et al., chapter "Starting Up Transition Management: A Closer View on the Systems Analysis and How It Initiated Transformative Thinking in Ghent and Aberdeen Cities")	Ghent (Belgium) (Maas et al., chapter "Starting Up Transition Management: A Closer View on the Systems Analysis and How It Initiated Transformative Thinking in Ghent and Aberdeen Cities")	Carnisse (Netherlands) (Wittmayer, van Steenbergen and Bach chapter "Transition Management in Urban Neighbourhoods: The Case of Carnisse, Rotterdam, the Netherlands")	Ludwigs-burg (Germany) (Hölscher and Wittmayer, chapter "A German Experience: The Challenges of Mediating 'Ideal-Type' Transition Management in Ludwigsburg")	La Botija (Honduras) (Silvestri and Frantzeskaki, chapter "Transition Management for Local Sustainability: A Case Study from La Botija Protected Area, San Marcos de Colón, Honduras")	Sydney (Australia) (Walsh, chapter "Translating Transitions' Thinking and Transition Management into the City Planning World")	Shenzhen (China) (Mai, chapter "Transition Management in Urban China: Co-creation of New Homes or a Monopoly Game?")	Urban settlements in Suriname (Lachman et al., chapter "Context-Driven Transition Management as a Necessary Vehicle for Sustainable Urban Futures in Suriname")	Cleveland (USA) (Schwarz, chapter "A Transition Management Approach for Shrinking Cities in the United States")	Rotterdam (Netherlands) (Jhagroe, chapter "Urban Transition Management as a Democratic Practice? The Case of Rotterdam's Waterfront Regeneration")
Size	210.000 inhabitants	240.000 inhabitants	11.000 inhabitants; neighbourhood in Rotterdam	87.000 inhabitants	La Botija is a protected area with 19 communities under the Municipality of San Marcos de Colon. There is no precise data on population.	5.070.000 inhabitants	113.787.000 inhabitants	500.000 inhabitants in Suriname	385.809 inhabitants	620.000 inhabitants
Transition management application	Empirical case study of operational TM process	Empirical case study of operational TM process	Empirical case study of operational TM process	Empirical case study of operational TM process	Empirical case study of operational TM process	Exploration of TM for future city planning	Exploration of TM for future city planning	Exploration of TM for future city planning	Exploration of TM for future city planning	Empirical case study of operational TM process

(continued)

Table 1 (continued)

	Aberdeen (UK) (Maas et al., chapter "Starting Up Transition Management: A Closer View on the Systems Analysis and How It Initiated Transformative Thinking in Ghent and Aberdeen Cities")	Ghent (Belgium) (Maas et al., chapter "Starting Up Transition Management: A Closer View on the Systems Analysis and How It Initiated Transformative Thinking in Ghent and Aberdeen Cities")	Carnisse (Netherlands) (Wittmayer, van Steenbergen and Bach chapter "Transition Management in Urban Neighbourhoods: The Case of Carnisse, Rotterdam, the Netherlands")	Ludwigs-burg (Germany) (Hölscher and Wittmayer, chapter "A German Experience: The Challenges of Mediating 'Ideal-Type' Transition Management in Ludwigsburg")	La Botija (Honduras) (Silvestri and Frantzeskaki, chapter "Transition Management for Local Sustainability: A Case Study from La Botija Protected Area, San Marcos de Colón, Honduras")	Sydney (Australia) (Walsh, chapter "Translating Transitions' Thinking and Transition Management into the City Planning World")	Shenzhen (China) (Mai, chapter "Transition Management in Urban China: Co-creation of New Homes or a Monopoly Game?")	Urban settlements in Suriname (Lachman et al., chapter "Context-Driven Transition Management as a Necessary Vehicle for Sustainable Urban Futures in Suriname")	Cleveland (USA) (Schwarz, chapter "A Transition Management Approach for Shrinking Cities in the United States")	Rotterdam (Netherlands) (Jhagroe, chapter "Urban Transition Management as a Democratic Practice? The Case of Rotterdam's Waterfront Regeneration")
Transition challenge	Energy efficiency and clean energy	Co-creation of climate neutrality by 2050	Co-creation of sustainable neighbourhood development	Co-creation of energy transition locally	Resource exploitation and dependency	Sustainable spatial development	Co-creating low-carbon transitions and co-creating visions for change	Sustainable urban planning and development	Shrinking cities	Urban regeneration in city ports area

Transition Management to Support Transformative Capacities for *Disruptive Innovation*

Transition management aims to influence transitions through the creation of space for actors to explore and build alternatives (e.g. ideas, practices, and social relations), as well as to challenge and change the status quo through experimentation and learning (Grin et al. 2010; Loorbach 2007, 2010; Frantzeskaki et al. 2012; Wittmayer et al., chapter "Transition Management: Guiding Principles and Appli cations", this volume). The transition management framework places much emphasis on enabling processes of niche and novelty creation: In the transition team and transition arena a problem is reframed and a transition agenda including a transformative sustainability vision and transition pathways are created (Wittmayer et al., chapter "Transition Management: Guiding Principles and Applications", this volume, Loorbach et al. 2015). The involved actors and growing transition networks implement innovative transition experiments, harvest from lessons learned, disseminate the transition agenda and adapt institutional frameworks and operational processes to embed innovation and transformative change (Roorda et al. 2014).

We analyse how the transition management interventions presented in this book's case studies have supported governance processes that enable novelty creation, support novelty in gaining ground and embed novelty in the specific contexts. All case studies testify the potential of transition management to spark the development of new ideas, discourses, strategic visions, practices and social relations. The extent to which these novelties have become taken up and integrated into broader local contexts differs across case study.

Enabling Novelty Creation

Major tools of the transition management framework are set around the creation of space for developing novelties in form of alternative ideas (e.g. reframing of actual challenges, new strategies, new discourses), practices (e.g. projects, experimentation, transformative action) and new social relations (e.g. between policy officers and citizens) to promote systemic change towards sustainability (Wittmayer and Loorbach 2016). In the transition team and transition arena settings new ideas and knowledge for a re-framing of cities' futures emerge through mutual and deep exchange, confrontation of multiple perspective and learning (Wittmayer 2016; Nevens et al. 2013; Roorda et al. 2014). New ideas and discourses are sown in strategic sustainability visions and transition pathways that provide a new orientation for action. The transition arena also creates space for new partnerships and actor relations that foster knowledge exchange and can result in collaboration in the implementation of transition experiments (Loorbach 2010; Frantzeskaki et al. 2014; Hölscher et al. 2017b). Finally, transition management promotes the remaking of the governance configurations in cities by fostering new relationships, for example in transition networks, between local policy officers and citizens or cross-departmental collaborations (Roorda et al. 2014; Roorda and Wittmayer 2014; Frantzeskaki et al. 2014; Wittmayer and Loorbach 2016).

Along these lines, the empirical case studies show a great variety of novelties that were created from the respective transition management interventions. In all case studies, the processes led to a reframing of the initial problem focus and goals. The chapter by Maas et al. (2018, chapter "Starting Up Transition Management: A Closer View on the Systems Analysis and How It Initiated Transformative Think ing in Ghent and Aberdeen Cities", this volume) particularly zooms in on problem (re-)framing as part of the initial process steps. The authors show how the applied system analysis methodology supported the involved actors to take system perspectives on problems and to identify complex interrelations between different problem domains. This created an urgency for change and a new view on systemic solutions that tackle the root causes of problems and open up opportunities for creating synergies through connecting diverse goals. Additionally, it helped to reveal innovative niches and positive assets on which new strategies and solutions can be built on. Starting from this broader perspective on problems, transition management has shown in all case studies to serve the formulation of new strategic visions and transition agendas. These can take up many different forms, from a more informal perspective on community life in Carnisse, Rotterdam (Wittmayer et al., chapter "Transition Management in Urban Neighbourhoods: The Case of Carnisse, Rotterdam, the Netherlands", this volume), to new policy-level strategies in Ludwigsburg (Hölscher and Wittmayer, chapter "A German Experience: The Challenges of Mediating 'Ideal-Type' Transition Management in Ludwigsburg", this volume).

In addition to new strategies and agendas, several case studies also indicate the development and implementation of new actions and concrete solutions. For example, in the neighbourhood Carnisse a new community centre has been initiated (Wittmayer et al., chapter "Transition Management in Urban Neighbourhoods: The Case of Carnisse, Rotterdam, the Netherlands", this volume). Similarly, the transition management process on urban regeneration in Rotterdam resulted in concrete pilot projects such as the floating pavilion (Jhagroe, chapter "Urban Transition Management as a Democratic Practice? The Case of Rotterdam's Waterfront Regeneration", this volume).

The arena setting, the connection of 'like-minded' frontrunners and process steps (e.g. visioning) thus seem to provide a space for actors to depart from business-as-usual thinking. However, the empirical experiences also indicate that moving from the identification of new initiatives towards their practical implications requires overcoming several hurdles. In Ludwigsburg, the move into action has never been achieved (Hölscher and Wittmayer, chapter "A German Experience: The Chal lenges of Mediating 'Ideal-Type' Transition Management in Ludwigsburg", this volume). One reason for this might have been the initial framing of the transition management process in terms of seeking citizens' input to the city government's strategies. Thus, rather than promoting a reflection on how different types of actors, including the participants themselves, could contribute to achieving the formulated transition vision, the city government remained the (sole) central actor. However, the city administration was not able to take up a strong role in implementing and supporting actions and projects.

Finally, in some of the case studies the transition management processes sparked a reflection on and re-definition of roles and responsibilities of actors. For example, in Carnisse, local community actors realised it was on them to improve their neighbourhood – rather than waiting for others to step in on their behalf (Wittmayer et al., chapter "Transition Management in Urban Neighbourhoods: The Case of Carnisse, Rotterdam, The Netherlands", this volume). This reframing of roles and responsibilities is supported by the open and co-creative process setting and a reflexive attitude that also questions existing actor behaviours (see also Hölscher et al. 2017b).

The exploratory case studies in Part III of the book similarly reflect on the potential values of transition management interventions for developing new types of solutions and approaches to address problems in the respective case studies. In this sense, the exploratory chapters also position transition management as a new type of governance approach that enables innovation and participation, thus complementing existing planning approaches that are unable to overcome unsustainability lock-ins. For example, Schwarz (chapter "A Transition Manage ment Approach for Shrinking Cities in the United States", this volume) suggests that transition management may prove relevant for shrinking cities in the US by providing an explicit focus on transformative change. Similarly, Lachman et al. (chapter "Context-Driven Transition Management as a Necessary Vehicle for Sustainable Urban Futures in Suriname", this volume) presents transition management as a tool to guide the creation of new urban settlements in Suriname, which requires breaking away from existing (e.g. tourism, planning, economic) paradigms. Interestingly, the authors point to the challenges of the developing countries context, which is often characterised by high power imbalances. Transition management interventions thus should take existing regimes and the role of status and hierarchy into account.

Increasing the Visibility of Novelties

To increase support and momentum for the newly created discourses, visions and practices, actors in transition networks communicate and disseminate the developed transition agenda, showcase pilot projects and establish supporting alliances (Roorda et al. 2014; Nevens et al. 2013; Frantzeskaki et al. 2012). Though these activities can be considered part of the 'activation phase' (Wittmayer et al., Chapter "Transition Management: Guiding Principles and Applications", this volume), transition management does not (yet) offer explicit tools to facilitate such lobbying for support, as this takes place after the actors leave the more structured arena setting. This is also because research on how transition initiatives can be coupled and rescaled are yet to be explored (Loorbach et al. 2015).

In the presented transition management interventions efforts were undertaken during and after the transition arena process to disseminate the developed transition agenda, engage additional actors and connect to on-going initiatives and processes. In Ludwigsburg, the transition agenda was presented to the city council and Lord Mayor and additional actors were invited to align their actions under the common umbrella of a long-term sustainability vision through for example broadening events (Hölscher and Wittmayer, chapter "A German Experience: The Challenges of Mediating 'Ideal-Type' Transition Management in Ludwigsburg", this volume). In Rotterdam, the transition management intervention was directly connected to the city government's waterfront regeneration process in the old city harbour area (Jhagroe, chapter "Urban Transition Management as a Democratic Practice? The Case of Rotterdam's Waterfront Regeneration", this volume). The showcasing of pilot projects also helped to attain a broader support base for the overall strategies.

Overall, the transition management case studies featured in this book – both empirical and exploratory – pay little attention to activities for increasing the visibility of the created novelties. One reason might be that these activities are commonly perceived as taking place after the initial transition management intervention, which is, as stated above, not yet explicitly conceptualised and operationalised in the transition management framework. These activities are also often not any more the focus of research activities – because of the set-up of such processes researchers usually withdraw from the processes after the first phases. Additionally, these activities require a large amount of time investment and the identification of actors and on-going initiatives willing to drive as well as to connect to the transition agenda. The initial system analysis supports the latter, as shown in Ghent and Aberdeen (Maas et al., chapter "Starting Up Transition Management: A Closer View on the Systems Analysis and How It Initiated Transformative Think ing in Ghent and Aberdeen Cities", this volume), as actors were identified that could be returned to at the later activation stage.

Anchoring Novelty in Context

The novelties that are created in transition management processes are to be anchored in local contexts by connecting them to on-going processes and strategies and by translating them into new operational and institutional frameworks (Roorda et al. 2014; Bos and Brown 2012). This contributes to scaling-up their innovative potentials and mainstreaming them into on-going practices and strategies (Roorda and Wittmayer 2014). Again, transition management does not offer explicit tools to guide actors in such processes of embedding but more implicitly expect actors in evolving transition networks to take them up. An important element that is implied in the methodology refers to the recognition of opportunities for change to boost the relevance of the new visions for existing processes (Rotmans and Loorbach 2010). This also requires reflexivity on how novelties fit to the context and taking up lessons learned from experimentation (Nevens et al. 2013; van den Bosch 2010).

The case studies feature few examples of such anchoring, which is largely due to their focus on what happened during the more immediate transition management interventions rather than on their follow-up. Additionally, the processes of anchoring novelties usually take time and longer-term outcomes of anchoring are more difficult to trace back to individual interventions. Nonetheless, in Rotterdam, for instance, the new waterfront regeneration strategy was connected to the strategic agendas and processes of the city government (Jhagroe, chapter "Urban Transition Management as a Democratic Practice? The Case of Rotterdam's Waterfront Regeneration", this volume). Similarly, in Ludwigsburg the transition agenda was connected to the existing urban development concept and energy strategies (Hölscher and Wittmayer, chapter "A German Experience: The Challenges of Mediating 'Ideal-Type' Transition Management in Ludwigsburg", this volume). Furthermore, in both case studies the initiated actor networks to coordinate and implement the strategies – such as in Rotterdam public-private partnerships and in Ludwigsburg the created cross-departmental working group on energy – have been established as more or less institutionalised bodies.

However, the case studies also point to several challenges for anchoring the novelties in the respective contexts, which relate to the need to connect the transition management intervention to on-going policy and planning processes in cities. Wittmayer et al. (chapter "Transition Management in Urban Neighbourhoods: The Case of Carnisse, Rotterdam, the Netherlands", this volume) reflect on the fact that according to most participants of the community arena in the neighbourhood Carnisse the arena process did not result in sufficiently tangible and obligatory actions – people thus did not recognise the impact of the intervention. A key problem was to connect the vision and process to existing policies and governmental structures – mostly because central decision-makers (e.g. housing corporations, city government) were absent, which left the participants with a feeling of insufficient leverage. Similarly, Jhagroe (chapter "Urban Transition Management as a Democratic Practice? The Case of Rotterdam's Waterfront Regeneration", this volume) criticises the failure to embed the developed experiments for urban regeneration in Rotterdam to on-going policy processes. This was also due to a disconnect between the transition management process and on-going policy and planning activities and resulted in a loss of momentum for the innovation impulse.

Interestingly, in this context Mai (chapter "Transition Management in Urban China: Co-creation of New Homes or a Monopoly Game?", this volume) points to the potential of transition management to contribute to anchoring of the novelty through the idea of 'reflecting'. The author states that comprehensive programme evaluation of niche implementation could capture the societal and environmental outcomes of the experimentation process, thus feeding back lessons. Also Lachman et al. (chapter "Context-Driven Transition Management as a Necessary Vehicle for Sustainable Urban Futures in Suriname", this volume) highlight that continuous monitoring and feedback on all levels of the transition are paramount in transition management interventions to learn about the progress of the transition and evaluate the objectives. However, the empirical applications show that thus far there seems to be too little attention to this as a concrete activity in transition management processes.

Transition Management to Support Orchestrating Capacities for *Coordinated Self-Organisation*

As stated in the introduction, sustainability transitions cannot be 'managed' or controlled, because of the complex and systemic interdependencies across problem domains and between problems and solutions. Rather, they can be realised by multiple societal actors who act in alignment towards common long-term goals by translating this transition perspective into their own operating context and concrete interventions (Loorbach 2014). A common orientation ensures that the diverse initiatives and actions are aligned and produce synergies between multiple priorities and goals across sectors and scales (ibid., Frantzeskaki et al. 2014). To promote such concerted action, transition management seeks to facilitate processes that align actors and networks under a common and long-term agenda, connect actors to mediate knowledge, conflict and resources and that establish boundary conditions for self-organisation and creation of synergies (Wittmayer 2016). For example, the transition agenda that is developed in the transition arena serves to orient the participants of the transition management intervention as well as of the emerging transition networks in their pursuit of an urban sustainability transition (Roorda et al. 2014; Loorbach 2010).

In the following, we review how the case studies describe the contribution of transition management to the coordination of self-organisation, including the strategic alignment of actors and networks, mediation of resources between actors and the creation of opportunity contexts for self-organisation. In all case studies, the transition team and transition arena manifested as settings that foster a sharing of knowledge, broadening of perspectives on problems and solutions and the development of a common, long-term orientation for a sustainability transition.

Strategic Alignment of Actors and Networks

Transition management creates space for actors to develop a common understanding of a transition challenge and a shared vision and agenda for navigating transitions towards sustainability (Wittmayer et al., chapter "Transition Management: Guiding Principles and Applications", this volume; Roorda et al. 2014). This serves to align the diverse efforts undertaken in transition networks towards a common long-term orientation (Loorbach 2014; Frantzeskaki et al. 2014). A critical element of such a shared system change perspective is the broadening of problem perceptions from singular issues and domains towards a systemic apprehension (Roorda et al. 2014; Frantzeskaki et al. 2012). A systemic long-term perspective on problems and solutions reveals that persistent unsustainability problems cannot be solved without recognising possibly detrimental repercussions on other domains and hence enables the creation of synergies across different problem domains, sectors and scales. It also provides shared reference points for concerted action that help diverse actors to relate to common broader ambitions and identify how they can contribute to a sustainability transition by connecting it to their specific interests (Hölscher et al. 2017b; Roorda et al. 2014). Following fundamental guiding principles of transition management like flexibility and learning, such a long-term orientation needs to be constantly reviewed (Loorbach et al. 2015).

All case study chapters highlight the development of a systemic, long-term and shared orientation for a sustainability transition as one of the most important contributions of transition management. For example, in La Botija, the transition agenda encompasses a comprehensive view on systemic problems and solutions that connect a variety of social, environment, economic domains (Silvestri and Frantzeskaki, chapter "Transition Management for Local Sustainability: A Case Study from La Botija Protected Area, San Marcos de Colón, Honduras", this volume). Participants were thus enabled to take more holistic and systemic thinking of the interconnection between diverse problems also using illustrations and systems maps. In La Botija, the ecosystem services framework was employed to systematise knowledge about resource use and ecosystem degradation and identify links between malpractices and ecosystems' deterioration. In Ludwigsburg, the initial view on energy has been opened up towards broader sustainability challenges through the systemic perspective (Hölscher and Wittmayer, chapter "A German Experience: The Challenges of Mediating 'Ideal-Type' Transition Management in Ludwigsburg", this volume). This integrated perspective on problems and solutions creates opportunities for connecting and aligning different actors with thus far individual agendas to a shared vision.

This broadening is achieved through the distinct process steps, such as the system analysis outlined by Maas et al. (chapter "Starting Up Transition Management: A Closer View on the Systems Analysis and How it Initiated Transformative Thinking in Ghent and Aberdeen Cities", this volume). The method facilitated an opening up of the view on problems and their interconnections from historical developments and current states of the system in Ghent and Aberdeen. For example, in Ghent the view on CO_2-mitigation was broadened towards considerations of social issues such as energy poverty. Also the heterogeneous arena composition supports a broadening of problems and goals, because it connects a variety of backgrounds and viewpoints (Hölscher and Wittmayer, chapter "A German Experience: The Challenges of Mediating 'Ideal-Type' Transition Management in Ludwigsburg", this volume).

The arena setting additionally creates a ‘safe’ space that fosters trust and promotes connections among the various participants. The openness of the participants towards each other promotes the willingness to align individual goals and cooperate for synergies (see also Hölscher et al. 2017b). Overall, the transition management interventions contributed to boosting new relationships between actors based on these new role definitions. In La Botija, Honduras, the participants forged new partnerships and started to self-organise in community groups. A highlighted learning outcome is the recognised need to develop own knowledge and community capacity to overcome local challenges deviating from the dependence on external aid (Silvestri and Frantzeskaki, chapter “Transition Management for Local Sustain ability: A case Study from La Botija Protected Area, San Marcos de Colón, Honduras”, this volume). In Ludwigsburg, transition management sparked cross-departmental collaboration within the Ludwigsburg city government, which has been important for developing and implementing co-beneficial projects in the future and pooling resources (Hölscher and Wittmayer, chapter “A German Experience: The Challenges of Mediating ‘Ideal-Type’ Transition Management in Ludwigsburg”, this volume).

The development of ownership over the process and goals seems important in this regard. In Ludwigsburg, the policy officers from the local government that led the implementation of transition management were not able to design a co-creative process setting (Hölscher and Wittmayer 2018, chapter “A German Experience: The Challenges of Mediating ‘Ideal-Type’ Transition Management in Ludwigs burg”, this volume). The participants, viewing the process as providing ‘input’ to the city government’s strategies, were hence not able to develop ownership. Similarly, this process framing did not spark a reflection on and (re-)definition of roles and responsibilities of the actors involved in the process, because the city government was regarded as main actor responsible to pursue the envisioned transition.

This strategic alignment that integrates and creates synergies and reveals trade-offs between diverse goals in the context of a systemic perspective on problems and solutions is also expected to be one of the key contributions of transition management in the exploratory case studies. Walsh (chapter “Translating Transitions’ Thinking and Transition Management into the City Planning World”, this volume) emphasises that transition management can contribute to integrate a systemic perspective focused explicitly on sustainability into urban planning. Interestingly, while referring to the contributions of conceptual models and frameworks like the multi-level perspective and multi-phase model for opening up avenues for radical change, he highlights that these abstract and complex notions need to be clearly translated into the city planning world to have an impact. Lachman et al. (chapter “Context-Driven Transition Man agement as a Necessary Vehicle for Sustainable Urban Futures in Suriname”, this volume) point out that developing countries’ context require more attention to the very rigid regime configurations by ensuring cohesion of the transition arena, continuous reflection and selection pressures on regimes. For example, it needs to be ensured that those within the arena with higher status will not overwhelm others. The authors suggest additional assessment tools such as Quinn’s culture model to guide the alignment processes and tailor the transition management intervention to suit the context, for exampling by adapting the communication practices.

While the case studies imply that transition management successfully achieves an alignment of actors that participate in the initial process phases of the transition team and transition arena, insights on how the framework contributes to a broader alignment in (constantly growing) transition networks are limited- for example, on whether the strategic vision gets connected to on-going policy goals and processes. This refers to lacking insights on the long-term outcomes of transition management interventions after the initial process phases as well as to limited anchoring of the new vision (due to lacking transformative capacity). Additionally, Jhagroe (chapter "Urban Transition Management as a Democratic Practice? The Case of Rotterdam's Waterfront Regeneration", this volume) highlights the need to critically reflect on who is involved in the processes and in broader transition networks that define and frame such an agenda. He criticises that only a group of already privileged actors took part in the process and in the later implementation. The agenda, hence manifesting a view on 'eco-gentrification', was not able to reach out to and align with more socially disadvantaged community actors.

Mediating Knowledge and Resources

In transition management processes actors can connect to each other, share knowledge and strike up collaborations and partnerships in the implementation of transition experiments (Roorda et al. 2014; Roorda and Wittmayer 2014; Loorbach 2010). The different actor settings – transition teams, arenas and networks – stimulate the mediation of knowledge and resources by creating a channel to express ideas and concerns, building trust, enabling integration between different institutions, city departments and policies, and pooling diverse resources for mutual benefits and coherent action (Frantzeskaki et al. 2014; Wittmayer 2016). While transition management provides rather clear guidelines on how such a mediation process can be organised in the transition team and transition arena settings (Wittmayer et al., chapter "Transition Management: Guiding Principles and Appli cations", this volume), there is no prescribed structuration of how to mediate between and across emerging transition networks, i.e. of how to mediate the longer-term transition process (Loorbach et al. 2015).

The development of new connections and partnerships both in the transition arena and beyond emerges as one of the most valuable outcomes of the transition management processes in some of the case studies. This is evident from the very beginning of the process: In Ghent and Aberdeen, for example, the system analysis and problem definition benefits from the various perspectives that open-up the view on the initially framed problems (in terms of energy) (Maas et al., chapter "Starting Up Transition Management: A Closer View on the Systems Analysis and How It Initiated Transformative Thinking in Ghent and Aberdeen Cities", this volume).

The different process steps of the transition arena showed to facilitate the further exchange of knowledge. This ignites a diversity of new collaborations and partnerships for the pursuit of the envisioned sustainability transition. The cross-departmental working group energy in Ludwigsburg is one example of such a new collaboration that serves to establish synergies between different goals and a pooling of resources across sectors and departments (Hölscher and Wittmayer, chapter "A German Experience: The Challenges of Mediating 'Ideal-Type' Transi tion Management in Ludwigsburg", this volume). In Rotterdam, strategic and operational partnerships emerged that coordinate and implement projects such as the Floating Pavilion by sharing knowledge and resources (Jhagroe, chapter "Urban Transition Management as a Democratic Practice? The Case of Rotterdam's Water front Regeneration", this volume).

Especially in view of the different types of actors and hence interests that are involved in transition processes, the creation of trust and cooperation is critical to stimulate the exchange of knowledge and resources (Loorbach 2014; Hölscher et al., chapter "Empowering Actors in Transition Management in and for Cities", this volume). The case of Ludwigsburg reveals some challenges relating to the necessary openness to be critical about existing problems, especially on the parts of the city government (Hölscher and Wittmayer, chapter "A German Experience: The Challenges of Mediating 'Ideal-Type' Transition Management in Ludwigsburg", this volume).

Similar to what we described in terms of the strategic alignment of actors, there is limited discussion of the on-going mediation of knowledge and resources in the later emerging transition networks. The outreach to broader groups seems to be limited (Jhagroe, chapter "Urban Transition Management as a Democratic Prac tice? The Case of Rotterdam's Waterfront Regeneration", this volume; Hölscher and Wittmayer, chapter "A German Experience: The Challenges of Mediating 'Ideal-Type' Transition Management in Ludwigsburg", this volume). Additionally, in many cases the involved actors from the city government seem to remain a central connection node for mobilising, accessing and mediating knowledge and (financial) resources. The limited involvement of key decision-makers in Carnisse, which resulted in a feeling of limited leverage of the arena participants (community actors), indicates that there is a need to mediate across diverse actors and resource bases (Wittmayer et al., chapter "Transition Management in Urban Neighbourhoods: The Case of Carnisse, Rotterdam, the Netherlands", this volume).

Similarly, the exploratory case studies indicate the value of transition management to establish connections across different actors and networks, which enables mobilising different power sources and capitals. Mai et al. (chapter "Transition Management in Urban China: Co-creation of New Homes or a Monopoly Game?", this volume) discusses on the contribution of transition management to address the disequilibria between the centralised state apparatus in China seeking to influence local transition dynamics top-down and localised urban governance systems that search for and experiment with innovative practices. In this sense, transition management could promote co-visioning across disequilibria by involving and connecting a variety of actors and resources. Lachman et al. (chapter "Context-Driven Transition

Management as a Necessary Vehicle for Sustainable Urban Futures in Suriname", this volume) highlight open, honest, clear and transparent communication at all levels as a critical condition for transition management interventions to ensure that actors can build trust and the willingness to cooperate.

Creating Opportunity Contexts for Synergies

Self-organisation requires opportunity contexts for facilitating the implementation of activities under the common agenda and the creation of synergies. Empirical insights from transition management processes show that especially in the longer term there is a need for political, financial and organisational contexts in which self-organised actors can develop and implement concrete initiatives and projects that collectively contribute to a sustainability vision and create synergies. For instance, actors need reference points on how to mobilise and combine resources for action and on how costs and benefits are distributed (Frantzeskaki et al. 2014). Transition management literature thus far does not specify when, how and by whom such context conditions are formulated. This again refers back to the lack of focus on the 'beyond the arena' processes as well as to the argued openness of what the concrete results of transition management can be. With regard to the latter, transition management does not seek to prescribe institutional changes but advocates a rather open-ended processes without pre-determined outcomes (Roorda et al. 2014).

While the established strategies serve to create an overarching orientation for coordinated action and for connecting a variety of different goals, the case studies do not explicitly discuss the creation of such opportunity contexts during or after the transition management process. One reason might be that there is so far limited anchoring of the vision into the cities' context, which would create a more institutionalised basis for seeking synergies. Another reason relates to the limited insights on the longer-term results of the transition management interventions. Overall, the question of resource availability is underlying the follow-up and 'activation phase' of transition management, which impedes the abilities to formulate new organisational and institutional context conditions. In Ludwigsburg, the follow-up hinged on the city government as the other participants were largely not willing and able to invest (more) time and financial resources into concrete activities (Hölscher and Wittmayer, chapter "A German Experience: The Challenges of Mediating 'Ideal-Type' Transition Management in Ludwigsburg", this volume).

It is evident from the case studies, however, that when experiments were implemented that created synergies between different goals and objectives, this was enabled because of ad hoc actor coalitions who pooled their different resources. Overall, we can conclude that there is a need to further advance thinking on what synergistic framework conditions could enable the implementation of the long-term vision in the short- and medium-term through activities that create multiple benefits but also trade-offs between different goals. Key questions in that regard are who should be expected to take up what kind of actions, who benefits and who bears the costs.

Concluding Remarks: Lessons for Transition Management as 'Transformative' Governance Approach in Cities

This book introduced and reflected on transition management as a new governance approach in and for cities. Transition management is explicitly oriented towards enabling actors to influence sustainability transitions. It builds on principles for governance that embrace complexity, uncertainty, radical change, experimentation, reflexivity and learning.

Our review of the case studies presented in this book shows that transition management is indeed apt for supporting new types of governance capacities for influencing sustainability transitions in cities. The case studies indicate that transition management supports urban governance capacities with a particular emphasis on enabling and coordinating transformative change towards sustainability. In particular, transition management supports processes of novelty creation (e.g. ideas, strategies, practices, actor relations, governance approaches) that provide seeds for more systemic change and for mobilising and aligning diverse actors and goals under a systemic and long-term orientation for a sustainability transition. Transition management also support the mediation of knowledge and resources between actors in the transition team, transition arena and in transition networks. Furthermore, new and shared strategic directions were in many cities more widely disseminated and connected to on-going processes and strategies. More indirectly these results of transition management ignite new role relationships between actors, reflection on existing governance approaches and institutional changes. These results further contribute to enabling cities to address sustainability challenges more comprehensively from systems perspectives, to identify synergies between different policy domains and to collaborate with actors across departments, sectors and societal spheres.

When implementing transition management to support these new types of governance capacities, different process design elements and tenets come to fruition. When devising transition management intervention it is important to keep these in mind.

- Creation of 'safe spaces' for communication and collaboration: Transition management creates different settings (e.g. transition team, transition arena) in which diverse actors can come together outside of their business-as-usual spaces to exchange ideas and perspectives and think creatively about problems and possible solutions while dreaming of a 'world they want'. This sparks new ideas, paradigms and practical interventions and contributes to an alignment of all these under a common sustainability umbrella. It also facilitates the

development of new forms of collaboration between diverse actors. It is critical to achieve an open setting that builds on trust between the members of the group.
- Systemic perspectives on problems and solutions: Through different process principles (e.g. systemic problem analysis, heterogeneous actor group, long-term visioning) transition management ignites systemic perspectives that connect different policy goals and problem domains with each other and reveal potential trade-offs and synergies. This contributes to the alignment of actors with different goals and it enables integrated approaches to tackling interconnected problems - such as climate change adaptation and social equity.
- Reflexivity and learning-by-doing: By going through the process steps suggested by transition management the participants embark on new learning trajectories that stimulate them to critically rethink their prior assumptions. On the one hand, transition management allows for trialling and experimenting new ideas without the fear of failure. This supports creative thinking and innovation. On the other hand, by working with diverse actors and revisiting existing problem frames and explicit and implicit desires for the future, also assumptions of existing actor roles and relationships are questioned. Transition management aids critical reflection on how everybody can contribute to achieving a desirable future.

The case studies in this book also reveal several gaps and points of attention for advancing the conceptualisation and operationalisation of the transition management framework to further contribute to building governance capacities. These largely require a more longer-term focus on the long-term follow-up of the initial transition management intervention. For example, this includes the positioning of the transition management intervention and its results within and their connection to ongoing policy and planning processes. Thus, we highlight as key challenge for transition management the connection of the novel visions, strategic agendas, initiatives and networks to on-going processes in the cities. This creates a need for activities that are yet to be made explicit within the transition management framework. For example, there is a need for activities that ensure broader outreach and mainstreaming of the outcomes of the transition management intervention to sustain longer-term momentum, involve growing actor networks and embed the novelties into institutional and organizational structures and processes. The case studies provide important insights on what types of activities could be relevant here, such as the lobbying for political support or the institutionalisation of a cross-departmental transition team within the city department. A key challenge is to ensure the financial basis and time availability for any follow-up activities. Additionally, in the spirit of reflexivity and learning that underpins transition management as a whole, the long-term follow-up requires continuous reflection on strategies, processes and goals, such as who is involved in defining the visions and how radical and effective the visions and solutions are for contributing to sustainability transitions. This also includes learning of what works and what does not in a specific context, and what are the implications of these insights for planning processes and concrete solutions. The diversity of case studies presented in this book can be instructive here–for example, there are different political cultures across regions and countries that might influence how a transition management intervention needs to be designed.

On the one hand, this book has introduced a step-by-step approach to how transition management can be implemented in cities and illustrated how these steps become translated in diverse case examples. On the other hand, it reflected on the implications and contributions of transition management for governance of sustainability transitions in cities. Ultimately, both go hand-in-hand: implementing transition management requires the creation of new types of governance capacities that enable flexibility, learning, collaboration and coordination. If implemented veritably, transition management can support new capacities by means of long-term and integrated visions and strategies, diverse and diffuse actor networks and innovative solution approaches including experimentation.

References

Adil AM, Ko Y (2016) Socio-technical evolution of decentralized energy systems: a critical review and implications for urban planning and policy. Renew Sust Energ Rev 57:1025–1037. https://doi.org/10.1016/j.rser.2015.12.079

Bos JJ, Brown RR (2012) Governance experimentation and factors of success in socio-technical transitions in the urban water sector. Technol Forecast Soc Chang 79(7):1340–1353

Castán Broto V, Bulkeley H (2013) A survey of urban climate change experiments in 100 cities. Glob Environ Chang 23(2013):92–102

Frantzeskaki F, de Haan H (2009) Transitions: two steps from theory to policy. Futures 41 (9):593–606. https://doi.org/10.1016/j.futures.2009.04.009

Frantzeskaki N, Loorbach D, Meadowcroft J (2012) Governing societal transitions to sustainability. Int J Sustain Dev 15(1):19–36

Frantzeskaki N, Bach M, Hölscher K, Wittmayer J (2014). Applications of transition management: taking stock of 13 years of transition management and lessons for future research, 5th International Sustainability Transitions (IST) Conference, 27–29 August 2014, Utrecht, The Netherlands

Frantzeskaki N, Bach M, Mungi P (2018a) Understanding the urban context and its challenges. In: Frantzeskaki N, Hölscher K, Bach M, Avelino F (eds) Co-creating sustainable urban futures. A primer on applying transition management in cities. Springer, Cham

Frantzeskaki N, Hölscher K, Bach M, Avelino F (2018b) Introduction to sustainability transitions thinking. In: Frantzeskaki N, Hölscher K, Bach M, Avelino F (eds) Co-creating sustainable urban futures. A primer on applying transition management in cities. Springer, Cham

Friend R, Jarvie J, Orleans Reed S, Sutarto R, Thinphanga P, Canh Toan V (2014) Mainstreaming urban climate resilience into policy and planning; reflections from Asia. Urban Clim 7:6–19. https://doi.org/10.1016/j.uclim.2013.08.001

Grimm N, Faeth S, Golubiewski N, Redman C, Wu J, Bai X, Briggs J (2008) Global change and the ecology of cities. Science 319:756–760

Grin J, Rotmans J, Schot J (2010) Transitions to sustainable development: new directions in the study of long term transformative change. Routledge, New York

Hölscher K, Wittmayer J (2018) A German experience: the challenges of adapting 'ideal-type' transition management in Ludwigsburg. In: Frantzeskaki N, Hölscher K, Bach M, Avelino F (eds) Co-creating sustainable urban futures. A primer on applying transition management in cities. Springer, Cham

Hölscher K, Wittmayer JM, Maschmeyer S, Frantzeskaki N (2017a). Transition Management als Meta-Governance-Rahmenwerk zur Gestaltung von Nachhaltigkeitstransitionen. Analyse von Governance- Kapazitäten durch Transition-Management-Interventionen in Gent und Montreuil. In: Kleinfeld R, Hafkesbrink J, Stuhldreier J (Eds.). Innovatives Regionalmanagement im demografischen Wandel. Springer, Berlin

Hölscher K, Wittmayer JM, Avelino F, Giezen M (2017b). Opening up the transition arena: an analysis of (dis)empowerment of civil society actors in transition management in cities. Technol Forecast Soc Chang (online first). https://doi.org/10.1016/j.techfore.2017.05.004

Jhagroe S (2018) Urban transition management as democratic practice? The case of Rotterdam's waterfront regeneration. In: Frantzeskaki N, Hölscher K, Bach M. and Avelino F (eds) Co-creating sustainable urban futures. A primer on applying transition management in cities. Springer, Cham

Kemp R, Loorbach D, Rotmans J (2007) Transition management as a model for managing processes of co-evolution towards sustainable development. Int J Sust Dev World 14(1):78–91

Lachman DA, Panday MRJ, Ferrier DJH (2018) Context-driven transition management as a necessary vehicle for sustainable urban futures in Suriname. In: Frantzeskaki N, Hölscher K, Bach M, Avelino F (eds) Co-creating sustainable urban futures. A primer on applying transition management in cities. Springer, Cham

Leichenko, R. (2011). Climate change and urban resilience. Curr Opin Environ Sustain, 3(3), 164–168, 10.1016/j.cosust.2010.12.015

Loorbach D (2007) Transition management. New mode of governance for sustainable development. Erasmus University, Rotterdam

Loorbach D (2010) Transition management for sustainable development: a prescriptive, complexity-based governance framework. Governance 23(1):161–183

Loorbach D (2014) To Transition! Governance Panarchy in the new transformation. Inaugural lecture. Erasmus University of Rotterdam, Rotterdam

Loorbach D, Frantzeskaki N, Huffenreuter L (2015) Transition management: taking stock from governance experimentation. J Corp Citizenship 58:48–66

Maas S, Fortuin K, Frantzeskaki N, Roorda C (2018) Starting up transition management: a closer view on the systems analysis and how it initiated transformative thinking in Ghent and Aberdeen cities. In: Frantzeskaki N, Hölscher K, Bach M, Avelino F (eds) Co-creating sustainable urban futures. A primer on applying transition management in cities. Springer, Cham

Mai Q (2018) Transition management in urban China: co-creation of new homes or a monopoly game? In: Frantzeskaki N, Hölscher K, Bach M, Avelino F (eds) Co-creating sustainable urban futures. A primer on applying transition management in cities. Springer, Cham

Malekpour S, Brown RR, de Haan FJ (2015) Strategic planning of urban infrastructure for environmental sustainability: understanding the past to intervene for the future. Cities 46:67–75. https://doi.org/10.1016/j.cities.2015.05.003

McCormick K, Anderberg S, Coenen L, Neij L (2013) Advancing sustainable urban transformation. J Clean Prod 50:1–11

Nevens F, Frantzeskaki N, Loorbach D, Gorissen L (2013) Urban transition labs: co-creating transformative action for sustainable cities. J Clean Prod 50:111–122

Rink D, Banzhaf E, Kabisch S, Krellenberg K (2015) Von der "Großen Transformation" zu urbanen Transformationen. Zum WBGU Hauptgutachten Welt im Wandel. GAIA Ecol Perspect Sci Soc 24(1):21–25

Roorda C, Wittmayer J (2014) Transition management in five European cities – an evaluation. DRIFT, Erasmus University Rotterdam, Rotterdam

Roorda C, Wittmayer J, Henneman P, van Steenbergen F, Frantzeskaki N, Loorbach D (2014) Transition management in the urban context. Guidance manual. Drift, Erasmus University Rotterdam, Rotterdam

Rosenzweig C, Solecki WD, Hammer SA, Mehrotra S (2011). Climate change and cities: first assessment report of the urban climate change research network (ed). Cambridge University Press, Cambridge

Rotmans J, Loorbach D (2010) Towards a better understanding of transitions and their governance: a systemic and reflexive approach. In: Grin J, Rotmans J, Schot J (eds) Transitions to sustainable development. New directions in the study of long term transformative change. Routledge, New York, pp 105–220

Rotmans J, Kemp R, van Asselt M (2001) More evolution than revolution: transition management in public policy. Foresight 3(1):15–31

Schwarz T (2018) A transition management approach for shrinking cities in the US. In: Frantzeskaki N, Hölscher K, Bach M, Avelino F (eds) Co-creating sustainable urban futures. A primer on applying transition management in cities. Springer, Cham

Shaw A, Burch S, Kristensen F, Robinson J, Dale A (2014) Accelerating the sustainability transition: exploring synergies between adaptation and mitigation in British Columbian communities. Glob Environ Chang 25:41–51

Silvestri G, Frantzeskaki N (2018) Transition management for local sustainability. A case study from La Btoija projected area, San Marcos de Colón, Honduras. In: Frantzeskaki N, Hölscher K, Bach M, Avelino F (eds) Co-creating sustainable urban futures. A primer on applying transition management in cities. Springer, Cham

Van den Bosch SJM (2010) Transition experiments: exploring societal changes towards sustainability. Erasmus University, Rotterdam. http://hdl.handle.net/1765/20714

Walsh P (2018) Translating transitions thinking into the city planning world In: Frantzeskaki N, Hölscher K, Bach M, Avelino F (eds) Co-creating sustainable urban futures. A primer on applying transition management in cities. Springer, Cham

Wittmayer, J.M. (2016). Insights and lessons for the governance of urban sustainability transitions. Loorbach, D., Wittmayer, J., Shiroyama, H., Fujino, J. and Mizuguchi, S.. Governance of urban sustainability transitions. European and Asian experiences. Springer: Tokyo/Heidelberg/New York/Dordrecht/London: 157–173

Wittmayer JM, Loorbach D (2016) Governing transitions in cities: fostering alternative ideas, practices, and social relations through transition management. In: Loorbach D, Wittmayer J, Shiroyama H, Fujino J, Mizuguchi S (eds) Governance of urban sustainability transitions. European and Asian experiences. Springer, Tokyo/Heidelberg/New York/Dordrecht/London, pp 13–32

Wittmayer JM, van Steenbergen F, Frantzeskaki N, Bach M. (2018a) Transition management: guiding principles and applications. In: Frantzeskaki N, Hölscher K, Bach M, Avelino F (eds) Co-creating sustainable urban futures. A primer on applying transition management in cities. Springer, Cham

Wittmayer JM, van Steenbergen F, Bach M (2018b) Sustainability transitions in urban neighbourhoods: Carnisse. In: Frantzeskaki N, Hölscher K, Bach M, Avelino F (eds) Co-creating sustainable urban futures. A primer on applying transition management in cities. Springer, Cham

Wolfram M, Frantzeskaki N (2016) Cities and systemic change for sustainability: prevailing epistemologies and an emerging research agenda. Sustainability 8(2):144. https://doi.org/10.3390/su8020144

Appendix: Exercises and Exam Questions for Lecturers Using This Book

Niki Frantzeskaki, Giorgia Silvestri, Matthew Bach, Katharina Hölscher, Chris Roorda, Rick Bosman, Flor Avelino, Julia Wittmayer, Roel van Raak, Marieke Verhagen, Derk Loorbach, Frank van Steenbergen, and Marijke de Pous

Abstract This appendix provide exercises, exam questions, group assignments and briefs on dialogue assignments that have been developed by the transition management research team of the Dutch Research Institute For Transitions (DRIFT) at Erasmus University Rotterdam, the Netherlands (www.drift.eur.nl) in more than 15 years of experience. The group assignments and exercises have been effectively used in lectures, trainings, courses and workshop sessions of transition management and urban sustainability transitions.

Introduction

This Appendix contains exercises, exam questions, group assignments and briefs on dialogue assignments that have been developed by the Transition Management Research team of the Dutch Research Institute For Transitions (DRIFT) at Erasmus University Rotterdam, the Netherlands (www.drift.eur.nl) in more than 15 years of experience. The group assignments and exercises have been effectively used in lectures, trainings, courses and workshop sessions of transition management and urban sustainability transitions.

N. Frantzeskaki · G. Silvestri · M. Bach · K. Hölscher · C. Roorda · R. Bosman · F. Avelino · J. Wittmayer · R. van Raak · M. Verhagen · D. Loorbach · F. van Steenbergen · M. de Pous
Dutch Research Institute For Transitions (DRIFT), Faculty of Social Sciences, Erasmus University Rotterdam, Rotterdam, The Netherlands
e-mail: frantzeskaki@drift.eur.nl; silvestri@drift.eur.nl; bach@drift.eur.nl; holscher@drift.eur.nl; roorda@drift.eur.nl; bosman@drift.eur.nl; avelino@drift.eur.nl; wittmayer@drift.eur.nl; vanraak@drift.eur.nl; verhagen@drift.eur.nl; loorbach@drift.eur.nl; vansteenbergen@drift.eur.nl; depous@drift.eur.nl

N. Frantzeskaki et al. (eds.), *Co-creating Sustainable Urban Futures*, Future City 11, https://doi.org/10.1007/978-3-319-69273-9

Questions to Make Lectures Interactive

- What are the four main phases of the Transition Management cycle? Elaborate on the characteristics of every phase and the activities taking place in every phase.
- What makes transition management different on
  - Neighborhood and city level?
  - Developed/undeveloped?
- Could you describe a historical transition and
  - Analyse the governance theory from a transition management perspective;
  - Name three characteristics of a transition.
- When do you recommend using transition management to a policy maker? Why?
- What is the role of cities in Sustainability Transition?
- Why would it be important to look at cities?
- What are the skills that change agents need to advance transition management processes in cities?
- What are strengths and weakness of transition management according to you?
- What are additional insights you have learned from transition management case studies to improve transition management theory?
- What are the similarities and differences of the Transition Management Cycle with the Policy Cycle (or Policy Process Cycle)?
- What are three main distinctions that can be drawn between transition management and regular policy processes? (+ examples)
- Which are the differences/relations between transition management and strategic planning/environmental governance?
- Give examples of a window of opportunity for institutional change.
- What role(s) can a policy maker have in a transition management processes?

Exam Questions

- What are the main phases of the Transition Management Cycle? Elaborate on the characteristics of every phase and the activities taking place in every phase.
- Fit of transition management to your research/case:

Please provide a brief description of your case study and discuss how you see transition management to fit it. You can address the following exemplary guiding questions:

- What are the boundaries of your case study?
- What are the (characteristics of the) main problems/challenges in your case study?
- What are opportunities for change in your case study?
- Why do you think transition management could benefit your case study? What is your rationale for applying transition management?

- How does transition management support addressing the problems/challenges?

  Briefly describe by reflecting on the guiding principles and characteristics of transition management how supports the problem-solving process in your case study. The following questions provide some guidance:

  - How does transition management help to address the characteristics of the problem?
  - What 'promised outcomes' of transition management relate to addressing the problem in your case?
  - What are (the limitations of) existing governance and planning processes and how do they need support?
  - What are potential limitations in applying transition management to your case?

- Please reflect on where you see potential limitations for applying transition management to your case. For example:

  - How does the scope of the problem relate to the transition management process?
  - How does transition management relate to on-going planning processes and incumbent actors/interests?

- How does transition management need to be adapted to be implemented in your case?

  Transition management requires reflexivity and tailoring to the specific context it is applied to. By reflecting on the guiding principles and process steps, briefly describe how you would adapt transition management to fit it to your case, including (but not limited to):

  - What are specific methods/process steps you would like to take up and how?
  - Who are the actors you would (not) like to involve, and why (not)?
  - What are preparatory and implementation steps?

- How are sustainability values represented and practiced in the Transition Management tools (transition arena or transition experiments)? Reflect upon the lessons learnt during the working sessions of the current module and the readings of transition experiments, transition scenarios and the transition arena.

- What is a transition scenario and how does it differ from other scenarios? What is a transition experiment and how does it differ from regular innovation projects? What is the relation between a transition scenario and a transition experiment? Illustrate your answers with an example.
- There is a paradox in the term "transition management", i.e. a tension between the concept of 'transition' and the concept of 'management'. Can you explain this tension? How does the transition management approach deal with this tension?
- In the transition studies we come across the paradoxical concept of radical change in incremental steps. Could you explain the concept and the relevance for the acceleration of a Transition?
- What are the characteristics of a sustainability transition?
- What is the role of different actors in transition processes? Give examples of role of state, market, community, etc.
- In what sense is the transition arena process different from regular participative processes? Name at least three differences.
- What are the main challenges in your area in urban (ground) water management (approx. 50–500 words)? Please think of issues related to:

  – Infrastructure and technology (e.g. water and sanitation infrastructures, etc.);
  – Environmental quality (e.g. water quality, soil quality, etc.);
  – Social, political and cultural aspects (e.g. value of water, access to water, community organization);
  – Governance and institutions (e.g. competences and policies, levels of collaborations).

  And think of big, fundamental questions for the next 5, 10 or 20 years.
- Please provide a demarcation of the physical, social and institutional system (approx. 50–250 words). Where are the boundaries of, amongst others:

  – Geographic boundaries;
  – Administrative boundaries;
  – Community/cultural boundaries;
  – Sectorial boundaries (e.g. boundaries between (ground) water system and other sectorial systems such as agricultural system, waste management system, etc.);
  – Time: how long to look back and forward?

- Please provide a short description of the structures that are the organisation and arrangement of the system in your context. For example what are the different governance approaches, institutional frameworks and policies related to the system in the context? What is the hydrological situation? Which infrastructure/networks/pumps are in place? Etc.

- Please provide a short description of the cultures that are the norms and values related to the system. For example: How do people perceive the quality of water in their daily life? Do people consider water quality as an important issue? Are people aware of sustainable water uses? Etc.
- Please provide a short description of the practices that are the choices, behaviors and habits related to of the system. For example: How is water used by residents? What kind of water sources people use? What are the behaviors related to promotion and maintenance of water quality? Etc.
- Change often starts on a small scale: individuals or groups doing things differently or using different technology. What are emerging alternatives (niches) promoting a transformative change of the system? Are there groups of organized citizens mobilizing for improving water quality (e.g. water committees, active community groups, local initiatives, etc.)?
- The regime refers to all dominant practices, rules and technologies providing stability and reinforcement to the prevailing socio-technical systems. The regimes can be characterised along different dimensions such as technology; user practices and application domains; physical infrastructures; policies; and culture and knowledge. For characterizing the regime the following questions can be asked: who holds the system in place? What institutions provides stability? What are the physical infrastructures of the system?
- Not only 'niches' but also changes in the broader context do have consequences for transitions. Please, list major, important, large scale changes happening now and are expected over the next 5–15 years that will affect the system? Think about physical, technological, social, economic and governance changes.
- Please list the main actors on a system level in the table below. Think about actors who would and/or could support a transition in different ways or who would stand in the way of a transition.

	Actor name	Role and main activities
Governmental actors (e.g. International and national institutes and departments)		
Business		
Research institutes and research projects		
NGOs, international aid programs, etc.		
Local initiatives		
Other relevant actors		

Exam Exercises

Transition Management and Complexity

During the course we have discussed the features of complex systems and how transition management is geared to deal with that.

(a) What are the four characteristics of complexity in complex systems?
(b) How is transition management geared to deal with complexity? It could help to refer to the principles of transition management in answering this question.
 Mention at least five principles.
(c) What are the two characteristics of transition experiments?
(d) Why would someone make use of transition experiments?
 Name at least three reasons.
(a) Mention

 a. society is not malleable
 b. effects & causes = uncertain & unpredictable
 c. societal change can not be planned
 d. we are part of the system under change

(b) Mention transition management principles and relate these to complexity

 a. long-term thinking as the basis for short term actions
 - relate to: envisioning/backcasting
 b. thinking in terms of multiple domains (multi-domain), different actors (multi-actor), different levels (multi-level)
 - relate to: system's analysis/ participants from different backgrounds, domains and levels (to generate diversity -> higher level of adaptation/ versatility)
 c. learning as an important aim in itself ('learning-by-doing' and 'doing-by-learning')
 - relate to: confronting different perspectives, expectations and agendas/ setting up experiments
 d. orient governance towards system innovation besides system improvement
 - relate to: radical experiments / out of the box thinking
 e. keeping options open, exploring multiple pathways
 - relate to: transition agenda, pathways, experiments
 f. selective participation focusing on frontrunners
 - relate to: actor analysis, participative process, change-agents/frontrunners

(c) Mention short term actions in line with arena narrative/sufficiently specific to be feasible AND enough radical to be symbolic for shift in structure, mind-set and practices of envisioned future.

System Analysis

Exercise 1

In class you have made a system's analysis.

(a) Explain the aims of such an analysis.
(b) Then explain what the three (possible) system boundaries are.
(c) Finally, choose your own functional societal system and demarcate it using at least two of these boundaries.

Mention: understanding of the current situation, long term change dynamics (how did the system evolve as it did?) and interaction between multiple domains. Distinguish between symptoms and deep-rooted problems, shifting focus from superficial solutions to systemic challenges. It encourages holistic thinking, supporting those involved to look beyond their own expertise, questioning their beliefs and values, making (the possibility of) different perspectives explicit.

Mention: geographical, temporal and thematic

- Define a system by its societal function
- Mention two out of three boundaries
- Demarcate your system using these boundaries

Exercise 2

In class you have made a systems analysis. Explain the system's analysis covering:

(a) Three lenses that could be used to make a systems analysis
(b) A concise systems analysis of the system 'sustainable housing in the Netherlands' using these three lenses.
(Use a minimum of 300 and maximum of 400 words).

Different lenses are:

- Historical sketch (macro)
- Landscape influences (macro)
- Description of stocks (meso)
- Relations between stocks (meso)
- Selection of main themes (meso)
- Trend analysis (meso)
- Dominant structure, culture and practices (meso)

- Identification of niches (micro)

Use the four different steps: system definition, system structuring, data gathering and data assessment & analysis.

Identifying Persistent Problems

There are several persistent problems related to the energy system

(a) mention two such persistent problems
(b) explain why these problems are considered persistent

(a) Environmental problems: pollution, earth quakes, increasing greenhouse effects
Health problems: e.g. respiratory diseases
Fossil fuels are limited and depleting
Monopoly position for fossil fuel companies – limits choice, transparancy and democracy
(Geopolitical) dependencies on foreign fossil fuels and instable regimes
(b) Persistent problems are ill-structured, deeply entrenched in societal structures, involve many stakeholders, that can have different perceptions of the problem at hand, making the problem extremely complex and contested.

From your answer it should become clear that you are not discussing symptoms (which is often the case in policy debates etc.) but the core underlying problems

Actor Analysis

In class we discussed the participants of the arena.

(a) Explain why actors are invited as individuals rather than as representatives. Mention at least two aspects.
(b) To ensure diversity of participants within the arena, which different characteristics would you look for among the actors? Mention at least three.
(c) Mention the five factors defining a relevant actor for a transition management process (transition arena).
(d) Why is it important to select participants with different backgrounds, competencies and interest profiles for your transition arena?

(a) Actors are approached as individuals and not as representatives of the organisations they work at, in order for actors to feel free to think out of the box and innovate.

By attending the arena as a person instead of representative, it is possible to let go of the restraints and protecting of interests that come with representing an organisation that has a stake in the changes at hand.

(b) Power: innovative, reinforcing, transformative.

Knowledge: practical, academic, policy, local, national etc.

Gender, Age, geographical background (where do you live), cultural background (where do you come from), possible income groups (what do you earn)

(c) Relevant are those actors:

- Whose interests are affected by the issue or those whose activities strongly affect the issue;
- Who have the potential to affect the issue but do not do so at present ('unexpected actors')
- Who possess information, resources and expertise needed for strategy formulation and implementation;
- Who control relevant implementation instruments;
- With new innovative ideas and initiatives regarding the issue.

Diversity in the arena is an important indicator for an innovative outcome. The more heterogeneous the group, the more social learning can take place and interesting ideas develop that could resonate with a large group of people outside the arena and it provides access to different networks.

The Role of the Frontrunners

During the course we have discussed the transition arena and the role of frontrunners.

(a) What are the four differences between a regular policy arena and a transition arena?

(b) For frontrunners, mention:

- Their position (for example in a regime or niche)
- What might drive them

(a) Differences between a regular policy arena and a transition arena:

- How they differ from regular stakeholders

Regular policy arena	Transition arena
Short term	Long term
Peloton	Frontrunners
Incremental change	System innovation
Problem- and goal oriented	Problem- and goal searching

(b) Frontrunners:

Position	Frontrunners can be innovative niche players but also change minded regime players
Drive	Driven by courage, vision and/ or frustration
Difference regular stakeholders	Not representatives; invited on personal & individual basis. So they are not stakeholders and not the usual suspects

Organisation of a Transition Arena

Transition Arena process: Imagine you are facilitating an arena together with other members of the transition team. Describe in a short essay (minimum of 20 lines) how this arena would look like.

Use the following aspects in your essay:

- What the subject is of the arena
- What the problem is that you are going to solve with this arena
- Which elements in the process will be your responsibility and why

During the course we have discussed the transition arena process elaborately.

(a) Explain the arena process and elaborate on the different phases.
(b) Then explain how an arena could contribute to (the emergence of) a transition. It could help to refer back to the principles of transition management in answering the second part.

  (a) Mention purpose, composition, set-up

    (i) Create safe-space in which 10–15 frontrunners from different backgrounds confront each other with their different perspectives, expectations and agendas, developing long term sustainability visions that inform short term actions.

    Mention seven phases
    Explain what happens in each phase (and why)

	Arena process phase	Process steps
I	Setting the scene for Transition Management	The Transition Team is formed to drive the TM process and start to explore the project dynamics (takes about 3–6 months)
II	Exploring dynamics in your system	
III	Framing the transition challenge	The Transition Arena group is formed and engages in a series of meetings to jointly structure the transition challenge, draft visionary images and develop transition paths and a transition agenda. (takes about 6–12 months)
IV	Envisioning a sustainable system	
V	Reconnecting long term & short term	
VI	Getting into action	Effort is put into disseminating the vision and transition agenda; initiating actions in line with the vision (transition experiments); and engaging more actors and networks
VII	Engaging & anchoring	

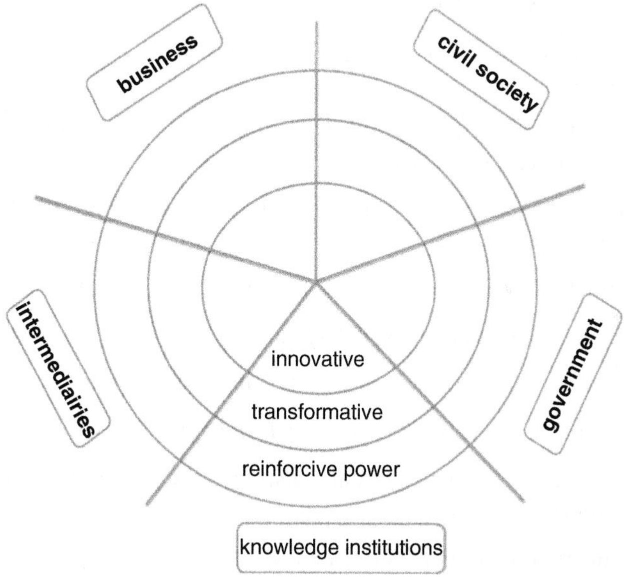

(b) Mention TM principles and relate these to arena process

(ii) long-term thinking as the basis for short term actions

– relate to: envisioning/backcasting

(iii) thinking in terms of multiple domains (multi-domain), different actors (multi-actor), different levels (multi-level)

Table B1 Competence profile of a frontrunner

	Process competences	Substance competences
Strategic	Networking skills	Systems thinking
	Communication skills	Creativity and imagination
	Decisiveness	Problem structuring skills
	Determination	General knowledge
	Leadership	Large network
	Vision	Abstract thinking
Tactical	Negotiation skills	Strategic thinking
	Communication and consensus building	Analytic ability
	Thinking in terms of co-production	Specific knowledge
	Open to new combinations	Innovative ideas
	Coalition building skills	

- relate to: system's analysis/ participants from different backgrounds, domains and levels

(iv) learning as an important aim in itself ('learning-by-doing' and 'doing-by-learning')

- relate to: confronting different perspectives, expectations and agendas/ setting up experiments

(v) orient governance towards system innovation besides system improvement

- relate to: radical experiments/out of the box thinking

(vi) keeping options open, exploring multiple pathways

- relate to: transition agenda, pathways, experiments

(vii) selective participation focusing on frontrunners

- relate to: actor analysis, participative process, change-agents/ frontrunners

Imagine you are on the transition team organising an arena process for housing corporation De Alliantie:

(a) What would be the criteria based on which you would invite people to join the arena?
(b) What should you pay attention to regarding the composition of the group?
(c) Propose three people (be as specific as possible) that you would invite for the arena, place them in the Power Domain map (see below) and explain why you place them there.

a. Mention: competencies (see skills Table B1), interest profiles (creative, out-of-the box thinking, intrinsically motivated, willing to commit time and effort), interests; types of power: innovative (new ideas), transformative (capacity to mobilize others for change) and reinforcive (position within dominant hierarchy)

b. Mention: different organizational backgrounds and domains; different types of power – majority of frontrunners
c. Show that you can apply the above by proposing specific people and adequately argue why you place them at a specific place in the power-domain map.

Envisioning

In class we developed a vision for a sustainable housing corporation. To guide the vision development process, we developed 11 sustainability principles.

(a) Give three examples of sharp and operational sustainability principles.
(b) Then, explain the purpose of co-creating a sustainability vision.

Provide three sustainability principles with following criteria in mind:

- Make them sharp and ambitious
- Relate them to your case of social housing
- Make sure the principle is generally applicable
- Formulate it positively ('100% renewable energy', instead of 'no fossil fuels')
- Make them operational → make sure to use a verb

E.g.:

- *'Sourcing all energy from renewables';*
- *'Becoming a neighbourhood where everyone feels welcome'*

a. Explain co-creation: developing something in a participative process, whereby the experts/designers/decisions makers have to shed their expert role and see eye to eye with 'normal' people

Link this to vision development from two angles: participants and vision itself

People: will feel (and take) ownership for vision
become ambassadors that are inspired to go beyond current interests and daily routines
common language

Vision: provides credibility and degree of aspiration to participants
becomes more tangible and inspiring, also for different kinds of people outside the process

Backcasting

In class we discussed that within the arena process there are several moments when the arena participants look back in time: either from now towards history or from an envisioned future towards the present moment.

(a) An important part of systems analysis is to look at the historical development of a system. In the context of an arena, why is history important?
(b) Why is the method of backcasting used in transition arenas?

Start your answer by giving a clear definition of backcasting.

1. Understanding how a system has developed historically provides information about how things got the way they are today. By looking back for decades or even centuries helps in questioning standards and ways of doing things that we now consider normal. This provides insight in dynamics and therefore where opportunities lie for interventions and which interventions have already been tried (and failed).
2. Backcasting is looking back from an envisioned future towards the current situation.

The choice to start from the envisioned future is essential to enable the formulation of future- oriented strategies that go beyond "business as usual" solutions and are not constrained by vested interests and stakes.

Interactive Exercises and Group Assignments

A 'Quick and Dirty' Transition Management Workshop Experience

In this interactive exercise we move from theory to practice by simulating a real world arena process. The students will take on different roles (e.g. transition manager and arena participant). In the first part of the exercise the participants will develop the system analysis. At a later stage the workshop will revolve around the development of a shared vision. In the third and final step we will reflect on the process and we will undertake the first steps to organise a 'real world arena' with 'frontrunners'.

This workshop is organised as an experimentation of Transition Management that has been practiced by DRIFT to urban areas and various service domains (e.g. energy, climate, health, food, social justice, etc.) as well as in different contexts. The goal is to run through a 'quick and dirty' problem analysis, visioning exercise and backcasting process with regard to a (pre-selected) sustainability challenge in question, so as to explore alternative futures and transition pathways.

The workshop aims to give a hands-on experience with the Transition Management methodology.

The workshop will consist of the following steps, which will each take 20 min:

1. Current situation: What are the challenges we face today? Why do these challenges exist?
2. Sustainability guiding principles: Which are the principles to guide future developments and solutions to the problems we face today? What does the desirable future system look like?
3. Transition pathways and experiments: What needs to change for achieving the vision? What are the long-term, mid-term and short-term goals that lead to achieving the vision? What actions and first experiments can help in achieving or stimulating a transition towards the vision?

Strategic Niche Management (SNM)

Below is a description of a simple SNM analysis that can be applied to improve the break through potential of innovative projects. The analysis helps you to identify chances and hindrances that the project could meet. It can be the basis for adapting your project in order to increase its potential to contribute to system innovation for sustainability.

- **Step 1: Define a project**

Define a 'niche-project'/'transition experiment' to analyze from a SNM perspective.

Describe briefly: what is the project about?

- **Step 2: Score quality of Vision**

Answer questions about the vision and score. Collect your scores in the table below

- To what extent is there a leading long- or medium-term vision guiding the project? What is it?
- How innovative or different is the vision from business as usual? (1: almost business as usual; ten quite different from business as usual?)
- Do you consider the vision to be an attractive vision, easy to convey? (1: no; 10 yes)
- Does the vision pay attention to non-technical aspects of the low-carbon innovation included (for example cultural, financial aspects)? (no = 1; yes = 10).

- Is the vision spread and shared? (1 = no; 10 yes, it is a vision that is shared by many, including major stakeholders. In case you need an example of possible stakeholders, see figure next question).

Scoring vision	Score (1–10)
Is there a leading, long or medium-term vision?	
How innovative is the vision?	
Do you consider the vision attractive, easy to convey?	
Are non-technical aspects included?	
Is the vision spread and shared?	
Total score(= average)	

- **Step 3: Score quality of Social Network**

Quickly determine: which are the relevant actors for your project and project ambition?

Jot down your results in the table below.

Actor group	Who? Present in the niche-project network? Yes or no?
Government	
Business	
Incl. suppliers, producers	
Community	
Incl. consumers, citizens	
Third sector	
Incl. NGOs, associations, research, education, activists	
Other/mixed	

Assess the quality of the social network around your project.

Answer the questions below and score.

Scoring actor-network	Score (1–10)
To what extent are (potential) users involved/represented?	
To what extent are (potential) partners involved/represented?	
To what extent are (potential) financers involved/represented?	
Are actors involved able to provide the necessary resources?	
To what extent does the network involved have the power to deal with resistance?	
Total score (= average)	

- **Step 4: Score Multi-Level Strategy**

What, in your case, are relevant landscape developments?

Determine major 'landscape' developments. Which landscape developments support your project or can be made supportive? Which landscape developments impede your project?

Landscape developments	Supportive, neutral or impeding?

What are the most relevant 'regime-factors' for your niche project?

What are the relevant regime-factors for your project? Think of factors from various regimes (e.g. fossil fuel regime, automobile regime, financial regime etc.)? Which regime factors are supporting your project? Which regime factors are impeding your project, i.e. which regime factors do you need/want to change? Remember to think of all three dimensions of the regime:

- Cultural: e.g. norms, expected behavior, scientific knowledge, routines
- Institutional: e.g. rules, regulations, financial arrangements
- Physical: e.g. infrastructure, spatial conditions

Regime-factors	Supportive op impeding?

Given your answers to questions and 6, score in the table the opportunities that your project ambitions come through (1, no opportunities; 10, very likely regime breakthrough)?

Scoring multi-level strategy	Score
To what extent is your project hampered by landscape trends (low score) or supported by landscape trends (high score)?	
To what extent is your project hampered by regime-factors (low score) or supported by regime factors (high score)?	
To what extent do you have a strategy to play into landscape trends?	
To what extent are there opportunities for regime break through?	
Total score (=average)	

- **Step 5: Score Learning**

What do you still need to learn for making your project successful? Think about lessons about technical aspects, cultural lessons, policy lessons, markets, new financial arrangements, legal conditions etc. Jot down your insights in the table below.

Learning aspects	Specify what there is to learn
Technical	
Cultural	
About markets	
About financial arrangements	
Juridical conditions	
..	
...	

Assess the present quality of the learning process in your project (score 1–10)? Answer the questions below and score.

Learning aspects	Score
To what extent is there space for learning in your project?	
To what extent is the learning in/about you project deepening?	
To what extent is the learning in/about your project broadening?	
To what extent is the learning in/about your project up-scaling?	
Are other actors involved in the learning process?	
Total score (=average)	

- **Step 6: Summary and Actions**

For a visual summary of your analysis, put the scores of steps 2, 3, 4 and 5 in the spider diagram. Based on the result, formulate follow-up actions for your project, in terms of dealing with weaknesses and/or further developing strengths.

SNM analysis diagrams:

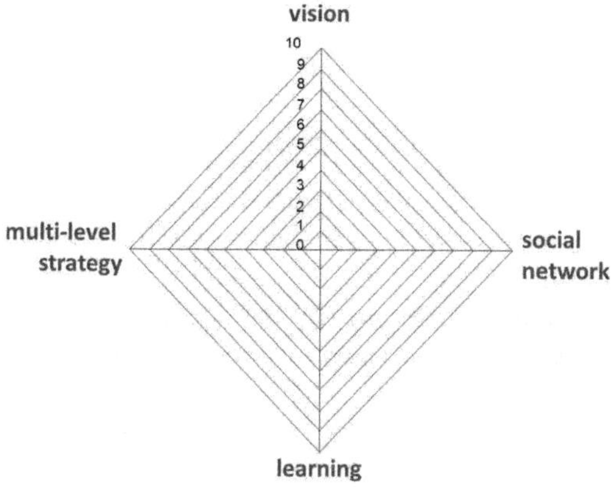

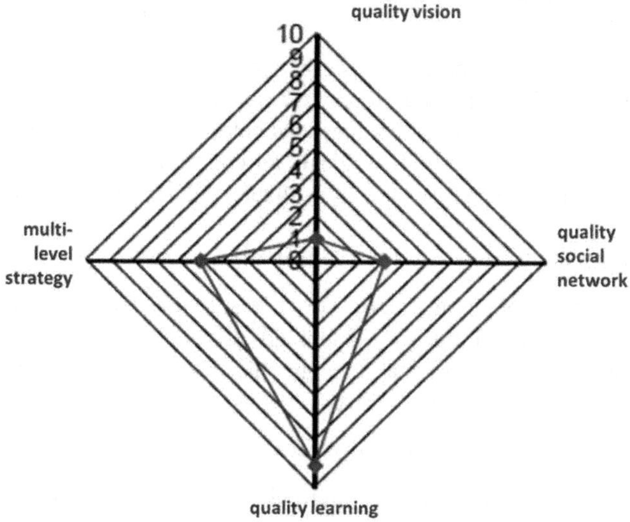

Strategic Niche Management (SNM)

The participants divide into different working groups and they pick a project to work on. Then the participants brainstorm about opportunities and needs for:

- Shielding
- Nurturing
- Empowerment

[Examples of four strategies for empowering & nurturing:

- Establish and strengthen relevant networks;
- Articulate and negotiate, with relevant actors, mid-or longer-term expectations/visions and interests;
- Try to align with supporting regime (fit-and-conform) or/and influence blocking regime (stretch-and-transform), using inspiring narratives, actor network, lobby, etc.);
- Stimulate collective learning.

Multi-level Perspective (MLP)

Every individual should choose one specific project, a specific innovative technology, or innovative idea. Each individual makes a "multi-level analysis" for his/her specific example, by answering the questions below. Collectively, the individuals fill in the Multi-Level Scheme. Individuals all contribute to filling in this Multi-Level Scheme by adding post-its and – if available – illustrative news paper cuttings to each level.

Step 1: What are the relevant landscape developments?

- Exogenous and autonomous trends – or at least trends that you as an individual or small group cannot influence – and major crises;
- Long-term developments such as demographic, environmental, macro-economic trends, trends in political culture, world views

Step 2: What are the relevant regime-factors?

- Established/mainstream practices and rules, culture, financial arrangements, powerful actors or infrastructure which provides stability and contributes lock-in of existing systems (e.g. fossil fuel lock-in, car-based mobility system)

Step 3: What are the (other) relevant niches?

- Small-scale protective spaces/projects for radical innovations
- Potential alternatives to regimes (path-breaking)

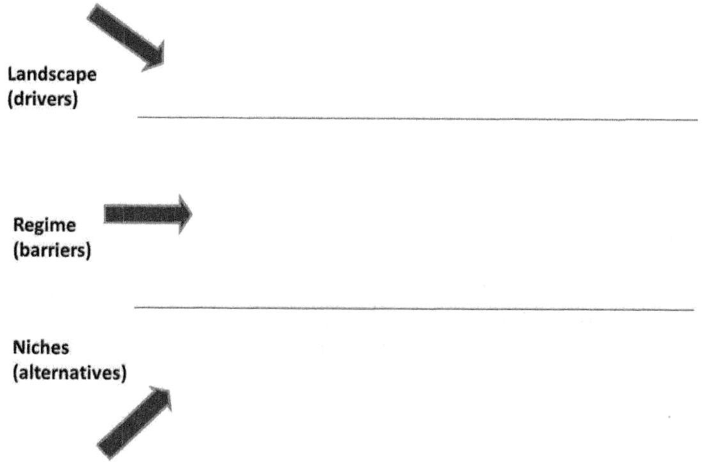

Multi-level scheme

Step 4: Discuss the resulting Multi-level scheme:

(a) What are the fundamental drivers for change (landscape level) for your project?
(b) What are the (most) critical regime barriers?
(c) What are the (most) promising niche alternatives?

Identification of Frontrunners

Identify frontrunners (intuitive exercise)

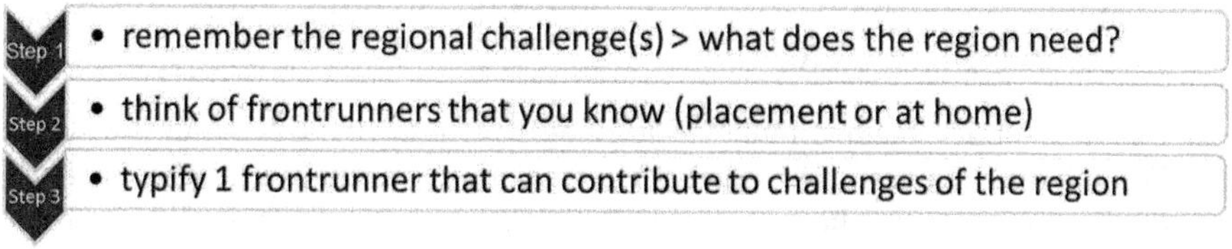

Material & tools:

- visualisation: pictures, scraps, drawings
- 1–3 words: expertise, characteristic, title

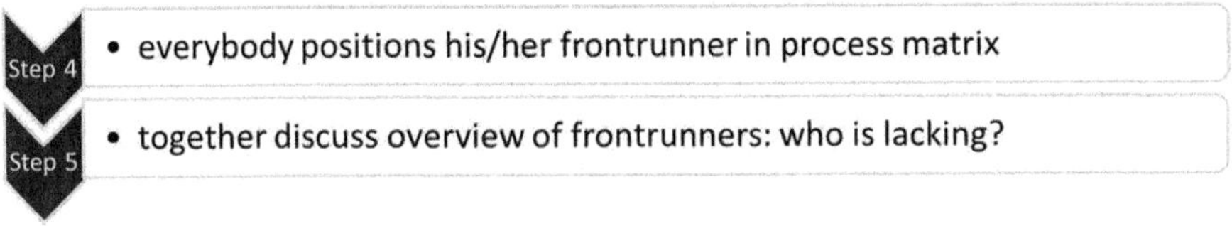

Material & tools:

- A0/A1 poster of matrix on the wall
- Glue the visualisations & words of frontrunners on the poster
- Draw relations between frontrunners & 'holes' in network

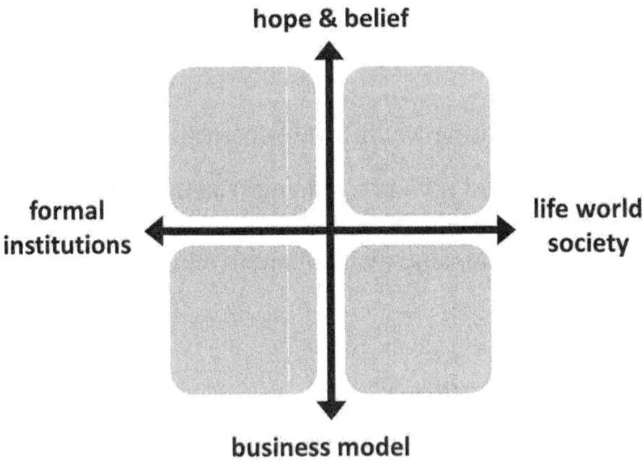

Simple matrix that distinguishes between different focus in what you are preoccupied with in your daily work:

(1) Business model (making money, surviving) versus (2) hope & belief (idealism)
(3) Formal institutions (rules, standards, regulations, formalities, organisational bureaucracy) versus the life world of society (real people living life, sensory experiences, feeling, smelling, doing, daily practices)

Position yourself

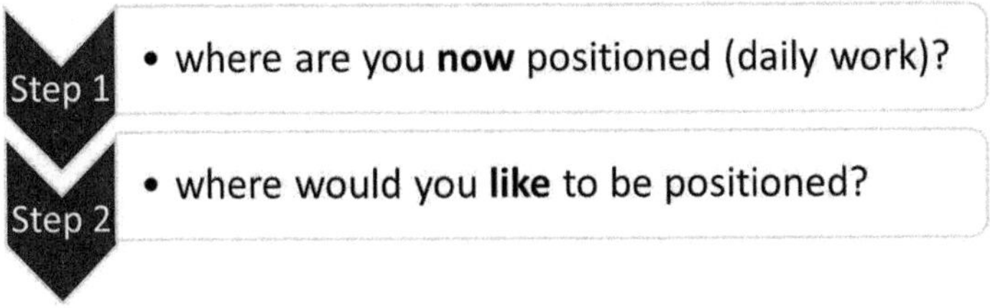

Position your host

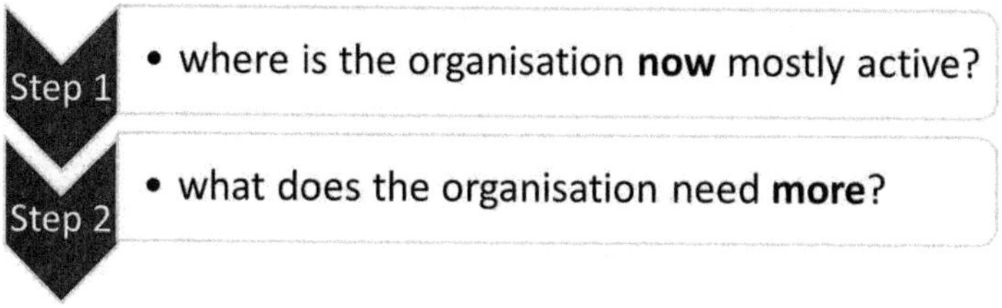

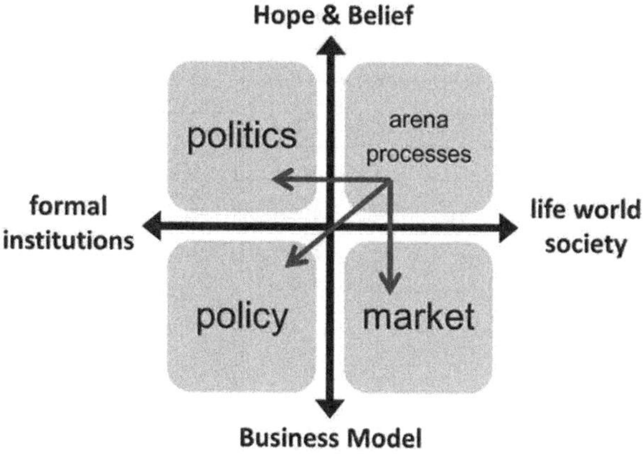

Formulate a Regional Challenge

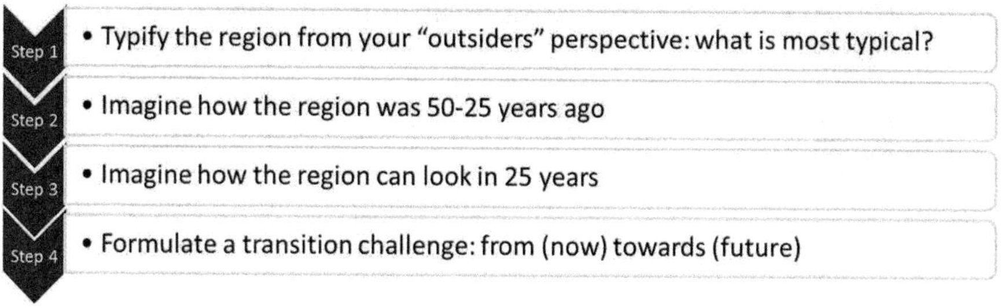

Material & tools:

- History: pictures/history books of region ±50–25 years ago
- If available: pictures/reports of future visions for region

typical regional traits:	25 years ago	now	(imagined) future 2040
water / sea port	nr. 1 world port	economic decline climate change	Bio-economy Port & Floating Port City

Actor Analysis

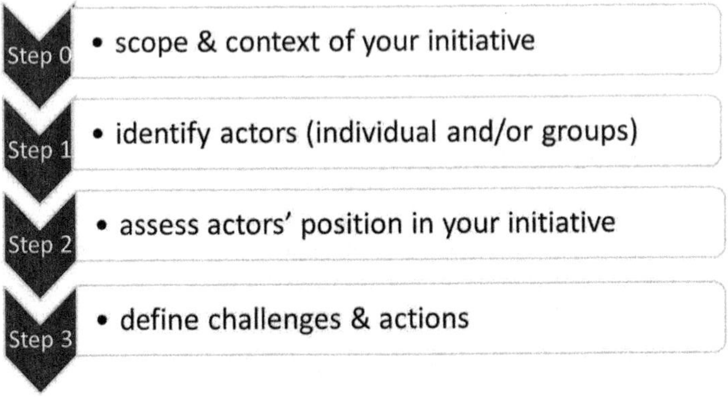

- **Scope of your initiative/project?**
  - What exactly are you developing? What should be the concrete result?
  - In what broader context should your product/ project/ programme/ initiative function?

  - What is the sustainability claim? And how does that relate to the broader context?

- **Why an actor analysis?**
  Exploring the actor network around your initiative:
  - Scope & context of your initiative
  - Identify relevant actors for your initiative: (a 'social map')
  - Assess the position of these actors regarding your initiative: interest, attitude and power
  - Their interests, position, and ideally their values, ideas on problems, solutions
- How do you do an actor analysis?
- Exploring the actor network around your initiative:
  - Scope & context of your initiative
  - Identify relevant actors for your initiative: (a 'social map')
  - Assess the position of these actors regarding your initiative: interest, attitude and power
  - Their interests, position, and ideally their values,ideas on problems, solutions

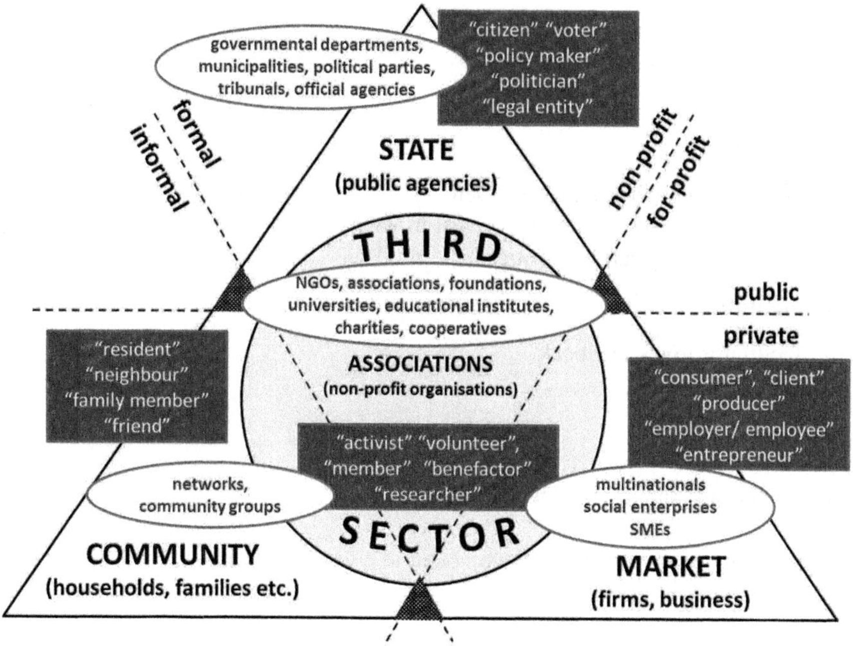

Avelino & Wittmayer 2014, Based on Evers & Laville 2004, Pestoff 1992

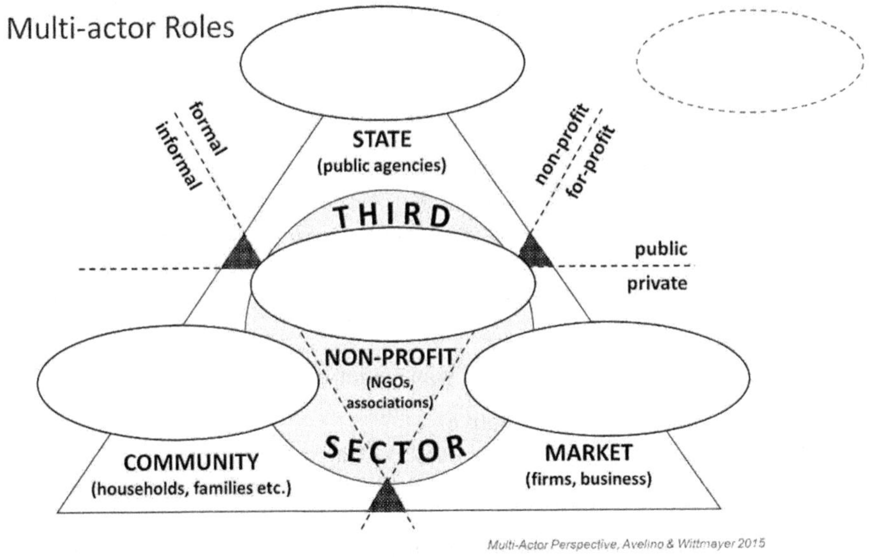

1. **Prepare role playing**
   - scope of project (very short)
   - ±10 relevant actors
   - volunteers
2. **Role playing**
   - set the scene: Nice Dragon's Den
   - pitch by project-owner: why invest in what?
   - actors = constructively critical
3. **Evaluation**
   - facilitator gives feedback
   - groups evaluates

Actor, Network or Stakeholder Analyses

Use for determining which actors you need to concentrate on.

- **Principles**

Producing a good ‘social map’ of the environment enables you to identify which actors you need to concentrate on.

- **What do you do?**

You produce a ‘social map’ containing information about the parties in the environment of a project or programme. A tip for the first step in producing a social map might be to break down the actors into the categories technology, science, users and regulators. For a technology-driven project or programme, this will produce at least an initial useful classification. Another starting point could be a general classification arranged by civil-society organisations, government, business, the academic community and the public. Alternatively, you could start with one or more key actors and make a map of their environment. For example, the organisers of an innovation project concerning telecommunication in the care sector started with the client. They then explored who the client had contact with and proceeded from there. In this way, the transition professionals produced a social map that included the general practitioner, the pharmacist, district nurse, neighbours, family and all the groups or organisations around them.

Drawing up a list of relevant actors is only the first step in a transition project. You will also have to explore the nature of their involvement and their views. This analysis can be superficial or in-depth. For example, a superficial assessment of commitment would be to identify actors with a positive interest, a negative interest or a neutral attitude. For an in-depth analysis, you would explore their views on the problem, the vision, their criteria for solutions and their proposed solutions. An in-depth actor analysis shows you what scope there is for innovation. On this point, see also the methods ‘Action theory’, ‘Cognition model’ or ‘ESTEEM’.

Multi-actor Roles in Transitions

- **Sustainability Transitions**

Which sustainability transition do you find most important: which sustainability transition do you want to contribute to?

- **Your roles**

What are your roles in this sustainability transitions? (Fill in the Multi-Actor Perspective map). Try to think of multiple roles that you play across all sectors. Be specific about HOW you contribute to sustainability transitions in each role.

Roles	How does this role contribute to the transition?

- **Share your insights**

Share your insights with each other. Then interview each other about question 3 and 4 (see other side).

- **Improving and creating roles**

What (other) roles would you like to take, improve and/or create to contribute (more/better) to your chosen sustainability transition? Are there gaps, challenges and/or tensions in your roles? Which roles would you like to integrate/connect more, and how?

Tensions & challenges in the roles you play:

Roles to improve/connect/integrate?	Improve/connect/integrate how?

Other (new) roles to take/create?	How and why?

- **Resolutions**

  Based on this exercise, do you have any resolutions about the roles that you (could) play in your chosen sustainability transition? Identify at least one resolution for the coming year.

Backcasting Transition Scenarios

- Identifying & specifying transition paths
  Backcasting transition paths
  - *We are in 2050 in Sustainable City XX*
  - *What happened since 2014 to get here?*

    Formulate milestones in 2035, 2020, 2015
    Which players have been involved and how?
    Which transition experiments were conducted?
    (How have you contributed?)
- Preparing group presentation: presenting backcasting stories

MIX
Papier aus verantwortungsvollen Quellen
Paper from responsible sources
FSC® C105338

Printed by Books on Demand, Germany